N1977

D1381244

TELEPEN

£18·60

External Readers | Staff & Research
Students

# *The Molecular*
# *Biology of Membranes*

David E. Green

# The Molecular Biology of Membranes

Edited by

**Sidney Fleischer**
*Vanderbilt University*
*Nashville, Tennessee*

**Youssef Hatefi**
*Scripps Clinic and Research Foundation*
*La Jolla, California*

**David H. MacLennan**
*University of Toronto*
*Toronto, Ontario, Canada*

**and**

**Alexander Tzagoloff**
*Columbia University*
*New York, New York*

*Plenum Press* · *New York and London*

Library of Congress Cataloging in Publication Data

Symposium on the Molecular Biology of Membranes, New Orleans, La., 1977.
    The molecular biology of membranes.

    "Proceedings of a Symposium on the Molecular Biology of Membranes held in New
Orleans, Louisiana, February 14, 1977."
    Includes bibliographical references and index.
    1. Membranes (Biology)—Congresses. 2. Molecular biology—Congresses. I. Fleischer,
Sidney. II. Title. [DNLM: 1. Membranes—Congresses. 2. Molecular biology—Con-
gresses. QH601 S992m 1977]
QH601.S925 1977                          574.8'75                          78-2207
ISBN 0-306-31114-3

Proceedings of a Symposium
on the Molecular Biology of Membranes
held in New Orleans, Louisiana, February 14, 1977

© 1978 Plenum Press, New York
A Division of Plenum Publishing Corporation
227 West 17th Street, New York, N.Y. 10011

All rights reserved

No part of this book may be reproduced, stored in a retrieval system, or transmitted,
in any form or by any means, electronic, mechanical, photocopying, microfilming,
recording, or otherwise, without written permission from the Publisher

Printed in the United States of America

# Contributors

*Harold Baum* • Department of Biochemistry, Chelsea College, University of London, London SW3 6LX, United Kingdom

*Helmut Beinert* • Institute for Enzyme Research, University of Wisconsin, Madison, Wisconsin 53706

*George A. Blondin* • Institute for Enzyme Research, University of Wisconsin, Madison 53706

*Gerald P. Brierley* • Department of Physiological Chemistry, College of Medicine, Ohio State University, Columbus, Ohio 43210

*Roderick A. Capaldi* • Institute of Molecular Biology, Department of Biology, University of Oregon, Eugene, Oregon 97403

*Yves M. Galante* • Department of Biochemistry, Scripps Clinic and Research Foundation, La Jolla, California 92037

*David E. Green* • Institute for Enzyme Research, University of Wisconsin, Madison, Wisconsin 53706

*David E. Griffiths* • Department of Molecular Sciences, University of Warwick, Coventry, United Kingdom

*Ruth M. Hall* • Department of Biochemistry, Monash University, Clayton, Victoria, 3168, Australia

*Youssef Hatefi* • Department of Biochemistry, Scripps Clinic and Research Foundation, La Jolla, California 92037

*Robert A. Haworth* • Institute for Enzyme Research, University of Wisconsin, Madison, Wisconsin 53706

*Douglas R. Hunter* • Institute for Enzyme Research, University of Wisconsin, Madison, Wisconsin 53706

*Saroj Joshi* • Department of Cell Physiology, Boston Biomedical Research Institute, Boston, Massachusetts 02114

*Annelise O. Jorgensen* • Department of Anatomy, University of Toronto, Toronto, Ontario, Canada M5G 1L6

*Vitauts I. Kalnins* • Department of Anatomy, University of Toronto, Toronto, Ontario, Canada M5G 1L6

*Giorgio Lenaz* • Istituto di Biochimica, University of Ancona, Ancona, Italy

*Anthony W. Linnane* • Department of Biochemistry, Monash University, Clayton, Victoria, 3168, Australia

*David H. MacLennan* • Banting and Best Department of Medical Research, Charles H. Best Institute, University of Toronto, Toronto, Ontario, Canada M5G 1L6

*Jerzy Popinigis* • Laboratory of Bioenergetics, Department of Physiology and Biochemistry, "Jedrzej Sniadecki" Academy of Physical Education, 80-336 Gdańsk, Poland

*D. Rao Sanadi* • Department of Cell Physiology, Boston Biomedical Research Institute, Boston, Massachusetts 02114; and Department of Biological Chemistry, Harvard Medical School, Boston, Massachusetts 02115

*Fariyal M. Shaikh* • Department of Cell Physiology, Boston Biomedical Research Institute, Boston, Massachusetts 02114

*A. E. Senior* • University of Rochester Medical Center, Rochester, New York 14642

*Alexander Tzagoloff* • Department of Biological Sciences, Columbia University, New York, New York 10032

*Garret Vanderkooi* • Department of Chemistry, Northern Illinois University, DeKalb, Illinois 60115

*Daniel M. Ziegler* • Clayton Foundation Biochemical Institute and Department of Chemistry, University of Texas at Austin, Austin, Texas 78712

*Elzbieta Zubrzycka* • Banting and Best Department of Medical Research, Charles H. Best Institute, University of Toronto, Toronto, Ontario, Canada M5G 1L6

# *Preface*

On February 14, 1977, a symposium entitled "The Molecular Biology of Membranes" was held in New Orleans in honor of Professor David E. Green, whose many contributions in mitochondrial structure and metabolism have influenced and guided research in this important area of biochemistry for many years. The symposium was attended by many former and present-day colleagues, friends, and interested scientists. The contents of this volume represent papers that were delivered at the symposium and other contributions from individuals who have been associates of Professor Green.

We wish to thank Plenum Press for their help in making the symposium and publication of this book possible.

Sidney Fleischer
Youssef Hatefi
David MacLennan
Alexander Tzagoloff

# Contents

# Impressions of David E. Green by His Colleagues

## 1930—1932

Events have shown that the brash, 20-year-old Dave Green, whom I met at the Marine Biological Laboratory during the summer of 1930, had a great deal to be brash about. The friendship which he gave me during the following year at Washington Square College, New York University, and during the summer of 1931 at the MBL, where we both worked with Professor Robert Chambers, played a great role in helping me maintain my morale during this most frustrating year of my professional life.

The following summer (1932), I learned from Dave that he had saved enough money for a year of graduate work at Cambridge University, England. With supreme self-confidence, he informed me that during that year he would make his mark to the point where Cambridge University would itself provide support to time of completion of work for his doctoral degree. His self-confidence was obviously not misplaced.

The Dave Green whom I knew in the early 1930s had a remarkable ability to read and critically evaluate pertinent literature with great rapidity and insight. In addition, he had a photographic memory—while I was groping for a reference, he would give me more than one, often complete with names of authors, journal, date, page numbers, and summary. Finally, it is my recollection that he had already arrived at the idea that to understand any particular biochemical activity of a living cell, it was necessary to elucidate the pertinent molecular structures.

Lyle V. Beck

## *1932–1940*

David Green joined the Department of Biochemistry of Cambridge University in 1932 as a Research Student for the Ph.D. degree. Sir F. G. Hopkins, the Head of the Department, suggested that he should work under my supervision on the enzymes and other catalysts concerned with biological oxidation processes and respiration. Research on that subject was then in a very interesting stage, when it was becoming possible to bring together the results of a number of then recent discoveries, and at Cambridge research was actively proceeding on a number of different aspects, e.g., study of separated enzymes, especially dehydrogenases, cytochromes and cell respiration, thiol compounds, "coenzymes I and II" (now NAD and NADP), flavins and flavoproteins, redox potential, respiratory inhibitors, and reactions in muscle, bacteria, plants, etc.

David threw himself into this research with great enthusiasm, energy, and enterprise. He was full of ideas which he expressed freely, and although not everybody agreed with all of them, they were always interesting and characterized by freshness and vitality. His research proved fruitful from the beginning, and he soon began the production of a continuous flow of important publications, which has continued now for about 45 years. He very soon obtained the Cambridge Ph.D. degree, and was enabled to continue his research in Cambridge for a further 6 years by the award of a Beit Memorial Fellowship, and later a Senior Beit Fellowship. During that time, he contributed very considerably to the growth of our knowledge of biological oxidation systems. He obtained a series of preparations of particular dehydrogenases, and although at that early date they could not be expected to be homogeneous, they were a great improvement on the extracts which had been available previously. He quickly realized the importance of starting with comparatively large amounts of material to obtain a reasonable yield, and this usually depended on teamwork. David had a gift of attracting other research workers, both senior and junior, to collaborate with him to form small research groups. This made it possible to carry out a series of important studies on the role of coenzymes, flavoproteins, cytochromes, etc., in the linking of dehydrogenases with $O_2$ (in respiration) and with other dehydrogenases (in fermentations). He also gave us much help in devising interesting experiments for our advanced practical classes.

His work, however, did not stop at systems of soluble enzymes. Our relations with Professor Keilin were close, and David soon became familiar with Keilin's particulate preparation of cytochrome oxidase from heart muscle mitochondria. Keilin drew attention to the possible importance of the physical interrelationships of the catalysts within the particle. David saw the importance of this, and it became the basis of much of his work for many years after his return to the United States, particularly with reference to the "cyclophorase system."

I must not fail to mention the book of essays *Perspectives in Biochemistry*, written in 1937 by over 30 distinguished writers, who were or had been connected with the Laboratory, in honor of the 75th birthday of Sir F. G. Hopkins. It was edited jointly by Joseph Needham and David Green, and attracted considerable attention at the time. It seems likely that it may have formed a model for the two later volumes *Currents in Biochemical Research*, edited by David in 1946 and 1956.

Malcolm Dixon

## 1940–1946

I met Green first in the spring of 1940 when he came from Cambridge to Harvard Medical School, an expatriate forced home by the war. I was a medical student being introduced to biochemistry. I later realized he must have experienced a kind of culture shock in coming from the fountainhead of biochemistry and enzymology that Cambridge then was, and being there one of the most enterprising of the younger men, to discover that the United States, and even Harvard, was a biochemical desert. Other refugees also experienced this, and their presence in the United States produced a flowering of the science by the end of the war that needs to be detailed. Green was a dominant influence among this now illustrious group. Because they called on him as they arrived, I saw that these future leaders of biochemistry placed a much higher value on Green and his work than did his colleagues at Harvard. This slanted my own evaluation of the subject. Green had been responsible for the famous Part II biochemistry honors course at Cambridge. He went through my Harvard course notes and assignments, with astonishment. "My God," he said, "at Harvard enzymes are the hair and fingernails of biochemistry!"

The balance began to change, but of course too slowly. His book, the first to present enzymes operationally, grouped by coenzymes, came in through the U-boats. It was to read this together that a group of eminent physicians formed the first Enzyme Club, and invited Green to New York City to preside at their learning, and to work at P & S. His trace substance hypothesis was published in Volume IV of *Advances in Enzymology*, the most stained and used volume in the series, which also contains Lipmann's high-energy phosphate paper.

Finally, there was some recognition. The Paul Lewis Award was created, and the first of these went to him without question. Wisconsin, never as enzymically desolate as most of the country, pioneered in starting an Enzyme Institute, and chose Green to organize it. I played a significant role at that time because I came from the Middle West. I reassured him that it was plausible to live west of the Hudson, a fate that Green, ever the New Yorker, looked at with some foreboding. When his appointment was announced in Madison, he brought me, rather quizzically, the newspaper clippings: "Enzymes," said the subheadline, "are in beer, cheese and wine."

W. Eugene Knox

## 1940–1946

During my last year at Harvard, I was required to prepare a research thesis to fulfill the honor requirements in biochemistry. Since I had become interested in enzymes, and since, in 1940, no enzymologist was on the staff in Cambridge, I made an appointment to see Professor A. Baird Hastings, at that time Chairman of the Department of Biological Chemistry at the Medical School at Brookline. At the appointed hour, I was ushered into the august and wood-paneled chambers of Hastings, and, after a brief series of questions, Hastings said that he no longer was active in this field but that a young chap just back from Cambridge, England, was downstairs and it would be a worthwhile experience to at least meet him. Hastings took me down to the high-ceilinged, darkish, cupboard-lined laboratory and introduced me to David E. Green, who promptly suggested, in a broad English accent, that I roll up my sleeves and go to work.

Thus, in September of 1940, I began a 6-year involvement with David E. Green. At the end of the 1940–1941 academic year, Green obtained a position as Instructor of Biochemistry at Columbia University College of Physicians and Surgeons. He suggested I apply for my graduate work at Columbia; so September 1941 found me in New York City with David as my major professor. I had the unique distinction of being Green's only graduate student in America. The period from 1941 to 1946 was an exciting one. With the war at full swing throughout this time, with research funds still severely limited, with supplies and equipment at a premium, today I am still struck by the highly successful operation that David ran in his two small laboratories on Floor "G." One of the two laboratories was occupied by Sarah Ratner, Marian Blanchard, and myself, and the other lab by David, his technician, Violet Nocito, Joe Dolan—a most unusual dishwasher—and visiting scientists who, during the war years, included Luis Leloir and later W. Farnsworth Loomis. The lab was the center of information on enzyme chemistry for all of New York City. David Nachmansohn, Konrad Bloch, David Rittenberg, and David Shemin were constantly wandering up for advice; S. Ochoa and E. Racker from N. Y. U. were frequent visitors, as were Fritz Lipmann, Otto Myerhof, and many others. At this time, Green organized a group of people who had interests in common and called it the "Enzyme Club" with monthly meetings at the Columbia University Faculty Club. After the war, this idea caught on all over the country, and for many years these "clubs" could be found at any large urban academic center.

Green had his desk in his small lab where he carried out all his own experiments, administered his laboratory, met people, ordered equipment, and wrote his papers. He was therefore a "lab" man. There is no question in my mind that Green was the best experimentalist I ever met. As we used to say, he had a "Green thumb," an intuitive sense to set up relevant experiments; he had a knack in isolating difficult enzymes. He was always enthusiastic, impetuous, perhaps stubborn, always available for advice and encouragement. We developed one of the first ultrasonic devices for the purpose of disintegrating bacteria; Green purchased one of the first Beckman DU spectrophotometers; he was the first, to my knowledge, to make use of Waring blenders for extraction of enzymes from tissues. One time, during the war, he needed a supply of xanthine oxidase. He managed to obtain about 10 liters of raw heavy cream—how he obtained this rare dairy product, I have no idea—and we extracted the milk xanthine oxidase and, more important, a very large amount of butter as a most desirable by-product.

P. K. Stumpf

## 1950–1953

I was fortunate in joining David Green's team in October 1950, at the start of a very exciting period of research. A few months earlier, the group had transferred to the new laboratory space on University Avenue, into what was then the new Enzyme Research Institute. David Green was now able to fully realize the dreams which he had developed during his time in Cambridge, England, that is, to build up a facility for large-scale enzyme preparative work, on a level that had not previously been attempted. He went about this with tremendous verve and enthusiasm, and I have fond memories of him, in a white coat, energetically supervising the installation of the 13-liter low-speed centrifuge and the Sharples centrifuge in the somewhat overheated boiler-room premises.

David had relatively few administrative duties at the time, and he was able to spend a good deal of this time at the bench, and, more importantly, with his younger colleagues. Discussion ranged widely; and although we mulled over the events of the day, the main topics concerned the science which was our chief interest. During the fall of 1951, discussion turned on the structure of the mitochondrion and of "cyclophorase." David insisted that his critics, who claimed the mitochondrion to be a "little bag of soluble enzymes," were entirely wrong, since the properties of the cyclophorase enzymes indicated the presence of a defined structure. I was so bold as to interpret David's theorizing in a diagram of a "cycloforasarus," which was photographed into multiple copies by Shirley Schweet, and which provided the Enzyme Institute group with an informal Christmas card for 1951.

Priscilla Hele


## 1958–1959

Those were the days when coenzyme Q had just been discovered by Fred Crane and his colleagues, which made Fred walk with his head a few feet higher. Absolute contrast was the case with another group in the same Institute, who were working on fatty acid synthesis. The carboxylation reaction was playing hide and seek with them, leading to unlimited gloom and frustration. Then, when one fine morning Wakil isolated malonate as an intermediate in fatty acid synthesis, the gloom suddenly disappeared and Wakil found himself firmly lodged in the hall of fame.

J. Ganguly


He was always a generous man to his young scientific colleagues in terms of publications—if anything, overgenerous; for I believe that he would have certainly been awarded the Nobel Prize for his work in the lipid field if the contribution that he made to fatty acid synthesis had been justly acknowledged. However, because of his generosity, this was not to be.

Anthony W. Linnane


## 1959–1962

The greatest aim of Dr. Green and myself, at that time, was the discovery and isolation of the hypothetical unit of the mitochondrial energy-transducing system. In the spring of 1961, Dr. Green and I visited Dr. Fernandez-Moran at Boston, carrying an ice-box with mitochondria in it, to ask him for the electron microscopic observation of mitochondria by negative staining with phosphotungstic acid. I cannot forget my dramatic emotion when we saw the repeating particles in the inner mitochondrial membranes (the headpieces of the elementary particles).

Takuzo Oda

## 1962 – 1967

I am very pleased to supply some personal recollections of the exciting scientific atmosphere at the Enzyme Institute when I was there. This was a period shortly following Hatefi's work on the isolation of the respiratory complexes and the demonstration by means of reconstitution and other criteria that these are genuine enzymes of the electron transfer chain. In my view, this was one of the major contributions that came out of Green's laboratory. The reality of the complexes as *bona fide* mitochondrial enzymes has withstood the test of time and is now generally accepted in the mitochondrial field.

My own exposure to these enzymes occurred when I first arrived at the Enzyme Institute and began working with David Wharton on the copper component of cytochrome oxidase. Later, David McConnell, our electron microscopist in residence, came, and the rather important discovery was made that the purified respiratory complexes were capable of forming membranes when appropriately supplemented with phospholipids. David Green, with his keen eye for ultrastructural details, was quick to recognize the significance of this phenomenon which became a strong experimental basis for his interpretation of the inner membrane of mitochondria.

One meeting I had with him at that time may be worth recounting, since it illustrates his analytical skill at connecting experimental observations with phenomena that at first glance might seem unrelated. On this particular occasion, we were discussing membrane formation by the complexes and Green quite suddenly realized that the conditions which had been found earlier by Hatefi to be necessary for the reconstitution of the electron transfer chain would obligatorily lead to membrane formation. Without spelling out all the details, he then suggested the idea that in order for two or more complexes to reconstitute an integrated activity, the sole structural requirement would be that they be present in the same membrane. In other words, Green was able to make the connection between Hatefi's conditions for reconstitution and a more abstract concept having to do with the assembly of the complexes into membranes. In fact, I was able subsequently to show that Green's idea was correct.

Alexander Tzagoloff


There are four essential things I learned from David Green. One is that before you can talk about the function of a biological component, you have to isolate it and characterize it as it functions *in vitro*, away from other components with whose functions its own could be confused. The second is that before you can appreciate the function of the component in its natural environment, you have to put it back together with other components to see how they work in concert. The third thing is that if you just run experiments without thinking carefully and at length about why you're doing them, they may all turn out to be trivial. The fourth is that when you make a mistake, you should make a quick and graceful acknowledgment of it, but be prepared to hear your critics bring it up the rest of your life.

David Green made numerous mistakes, almost always *en route* to a new and better synthesis of ideas. His energetic mind continually leaped from crag to crag in the cloudy heights of biological phenomenology. His leaps were intrepid and magnificent. The timid were sometimes left behind in the well-cultivated valleys of more pedestrian ideas. The uncritical were sometimes plunged into the abyss, from which Green himself seemed to escape, although not without scars. There was a sense of heady, almost extravagant excitement about research with this extraordinary scientist.

Among the many significant contributions he had made to biology, one that I recall most personally was his early realization and demonstration that the electron microscope was not the exotic plaything of a scientific elite, but a powerful interpretive tool to be put at the disposal of every working biochemist. In the early 1960s, to bring this about, he traveled repeatedly—thermos jug in hand to transport the ever-present mitochondria—to the laboratory of a world-renowed electron microscopist for instruction and collaborative research. Soon thereafter, with the merest of credentials in this new field, he was calumnied by morphologists and biochemists alike for venturing into the structure–function arena. Yet he persisted and triumphed in his goal. Characteristically, it was of less interest to David Green that his inductive generalizations fit completely the available facts than that he enter forbidden temples with searchlight and rapier in hand, his mop-up troops to follow.

David G. McConnell

Having spent 5 years in the old quarters on University Avenue, I remember particularly the rise and fall of several first-class theories. These were initially tested on his Institute fellows before they were voiced on an international stage. But I shall never forget the almost reckless courage of the man to state his most recent brilliant, intuitive synthesis from fragmented data. Although many of his concepts were certainly not sustained in detail, some nidus of these was always fixed in the subconscious of his wide audience. Everyone had to look at things his way, at least once. Every subsequent theory, by whomever, always had an element of Green in the fabric.

David M. Gibson


Weekly Saturday morning work report sessions were of course solely devoted to biochemistry, to research in progress. There, David Green could be scathingly critical of work which he considered not to be up to the highest possible standard, or not moving ahead fast enough. There also, he revealed his unique talent for collecting the bits and pieces and integrating them into an alluring hypothesis with a new dimension. And if in the course of events further evidence collected toppled this hypothesis, that was no problem: a new and better one was immediately erected in its place.

He must have spent a good deal of his working life thinking, arranging, and rearranging his bits and pieces until a structure began to emerge which made sense to him. In those days I am talking of, he would occasionally emerge from a thinking session, in search of an audience to try out his ideas on. He would pop his head around the door, asking if one were free and take one to a drugstore for a cup of coffee and an exposition of his latest ideas. These occasional outings were the moments when one had the impression of meeting the real David Green: an extraordinary man, using his gifts in an unconventional manner, with perhaps a whiff of arrogance, but simple at heart.

Elizabeth P. Steyn Parvé


My recollection is of a conversation with him during lunch at the Breeze Terrace Cafeteria. I don't remember the date, but Dr. Green had recently given up smoking. After lunch, when I lighted a cigarette, he asked me for one. I was too polite to remark on his recent resolution, but he himself volunteered an explanation. "My wife says that I am difficult to get on with after I stopped smoking." I said that I too became irritable with my colleagues when I stopped smoking. He then remarked: "But, you know, Jack, I have worked with many I liked and a few I disliked; but even if I dislike a person intensely, I would go all out to support him if he is good. My likes and my work are in separate watertight compartments."

It occurred to me then, as it does now when I write these lines, that during my meetings with him over a long period I have often heard sharp criticisms from him of the scientific work of others, but never of a personal nature. The compartments have always remained watertight and separate.

V. Jagannathan

As a former postdoctoral associate of David Green, I saw a somewhat more intimate side of him. He was, at social gatherings, a most delightful, charming, and gracious host. His gatherings for his associates were always lively and interesting. In contrast, his research was always a serious business. He was always in charge and he never lacked ideas for new experiments or interpretations of completed studies. He was one of the very best of individuals in directing large groups and in appraising the status of research being carried out by each one. He was also a penetrating thinker. He was able to discern the defects in other people's work, including that of his associates, very quickly, and to devise experiments to overcome these defects.

John W. Porter

Once, as a willing newcomer, I had embarked upon research on "Q-275" aided by the catalysis of his enthusiasm, he rarely, if ever, contacted me for a report of progress, but he was always receptive to hear about what was happening when I made the overtures. One may analyze this personal characteristic of David to say that, in the best sense of modern management, he instinctively delegated responsibility with trust and confidence.

Karl Folkers

As I had been accustomed to research as commonly practiced, namely, the insistence on unbiased attitude, it was a refreshing experience to interact with Dr. Green, who reminded me of the importance of a strong drive, enthusiasm, or even bias, at the initial phase of research. Detached objectivity may come later, but you have to have a strong conviction—even if it turns out to be wrong—to get a project started and to fight discouragement. Along with this is his insistence on clarity, focus, to the extent that he said to me once, "You become articulate when you get angry," seemingly enjoying my clarity and not being annoyed by my anger.

Hirochika Komai

We had our good days and we had our bad days. While the good days defied description, the bad days never lacked recognition or identification. But they are all lumped together now and I'll always remember them and make constant reference, saying "Those were the days, my friend." Truly an unforgettable decade of my life.

Howard Tisdale

# Enzyme Institute Days

It has been said that "those who make history are often unaware of it at the time." This judgment may be particularly relevant for those of us who were privileged to spend a period of time in the laboratory of Professor David E. Green but who have seldom had an opportunity, such as afforded by the present Symposium, to reflect upon the broad contributions that his laboratory has made to enzymology and related areas of biochemistry. For nearly four decades the research contributions of this group have been recorded in hundreds of journal articles and reviews, and several generations of Dr. Green's colleagues now occupy major positions in the scientific community both here and abroad. Subsequent presentations on this program will provide a cross-section of these research interests, since each speaker's work has had its origin in some problem that he undertook earlier in David Green's laboratory.

But in this early part of the Symposium we are not looking at the science *per se* but rather at the framework in which the science was set. Dr. Eugene Knox has described David Green's activities at Columbia University, where he set up a burgeoning research unit that was to become the precursor of the Enzyme Institute at the University of Wisconsin. I would like to take up the thread of the narrative at this transition and give you a brief account of the early years of the latter Institute, which was soon to become a renowned center for research in enzymology.

In the mid-1940s, the far-sighted University of Wisconsin had decided to create an Institute devoted entirely to research in enzymology. The selection of this particular field as the focal point for the new Institute stemmed largely from the fact that several distinguished members of the Wisconsin faculty, such as Henry Lardy, Van Potter, and Philip Cohen, were among the leaders in this rapidly developing discipline. David Green was one of the obvious choices as Director of the new Institute, and he accepted the post, arriving from Columbia together with some key personnel and equipment in the early part of 1948.

My own acquaintance with David Green and with the Enzyme Institute came in a rather roundabout fashion. During the final few months when I was completing my Ph.D. work in the Department of Chemistry at the University of California, Berkeley, under the direction of Professor Melvin Calvin, I had become interested in oxidoreduction reactions in biological systems. Along with several other graduate students and Dr. Calvin, I organized a small group which met weekly to discuss this fascinating extension from classical physical-organic chemistry. We soon became aware of David Green's book, *Biological Oxidations,* and were impressed by the author's insight into the complex processes of electron and hydride ion transfer among pyridine nucleotides, flavins, and cytochromes. By chance, Melvin Calvin and David Green happened to meet on a plane trip and, as one would expect, they were soon immersed in a lively discussion of biological problems. During the course of their conversation, David Green mentioned that he was setting up a new Institute at the University of Wisconsin and asked for suggestions concerning prospective postdoctoral fellows. This information, in turn, was passed on to me; I wrote to David Green and was subsequently accepted as a member of his laboratory.

Upon arriving in Madison in the summer of 1948, I found that the Enzyme Institute was housed in a small wooden building located literally "across the tracks" from the University campus. David greeted me with his usual warmth and enthusiasm, introduced me to my new colleagues in the laboratory, and immediately gave me some reprints to read. Lacking any formal training in biochemistry, I anticipated settling down to several weeks of preliminary reading in order to be able to discuss the current research problems. Within the hour, however, David had outlined a project to start my on-the-job training, and he was soon teaching me the art of manometry. To a chemist, the visual evidence of oxygen uptake by a tissue homogenate was truly impressive.

We had a small research group in the old Enzyme Institute building. In addition to Jack Still from Australia, who had been one of David's colleagues at Columbia, the other postdoctorals included Jo Nordmann from France, Lester Teply, and Rao Sanadi. Our capable research assistants were Irene Rechnitz and Muriel Feigelson. It is of interest that Muriel subsequently obtained her Ph.D. degree at Columbia and now supervises her own research group. The cyclophorase concept was just being developed at this time, and we all worked on various aspects of this problem. David Green had reasoned that enzymes which comprise linear or cyclic sequences (e.g., the citric acid cycle) probably do not occur as discrete, soluble entities but instead are fixed in a matrix in order to function with maximum efficiency. It was beginning to be recognized at the same time that the mitochondrion was the subcellular site of the citric acid cycle and oxidative phosphorylation, but the status of individual enzymes within the mitochondrion was not at all well understood. In particular, I was looking at the problem of bound pyridine nucleotides, and I found that the properties of pyridinoproteins (e.g., malate dehydrogenase) in the mitochondrion were quite different from those of their isolated counterparts. These observations, and their implication, are now well accepted; however, at the time there was some reluctance to accept our conclusions. David persevered, and the data were eventually published. I recall his strong conviction that theories are transient, and that at each point in time an interpretation needs only to be consistent with the available data and with the general concepts of chemistry. It did not matter, he felt, if conclusions had to be modified as new and additional data were obtained.

In 1950, our group, which had outgrown the original structure, moved into a spacious new concrete and brick structure at 1710 University Avenue. This was the core building of the "new" Enzyme Institute onto which wings have been added in subsequent years. A photograph of Institute personnel taken just after the new building had been occupied shows the youthful enthusiasm of the postdoctoral contingent—Henry Mahler, Rao Sanadi, Jesse Rabinowitz, Harold Edelhoch, Osamu Hayaishi, Richard Schweet, Venkataraman Jagannathan, Lester Teply, Arthur Tomisek, Richard Potter, Frixos Charalampous, Gerson Jacobs, Joseph Betheil, Ephraim Kaplan, Abner Schepartz, and Bernard Katchman. Visiting professors included Vernon Cheldelin, Mary Buell, and John Harman. Several of these colleagues have since died, and we remember them with special affection.

The scope and extent of the research work expanded markedly in the new Institute—first into fatty acid oxidation, then into fatty acid synthesis, and on to the mitochondrial electron transport system and oxidative phosphorylation. In turn, these efforts paved the way for the present, multidisciplinary attack on membrane structure and function. Time does not permit me to review the many discoveries that resulted from these intense research efforts, or to identify the many postdoctorals and visiting professors from various institutions in the United States and abroad who received training at the Enzyme Institute. Few centers have more numerous or more loyal alumni.

I could entertain you with numerous anecdotes that are illustrative of those early days at the Enzyme Institute. There was, for example, a visit by the legendary Otto Warburg. While being shown through the laboratory, he noticed the Lardy modification, i.e., the circular form, of the Warburg manometric apparatus. The maestro carefully inspected one of the instruments, and, when asked by David what he thought of it, replied, "It isn't necessary."

On another occasion, when we were still in the old building, David had purchased a large refrigerator-freezer for the second floor laboratory. When delivered, the ponderous box could not even pass through the front door, let alone negotiate the stairs. Undaunted, David had an entire wall of the building removed and the box swayed up into position with a crane. It happened, though, that a University official came to see David on some other business just as this engineering feat was taking place. I am sorry that no picture was taken to record the look of consternation on the man's face as he witnessed this apparent destruction of University property. I may add that this episode is entirely in character for David Green. He has always dealt briskly with administrators and logistical problems.

And, finally, there was the Great Avian Confrontation. Appalled to find one day that there were no pigeons available for a breast muscle preparation, David overcompensated somewhat and ordered several truckloads of pigeons, which then had to be housed in a hastily constructed coop located behind the Institute. Within minutes, an apoplectic gentleman rushed into David's office, identified himself as the caretaker of the University's poultry farm immediately adjacent to the Institute, and then sarcastically asked David if he had any bright ideas for preventing his prize birds from being infected by the pigeons, which are, in fact, notorious disease-carriers. Cutting across this tirade, David looked the fellow squarely in the eye and said, "I suggest that you move your chickens!"

I would like, if I may, to close on a personal note. David Green needs no testimonials and I would not wish to embarrass him by providing one. However, it is in keeping with the intellectual tradition to honor our teachers, and I have always considered it a great privilege to have received my postdoctoral training under David Green's tutelage. More than just becoming acquainted with the mechanics of research, I learned from him the art of using today's results to formulate tomorrow's approach to a problem and, most important, a philosophy of scientific investigation. For these and many other reasons, I have always had great respect for David as a scientist and affection for him as a person.

F. M. Huennekens

*La Jolla, California*

# Perspective for the Mitochondrion

I intend to provide you with a perspective of the mitochondrion that reaches back to the earliest beginnings and extends to the here and now. But before I become immersed in this historical and overview sweep, I think it appropriate to say that this symposium is an occasion to do honor to the group of investigators who collectively laid the foundations of mitochondriology. Many of this group are here today at this Symposium. I am proud that many of this group have been my colleagues at the Enzyme Institute.

As I see it, the history of mitochondriology can be divided into four periods: the period from 1945 to 1950, during which the mitochondrion was first recognized as an organelle; the period from 1950 to 1970, when the mitochondrial system was systematically defined and explored; the period from 1965 to 1975, which we may describe as the period of conceptual crisis; and, finally, if you allow me to extrapolate, the period from 1976 to 1980, when the principle of energy coupling was recognized and classical mitochondriology was brought to a definitive conclusion.

## 1945–1950

In the 1940s we were aware of an electron transfer chain from the monumental studies of David Keilin, of the citric acid cycle from the work of Sir Hans Krebs, of the phenomenon of oxidative phosphorylation from the studies of Herman Kalckar and E. T. Tsybakova, and, finally, of the fatty acid oxidation system from the work of Luis Leloir. No one suspected or found evidence for the intimate interrelationships of these four major metabolic systems. When we studied at Columbia University School of Medicine the pyruvate oxidation system in homogenates of heart muscle and kidney, we were amazed to find that

17

this oxidation system was inextricably linked to readily sedimentable particles and that the same particles had the capacity not only for pyruvate oxidation but also for electron transfer, citric cycle oxidations, fatty acid oxidations, and oxidative phosphorylation. After trying unsuccessfully in many trials to separate these different functionalities from the particles, we finally realized that we had stumbled on a momentous discovery. All these functionalities were intrinsic to one integrated system—a system we elected to call the cyclophorase system. Shortly afterward this particular system was identified as the mitochondrion by A. L. Lehninger and E. Kennedy, then at the University of Chicago, and by V. R. Potter and W. Schneider of the University of Wisconsin.

What amazed and intrigued us as we studied the cyclophorase system was its completeness. It had all the enzymes and coenzymes required to oxidize citric cycle intermediates, fatty acids, and a group of amino acids to $CO_2$ and water, and all the components requisite for linking these oxidations to the synthesis of ATP.

It was soon obvious that the mitochondrion contained the metabolic systems that provided fuel for the ATP-generating dynamo. The oxidation of fatty acids, amino acids, and citric cycle intermediates generated the ultimate fuel—namely, NADH; and it was this common fuel that linked all three oxidizing systems. Moreover, the citric cycle was the common ultimate pathway by which fatty acids and amino acids were oxidized to $CO_2$ and water.

What we did not perceive in these early years was the true meaning of a particulate organelle. It was only later that we recognized that the organelle represented the marriage of enzymology to the membrane—a marriage prerequisite for the execution of energy coupling.

A few of us—A. L. Lehninger, B. Chance, E. C. Slater, and our group—were quick to perceive that the mitochondrion was the powerhouse of the cell and as such deserved the most intensive and close study. This perception and the events that soon followed marked the beginning of classical mitochondriology.

John Taggart, Eugene Knox, Jack Still, Frank Huennekens, Allan Grafflin, Mary Buell, and Victor Auerbach were the pathfinders in our group that explored the cyclophorase system and thus laid the foundations for the second phase of mitochondriology.

## 1950–1970

We had just moved into our new Institute building at 1710 University Avenue in Madison from the temporary quarters in the frame houses astride the railroad track just opposite the Badger Tavern. Through the generosity of the Rockefeller Foundation the Institute had unique facilities for large-scale processing of mitochondria. It was also a boon to have the Oscar Mayer Packing Company around the corner, so to speak. We were, in short, all set to explore the mitochondrion from one end to the other.

*Localization of Coupling Function in the Inner Membrane.* The mitochondrion is a two-membrane organelle. After an extended search in several laboratories, the inner membrane was found to be the locale of the electron transfer chain, and of the mechanism for coupled ATP synthesis and hydrolysis. Moreover, in the space enclosed by the inner membrane were contained the citric cycle enzymes as well as the enzymes for oxidation of fatty acids and amino acids (Lars Ernster).

*Unit of Mitochondrial Function.* One of the most fascinating problems that early absorbed our attention was that of the unit of mitochondrial function. We found that the capacity for electron flow or coupled ATP synthesis did not require the intact mitochondrion. The mitochondrion could be fragmented by several means, notably sonic irradiation, into relatively tiny membranous vesicles that fully retained these two functionalities. The unit of function thus turned out to be the inner membrane and not the whole mitochondrion. Particles such as ETP, $ETP_H$, PETP, and EP were the submitochondrial particles that a whole generation of mitochondriologists used to advantage in their studies. Dan Ziegler, Anthony Linnane, Youssef Hatefi, Archie Smith, Robert Lester, Fred Crane, and Marc Hansen were the pathfinders in the tactics of fragmenting mitochondria to functional submitochondrial particles.

*Unit of Electron Transfer Function.* The problem of determining the unit of electron transfer activity turned out to be the most baffling and formidable of all the problems that had to be tackled in the early years of mitochondriology. Finding the solution required the development of a new concept—the concept of a transmembrane complex. The electron transfer chain was found to be a composite of four such transmembrane complexes, each capable in its own right of generating a membranous vesicle. Once this organizational pattern was recognized, the one-by-one isolation and characterization of the four complexes could be undertaken and completed. In this search for the unit of electron transfer activity, Youssef Hatefi was the acknowledged leader and innovator; Dan Ziegler carried out parallel studies on complex II and David Griffiths and David Wharton parallel studies on complex IV.

The unit of electron transfer activity—the transmembrane complex—was shown by Alex Tzagoloff, David MacLennan, and David McConnell to be also the unit of membrane construction. The importance of this discovery was enormous, for it established not only that protein was as important as lipid in membrane construction but also that electron flow was vectorial—one of the principal tenets of the chemiosmotic model of P. Mitchell.

*The Tripartite Unit.* The electron microscope was the tool that revealed for the first time the tripartite unit (TRU) of the inner membrane. H. Fernandez-Moran of the University of Chicago and T. Oda of our laboratory teamed up in this discovery. E. Racker, then at the Public Health Research Institute, identified the headpiece of the TRU as the ATPase; David MacLennan and Junpei Assai identified the stalk as the determinant of oligomycin sensitivity; and Alex Tzagoloff and Krystyna Kopaczyk identified the basepiece of the TRU as the membrane-forming element.

Shortly after the TRU was discovered, it was extracted from the inner membrane as a membranous vesicle and separated from the electron transfer chain. In this way it was established by David MacLennan and Alex Tzagoloff that the TRU was the unit of ATP hydrolysis and synthesis and also a unit of membrane construction.

*Phospholipids of the Mitochondrial Membrane.* In all these studies directed at the unit of electron transfer and the unit of ATP hydrolysis or synthesis, the membrane was flitting in and out like Hamlet's ghost. We early recognized the imperative of taking the membrane seriously. It was Sidney Fleischer who took the lead in this endeavor. He developed the methodology for extracting and identifying the characteristic phospholipids of the mitochondrion—phosphatidylcholine, phosphatidylethanolamine, and cardiolipin. Together with Gerald Brierley and David Slautterback, he established that phospholipids were essential for electron transfer and that the ratio of phospholipid to protein in the inner membrane was constant—a token of the precise fitting of lipid to protein in the complexes of the electron transfer chain and in the TRU.

*Intrinsic Proteins of the Inner Membrane.* An important consequence of the capability of electron transfer complexes and of the TRU to form membranous vesicles was the recognition that proteins like phospholipids are the building blocks of the membrane continuum. The theoretical and experimental studies of Rod Capaldi and Gary Vanderkooi established the concept of intrinsic proteins—the specialized hydrophobic proteins in the electron transfer complexes and in the TRU that underlie the marriage of protein and phospholipid in the membrane continuum. These proteins are readily distinguishable from the water-soluble extrinsic proteins that are external to the membrane continuum. It is the intrinsic proteins that are the vehicles of energy coupling.

*Structure of the Electron Transfer Complexes.* An incisive study of complex III by Harold Baum, John Rieske, Israel Silman, and Sam Lipton was the first of such studies that defined the characteristics of an electron transfer complex. Complex III contained one molecule each of cytochrome $b$ (a dimer), cytochrome $c_1$, and an iron protein. These three oxidation-reduction proteins were embedded in a matrix of phospholipid and so tightly associated that dissolution of the complex and separation of these proteins required high levels of bile acids such as taurodeoxycholate or deoxycholate. When the oxidation-reduction proteins were in the oxidized state, dissolution of the complex by these bile acids took place rapidly, but when the proteins were in the reduced state, the same detergents at the same concentrations were ineffectual in this dissolution.

Antimycin was found to act in the same fashion as did reduction; i.e., it prevented the dissolution of complex III in the oxidized state by bile acids. Thus the antimycin-binding site was found to play a controlling role in the fragmentation of complex III.

The discovery by Israel Silman that some 50% of the protein of complex III was accounted for by a colorless protein (core protein) came in the nature of a bombshell, for there was no rational explanation in terms of theory for the presence of such a protein in the complex. Some 10 years later, complex III was identified as a duplex of an electron transfer complex and an ionophoroprotein of which the core protein is the major component.

*Structural Proteins of the Mitochondrion.* In the life of a laboratory there is always one problem that sears the souls of all the investigators who become involved. The structural proteins of the mitochondrion represented one such problem. If this were merely an instance of studying one's own brain, it could be conveniently forgotten. But there was something much more important at stake in this problem, and for this reason it deserves to be told.

Richard Criddle, then a graduate student working with Robert Bock of the Department of Biochemistry, became intrigued with the nature of the colorless proteins that were found to be intimately associated with intrinsic hemoproteins, such as cytochromes $b$ and $c_1$. These colorless proteins turned up whenever one tried to fractionate complexes of the electron transfer or hemoproteins derived from these complexes. Criddle, together with Howard Tisdale, developed methods for the large-scale isolation of these proteins and with the use of sodium dodecylsulfate succeeded in isolating a water-soluble species of low molecular weight that could combine in 1:1 molar proportions with cytochromes $b$, $c_1$, and $a$, and could bind phospholipid. The protein free of detergent and phospholipid was utterly insoluble in water and later was found to be the most hydrophobic protein in the mitochondrion as judged by the ratio of nonpolar to polar amino acid residues.

The particular protein isolated by Criddle was perceived to be a representative of a class of proteins designated as "structural proteins" because at the time no functional role could be assigned. Later H. Hultin, S. Fleischer, and S. H. Richardson introduced a milder method for preparing the structural protein fraction which greatly facilitated its systematic study. This fraction was found to be capable of binding a wide variety of molecules—phosphate, ATP, citrate, etc. Moreover, it bound phospholipid to the degree characteristic of other inner membrane proteins.

Then the crisis arose. Were we dealing with proteins denatured by the relatively severe procedures required to isolate intrinsic proteins, or did the structural protein fraction correspond to a class of proteins that served some important role in the mitochondrion? The protein specialists, Allan Senior, David MacLennan, Giorgio Lenaz, and Norman Haard, found unmistakable evidence of gross inhomogeneity in the structural protein fraction and on the basis of this inhomogeneity were inclined to question the reality of the structural protein family of proteins. On the other hand, Krystyna Kopaczyk found clear evidence that all the complexes of the electron transfer chain were intimately associated with colorless proteins of the genre of the structural proteins and these proteins could be separated from the catalytic oxidation-reduction proteins of these complexes. Moreover, the catalytic proteins could be recombined with the structural proteins to reconstitute the original complexes.

I think we can finally specify the role of structural protein. Complex IV has been shown to be a duplex of an electron transfer and an ion transfer complex. Common to both of these resolvable complexes is a set of three subunits that are highly hydrophobic, that combine with phospholipid, and that are synthesized by mitochondrial DNA. The properties of these subunits correspond to those of Criddle's structural protein. Subunits with similar properties are present in other mitochondrial coupling complexes.

*Role of Metals in Electron Transfer.* Several years prior to the study of mitochondrial electron transfer, we had been aware of the intimate association of flavoproteins with metallic components. Xanthine oxidase had been shown to be a molybdeno-iron flavoprotein enzyme and aldehyde oxidase a molybdeno-flavoprotein enzyme. It was this association established by Helmut Beinert, Bruce Mackler, and Henry Mahler that induced us to examine the electron transfer system for further evidence of metal involvement. Dan Ziegler and Ken Doeg discovered the presence of substantial amounts of nonheme iron in the electron transfer chain. Later Youssef Hatefi demonstrated the presence of nonheme iron in the complexes of the chain except for complex IV. David Griffith and Helmut Beinert established copper as the nonheme metal in complex IV. Thus at a relatively early stage we were aware that the electron transfer chain was a symphony of metal atoms either associated with heme groups or not associated with such groups. The meaning of this involvement of metal atoms in the electron transfer process was not clear at the time. But now we know that metals are the molecular instruments of charge separation and charge elimination as well as the mediators of vibronic coupling (electron tunneling)—the most efficient form of electron transfer according to J. J. Hopfield of the Bell Laboratories and D. S. Chernavskii of Moscow State University.

*Mitochondrial Control Mechanism.* The control mechanism was discovered more by accident than design. Mitochondria were found to fluctuate between three configurational states—the aggregated state in which oxidative phosphorylation took place, the orthodox state in which uncoupled respiration took place, and the twisted state in which active transport of anions took place (phosphate, acetate). These profound configurational states were interpreted in terms of the energy coupling mechanisms. Some 10 years later Douglas Hunter and Robert Haworth showed that the configurational changes were expressions of a control mechanism that regulated the pattern of coupling options. In my chapter on the molecular mechanism of energy coupling that comes later in this volume, I shall consider in greater detail the nature and properties of the control mechanism. But one must salute the pioneers who first discovered and described this mechanism: David Allman, Robert Harris, John Penniston, Junpei Assai, and Takashi Wakabayshi.

*Isolation of Mitochondrial Coenzymes and Hemoproteins.* Our large-scale facilities made it possible to undertake the one-by-one isolation of the important coenzymes and oxidation-reduction proteins of the electron transfer chain. We were able to start with kilograms of purified mitochondria and thus could avoid the pitfall of ambiguity with respect to the source of the isolated species.

Fred Crane discovered a yellow reducible component in heptane extracts of mitochondria and in collaboration with Y. Hatefi and R. Lester undertook its large-scale isolation. The reducible component was soon identified as a quinone, and its structure as $Q_{10}$ (ubiquinone-10) was established by Karl Folkers and his group at the Merck Laboratories. Coenzyme $Q_{10}$ was found by Y. Hatefi to be the reductant for complex III and the oxidant for complexes I and II. It is one of the ironies of mitochondriology that 10 years had to elapse before the participation of $Q_{10}$ in electron transfer was generally accepted.

Robert Goldberger and Ralph Bomstein broke new ground in isolating for the first time cytochromes $b$ and $c_1$—the two hemoproteins of complex III. The technical problems of converting these two hydrophobic intrinsic proteins into water-soluble species can only be described as horrendous. Yet despite these formidable difficulties, both hemoproteins were brought to a stage of homogeneity without modification of the spectrum of the native protein.

John Rieske has the distinction of recognizing and isolating the first nonheme iron protein from the electron transfer chain. This protein was isolated from complex III after exposure to mercurials such as mersalyl. By virtue of this isolation it was possible for Helmut Beinert and Ray Hansen to study in some depth the electron spin resonance properties of iron in an iron sulfur protein, and these studies opened the door to the whole spectrum of iron sulfur proteins found not only in the mitochondrion but also in other organelles.

When we came to grips with the problem of the molecular mechanism of fatty acid oxidation, we soon recognized that the key to its solution was the availability of coenzyme A in massive amounts. The then available methods for its preparation produced milligram quantities whereas gram quantities were needed. We introduced the technique of precipitating coenzyme A as the copper salt of the mixed disulfide of coenzyme A and glutathione (glutathione was added to the yeast extract) and by this stratagem were able to obtain relatively pure coenzyme A in gram quantities. Helmut Beinert and Richard Von Korff in collaboration with Frank Strong of the Biochemistry Department of the University of Wisconsin were the key individuals in this isolation. It was this large-scale isolation of coenzyme A that paved the way to the solution of the mechanism of fatty acid oxidation and synthesis.

*Molecular Mechanism of Mitochondrial Fatty Acid Oxidation.* When Richard Schweet established the participation of coenzyme A in the oxidation of pyruvate catalyzed by a soluble pyruvate dehydrogenase complex and D. Rao Sanadi isolated succinyl CoA as the product of oxidation of $\alpha$-ketoglutarate catalyzed by a soluble $\alpha$-ketoglutarate dehydrogenase complex, we deduced that a similar role would be found for coenzyme A in fatty acid oxidation. By virtue of the ready availability of coenzyme A, we could form the CoA esters of fatty acids and fatty acid derivatives using the ATP-dependent fatty acid activating enzyme. Given ample amounts of these esters, it was relatively simple and straightforward to isolate the other four enzymes of the fatty acid oxidation cycle and thus establish the molecular sequence by which fatty acids were oxidized to a $\beta$-ketoacid and then decarboxylated to a fatty acid with two carbon atoms less than the parent fatty acid. The entire enterprise was over in about a year's time. Helmut Beinert, Salik Wakil, Henry Mahler, Dexter Goldman, Robert Bock, and Sanae Mii were the principals in this dramatic resolution of the mechanism of fatty acid oxidation.

*Molecular Mechanism of Fatty Acid Synthesis.* There was an interesting sequel to the story of fatty acid oxidation. We were interested to know whether fatty acid synthesis followed the same path as the oxidation of sequence, only in reverse. Feodor Lynen of Munich felt that it did. We were unable to show synthesis in our system for oxidation of fatty acids and decided to go at the problem systematically—looking for a system in liver that could mediate this synthesis. We finally found this sytem in extracts of pigeon liver following the lead of Sam Gurin of the University of Pennsylvania and showed that this nonmitochondrial system was entirely different from the fatty acid oxidizing system. It was a biotin-containing particulate system, required $CO_2$ and NADPH, and generated malonyl CoA. The rest of the story is by now well known. Salih Wakil, David Gibson, John Porter, and Alisa Tietz were the heroes of this particular sweep.

## 1965–1975

The decade 1965–1975 encompasses the third period in the development of mitochondriology—a period we may describe as the period of crisis and bewilderment. It was generally assumed, and we were among those who held this mistaken view, that once the basic systems of the mitochondrion were defined, all the remaining uncertainties would be resolved; the many pieces would nicely fall into place. Of course, that did not happen. The more we learned about the mitochondrion, the more confused the investigators became. Data were pouring in from all sides on the structure of the electron transfer chain and the TRU, on the spectroscopic behavior of the oxidation–reduction components of the electron transfer chain, on the multiple options for energy coupling, and on the myriad ways in which mitochondrial coupling could be modulated. A variety of specific reagents such as uncouplers, ionophores, oligomycin, aurovertin, and dicyclohexylcarbodiimide came into general use. The mind reeled with the avalanche of new information, but the means of encompassing this information within a simple consistent framework was conspicuously lacking.

What was wrong? It is now easy to see what was wrong. The mitochondrion is a device or machine designed for the execution of energy coupling. To understand this machine, we had to know and understand the principle of energy coupling. Without a knowledge of that principle, it became impossible to impose order on the accumulating information dealing with coupled mitochondrial processes.

The high-energy intermediate hypothesis—the hypothesis that has dominated thinking about oxidative phosphorylation for the past two decades—probably was one of the main contributing causes of the stagnation that beset mitochondriology in its third period. This hypothesis is a negation of the necessity for energy coupling. It simply bypasses the problem of energy coupling. Curiously enough, there has never been any evidence for this hypothesis despite many attempts to find the postulated intermediates. But the lack of evidence proved no obstacle to the general acceptance of this hypothesis. Little wonder that with the millstone of this hypothesis around the necks of mitochondriologists progress in understanding how the mitochondrion works all but ground to a halt during the past decade. The first major break in the impasse came with the introduction by Peter Mitchell of the chemiosmotic hypothesis. This hypothesis incorporated three powerful new ideas—the vectorial nature of electron flow, charge separation as a prerequisite for energy coupling, and the formation of transmembrane gradients as a necessary consequence of charge separation. These ideas introduced in the early 1960s were strongly resisted by other workers, but in the early 1970s they became increasingly more popular and dominant. It was the Mitchell model that finally broke the back of the high-energy intermediate hypothesis and thus opened the door to realistic thinking about energy coupling.

The Mitchell model accounted elegantly for proton gradients and for the simple stoichiometric relations of the moving ions in energy coupling. But that was as far as it could go. It provided no insights into the mechanism of uncoupling; it provided no rationalization for an elaborate electron transfer chain; it shed no light on the structure of the transducing assembly. More seriously, the Mitchell model was found to be incompatible with several crucial experimental observations. Henry Tedeschi failed to find any experimental evidence for the transmembrane potentials that the model requires as a precondition for energy coupling. As we shall present evidence for in another article in this volume, energy coupling can be shown to be maximal when the protonic gradient is virtually zero and uncouplers can be shown to induce coupled cyclical cation transport. If gradients are the driving force in energy coupling, as the Mitchell model assumes, then the maximization of coupling by uncouplers under conditions in which the gradient is close to zero pretty effectively invalidates the assumption of a protonic gradient as the driving force for energy coupling. Finally, there is the phenomenon of energized contraction of swollen mitochondria in which both $K^+$ and $Cl^-$ are extruded from the mitochondrion. This process has been shown by Douglas Hunter in our laboratory to be accelerated by uncoupler. The assumptions of the Mitchell model are totally at variance with the experimental observations on energized contraction. How can a coupled process be accelerated by uncoupler if the uncoupler erodes all gradients? How can the direction of ion flow be reversed if the direction of the membrane potential is invariant?

These and others not mentioned were among the considerations that led us to the conclusion that the Mitchell model was only the steppingstone to the correct model. The hallmark of the correct model would have to be the complete rationalization of all energy coupling phenomena and the opening of the doors to a grand synthesis of all we know about the mitochondrion. As we have pointed out above, the Mitchell model has been unsuccessful in achieving either the rationalization or the synthesis.

If there is a lesson to be learned from the trauma of the past decade, it is that experiment without theory is just as sterile as theory without experiment. The greatest advances in biological science have been made when experiment and theory are coequals.

## 1976–1980

The key question in energy coupling is whether charge separation is unpaired as Mitchell postulates, leading to a transmembrane potential, or paired as we postulated, in which case no transmembrane potential is generated. The principle of paired charge separation requires the participation of ionophores in all coupled processes. Evidence is accumulating that the mitochondrion is a symphony of ionophores and ionophoric systems. This is only one of the more dramatic pieces of evidence that the mitochondrion is constructed to implement energy coupling via the principle of paired charge separation. In the chapter on the molecular mechanism of energy coupling appearing later in this volume, we have developed this principle of energy coupling and the model based on this principle. Evidence has been presented to show that the model rationalizes the mechanism of uncoupling, the mechanism of the control of coupling function, the emergence of respiratory control in cytochrome oxidase, the structure of the transducing assembly, and the uncoupled state of mitochondria in the orthodox configuration. No other model has hitherto been successful in rationalizing any of these phenomena. Moreover, evidence has been presented for the presence of multiple ionophores in the mitochondrion and for the presence of ionophoric systems closely associated with the electron transfer complexes and the TRU. On the basis of this accumulated evidence, we now feel confident that the principle of energy coupling has finally been recognized and that with the insight provided by this principle the task of resolving the still outstanding questions of mitochondrial energy coupling should be completed certainly before 1980.

## Epilogue

As I see it, we are now close to the stage when classical mitochondriology will disappear and merge into bioenergetics. The principle of energy coupling is universal. It applies to enzyme catalysis, nerve action, muscle contraction, active transport of $Ca^{2+}$ by the sarcoplasmic reticulum, and regulation of membrane function by hormones. It applies in fact to all biological activities that are energy dependent. The role of bioenergetics in the study of physiological processes and disease will become, I am certain, the cutting edge of biomedical research. To my former colleagues and friends, I say prepare yourself for this renaissance of bioenergetics and for the fascinating research opportunities that inevitably will unfold.

No account of the development of mitochondriology in this country would be complete without mention of the central role of the National Institutes of Health (first the National Heart Institute and later the National Institute of General Medical Sciences) in providing a high degree of support for this enterprise during the past 25 years.

David E. Green

*Madison, Wisconsin*

*Chapter 1*

# Organization of Protein and Lipid Components in Membranes

## Garret Vanderkooi

### Introduction

The essential structural characteristics of biological membranes are by now well known. Membranes consist largely of amphipathic molecules (lipids and proteins) which spontaneously aggregate in such a way as to expose their polar parts to water while at the same time creating a nonaqueous interior for the nonpolar parts.

The rapid progress being made in the field of membranology has been possible only because of the simple but believable molecular models which were proposed at the beginning of this decade.[1-6] These models differed in their details but were sufficiently similar in their essential features to show that a consensus of opinion on the fundamentals of membrane structure was finally being arrived at. This was in marked contrast to the situation during the early to middle 1960s, when widely divergent models were being proposed and defended.

The fact that a unifying model for membrane structure is now available does not mean that studies on the principles of membrane structure are finished. Really, it is only the beginning; it may be looked upon as the arrival at the first milestone on a long and arduous pathway. There are two interdependent routes which may be followed in seeking to further elucidate membrane structure. One of these is the experimental route, the objective of which is to determine, in as great detail as

*Garret Vanderkooi* • Department of Chemistry, Northern Illinois University, DeKalb, Illinois 60115.

possible, the molecular structures of particular membranes. The other route is to continue to consider the general characteristics of membranes; the rationale behind this approach is that if the physical laws which govern the membranous state of matter can be deduced, the properties of individual membranes will then be understandable as particular manifestations of these general principles. To put this in more operational terms, if the experimentalist knows the range of phenomena *theoretically* possible, he will be less likely to overlook important features of his system which he might otherwise not have noticed.

Our laboratory has followed the second of these two routes. It is our objective to understand membrane structures in terms of the thermodynamic and statistical thermodynamic requirements which give rise to these structures. The expected fruit of this undertaking is not an esoteric mathematical description of a membrane, although that may be a necessary way station, but rather a pratical understanding which can be used to interpret such seemingly diverse characteristics as protein aggregation in membranes, the temperature dependence of membrane-bound enzyme kinetics, and the effect of foreign agents, e.g., drugs and anesthetics, on the physiologically important properties of membranes.

An evidently important but as yet inadequately understood area of membrane structure concerns the relationship between the mode of distribution of the proteins in the plane of the membrane and the functional properties of the membrane. It is well known that in some membranes the proteins are seemingly distributed at random, or at least have a high degree of motional freedom, e.g., retinal rod outer segment (ROS) disk membranes, whereas in other membranes the motional freedom is much less or even, for practical purposes, nonexistent, e.g., *Halobacterium halobium* purple membrane.[7] The mere observation that the *H. halobium* membrane is crystalline whereas the ROS disk membrane is fluid cannot be used as evidence that in the one case fluidity is *required* for proper functional activity, but not in the other case, and *vice versa*, but one is certainly led to suspect that a relationship does exist between the motional freedom and function. (There are of course other cases where a direct relationship between motional freedom and enzymic function has been inferred, with the cytochrome $b_5$–cytochrome $b_5$ reductase system being a case in point.[8])

The modulation of the motional freedom or state of aggregation and dispersal of proteins within a given membrane may prove to be an important mechanism for metabolic control. Two possibilities along these lines immediately come to mind. First, when it is necessary for two proteins to come into contact with each other by lateral diffusion in the plane of the membrane in order for a given enzyme reaction to occur, then the membrane viscosity will directly influence the kinetics of the reaction. If this lateral diffusion is rate limiting, any physical or chemical changes in conditions which tend to decrease the membrane viscosity should enhance the reaction rate. Second, the membrane proteins may be involved in a dynamic equilibrium between an enzymically active associated state and an inactive dissociated state (or *vice versa*). Systems of this type are very common among the allosteric regulatory enzymes found in the soluble phase of cells; it seems eminently reasonable to suppose that similar control mechanisms may be prevalent in membranes, and, indeed, a theory along these lines has recently been proposed to explain the interaction between adenylate cyclase and the hormone receptor protein.[9] Allosteric control of the enzyme activity could then be brought about by membrane structural changes affecting the association-dissociation reaction.

One of our major objectives, within the general framework of elucidating membrane thermodynamics, has been to identify the determinants of the mode of distribution of proteins in membranes. Once these determinants are understood, we will be in a position to intelligently modify the actions of membrane-bound enzymes whose activities are found to be dependent on their state of association or dissociation. Much of the remainder of the present chapter will be devoted to this subject. I will consider the several types of generalized interactions between proteins, as well as between lipids and proteins, and will relate each of these to the expected effects on the equilibrium between the associated and dissociated states of membrane proteins.

## Protein Distribution in Membranes

Lateral mobility is by now a well-established characteristic for many membranes. The ability of the proteins in a membrane to move by no means implies that their distribution in the plane of the membrane will be random, however. Freeze-fracture electron microscopy has demonstrated that change of pH or temperature can cause changes in the state of distribution of the intrinsic membrane proteins in several membranes, including erythrocytes,[10] recombined rhodopsin membranes,[11] and mitochondria,[12] among others.[13,14] In these cases, nonphysiological conditions were employed to induce a gross change from a state in which the proteins are fairly randomly scattered throughout the available membrane area to one in which they have become aggregated into clusters, leaving large areas of protein-free lipid bilayer. Gross changes of this type may have little or no physiological relevance, considering the abnormal conditions employed to induce the changes.

However, the same kinds of forces which control the observable aggregation or dispersal of the proteins on a gross level may also be expected to influence the equilibrium between the proteins in specific associating systems within the membrane. As I have already mentioned, such associating systems may well play an important role in the control of cellular metabolism.

In the following sections, I will outline the several thermodynamic factors which may be expected to influence the state of distribution of proteins in a membrane. In the course of doing so, several ways will be pointed out by which the gross aggregated-dispersed equilibrium and the hypothesized specific association-dissociation equilibria may be shifted.

The three major thermodynamic categories to be considered in regard to protein distribution are as follows:

a. Entropy of mixing.
b. Interactions among proteins.
c. Lipid–protein interactions.


### Entropy of Mixing

It is intuitively obvious that the ideal entropy of mixing should favor a random distribution of the proteins throughout the available membrane area. It is not at all obvious, however, how strong this randomizing tendency is relative to the other forces which may be present.

The ideal entropy of mixing for two liquids in which the molecules have approximately the same size and are noninteracting can be easily calculated using the following formula:

$$\Delta S/R = -x \ln x - (1 - x) \ln (1 - x) \qquad (1)$$

This formula can be found in any book on the thermodynamics of solutions. $\Delta S$ is the ideal entropy of mixing per mole; $x$ and $1 - x$ are the mole fractions of the substances being mixed, and $R$ is Boltzmann's constant. This same equation could also be used to calculate the ideal entropy of mixing of lipids in the plane of a bilayer, since the surface areas of lipids are all of the same order of magnitude. It cannot be used to compute the $\Delta S$ for the mixing of proteins and lipids in the plane of a membrane, however, because of the considerable difference in surface area between proteins and lipids. The assumptions employed in deriving equation 1 break down down if the molecules involved differ greatly in size. We have therefore derived another expression for the mixing of lipids and proteins in the plane of a membrane, taking account of the size differences, but still retaining the (unrealistic) assumption of noninteraction, by which is meant that the enthalpy of mixing is zero. The derivation of this equation is written out elsewhere[15]:

$$\frac{\Delta S}{N_s k} = \frac{-1}{r} [\phi \ln \phi + (1 - \phi) \ln (1 - \phi)] \qquad (2)$$

where $\phi$ is now the fraction of total surface area occupied by protein and $1 - \phi$ is the remaining area which is filled with lipid, $N_s$ denotes the total surface area expressed in units of area per lipid molecule, $k$ is the molecular Boltzmann's constant, and $r$ is the ratio of protein to lipid surface area per molecule. Equation 2 readily reduces to equation 1 for molecules of the same size, since in that case $r = 1$ and $\phi = x$, and if $N_s$ is equated to Avogadro's number, then $N_s k = R$.

The entropy of mixing of lipids and proteins in a membrane is proportional to the area fraction of protein, and not to the mole fraction as in conventional ideal solution theory. $\Delta S$ also varies inversely with the size of the proteins. This means that if the area per protein is increased by any means the value of $\Delta S$ per unit area will decrease, for the same value of $\phi$. In this context, the term "protein" must be understood to mean the smallest freely diffusing unit, which may in fact be a complex of several polypeptides. Thus cross-linking of intrinsic membrane proteins with bifunctional reagents, or by antibodies or lectins, will cause a decrease in the entropy of mixing, which in turn will favor a generalized, nonspecific aggregation of the proteins. These qualitative conclusions are expected to be true not only in the idealized case for which the equation was derived but also for real membranes.

## Interactions among Proteins

Since one-third or more of the surface area of membranes may consist of intrinsic proteins,[7] it is inevitable that there will be interactions between these proteins, since in effect they form a highly concentrated two-dimensional solution. The interactions may be either specific or nonspecific. The specific interactions will give rise to the formation of multimolecular enzyme complexes, whereas the nonspecific interactions must necessarily be of a more generalized nature, and will affect all of the proteins present.

The same kinds of forces which hold the subunits of water-soluble multiprotein complexes together may also be expected to be operative within membrane-bound protein complexes, except that the relative importance or strengths of the hydrophilic/hydrophobic forces may be altered on account of the partially lipoidal environment in which the membrane proteins are found. The complexes may be held together very firmly, as for example the mitochondrial electron transport complexes, or else they may be more loosely associated. The latter class would include the reversible associations which may occur among the several electron transport complexes. The nonspecific, generalized intermolecular forces may be expected to have an effect on weakly associating systems of the latter type but not on the strongly associating systems of the former type. The generalized, nonspecific interactions are the major subject of interest in the following sections. These may be modulated by environmental factors such as temperature and pH or else by the addition of small lipid-soluble molecules, including drugs and anesthetics.

Formation of specific complexes involving intrinsic proteins may occur either as a result of nearest-neighbor contacts between these proteins, or else two or more intrinsic proteins may be linked together at a distance through the intermediate of extrinsic protein binding. This type of role has been proposed for spectrin in erythrocyte membranes.[16] The interaction of extrinsic proteins with membranes is undoubtedly of great importance in determining not only the chemical, enzymatic, and transport properties of membranes, but even the macroscopic shape. Figure 1 illustrates these major categories of mechanisms for protein association or aggregation.

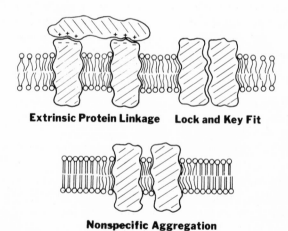

**Extrinsic Protein Linkage    Lock and Key Fit**

**Nonspecific Aggregation**

Fig. 1. Possible mechanisms for the aggregation of proteins in a membrane. Upper left: Cross-linking of intrinsic proteins by an extrinsic protein. The possible involvement of charge–charge interaction in this type of interaction is suggested in the figure. Upper right: Direct, specific association between intrinsic proteins. Lower: Nonspecific aggregation, involving generalized protein–protein interactions and possibly unfavorable lipid–protein interactions.

There are two major types of nonspecific forces between intrinsic proteins which need be considered:

a. Electrostatic forces.
b. Nonbonded, or London–van der Waals, dispersion forces.

One might ask why hydrophobic forces were not included in this list. The hydrophobic effect is without a doubt the major driving force which determines the bilayer structure of membranes, but once the bilyaer arrangement has been achieved, lateral movement of the lipids or proteins within that bilayer would have little or no effect on the hydrophobic energy, since lateral movements would not be expected to significantly change the degree of exposure of membrane components to water. Thus the hydrophobic effect is all-important for the geometric disposition of molecules in a direction normal to the plane of the membrane, but it is to a first approximation blind to the two-dimensional distribution of molecules within the membrane plane.

*Membrane Electrostatics*

The electrical properties of membranes constitute a broad and difficult subject. I will at this point be concerned only with electrostatics, and of that only with a few qualitative aspects which are directly related to our principal theme of protein distribution in membranes.

Lipid bilayers and membranes are a special class of polyelectrolytes, since the lipids and proteins possess ionizable groups in close proximity to each other. The surface charge density which results is dependent on the pH and ionic strength of the medium, as well as on the chemical composition. The feature of polyelectrolytes which makes their mathematical analysis so difficult is that the entire system of polyelectrolyte plus ions in the medium must be treated simultaneously. The charged surface of the polyelectrolyte induces the formation of a diffuse ionic "double layer" in the medium, and conversely the presence of the ionic atmosphere affects the degree of ionization of the polyelectrolyte. The electrostatic free energy of a polyelectrolyte is thus intimately related to the ionic composition of the medium.

The Poisson–Boltzmann equation provides one theoretical approach to the calculation of the electrostatic free energy of membranes.[17] For mathematical tractability in using this equation, it is ordinarily necessary to assume that the surface charges are uniformly "smeared," so as to give a constant surface charge density over the region of interest. This approximation is generally deemed to be acceptable in studies of the effects of the polyelectrolyte on the ions in the medium as, for example, in the analysis of pH titration curves. This approximation may be seriously questioned, however, when the objective is to learn the effects of electrostatics on the distribution of proteins in the plane of the membrane. Suppose that proteins bearing a net charge are dispersed at fairly high dilution in a zwitterionic lipid bilayer. It would then be unreasonable to assume that there is constant charge density over the entire surface. The smeared charge model would be more appropriate if the proteins are aggregated together, and might actually be used to calculate the electrostatic repulsion between aggregated proteins which will develop when the pH is moved away from the isoelectric point. In light of these considerations, it appears that membrane electrostatics deserves much more attention by theoreticians than it has been given in the past.

*Coulomb Interactions.* I will for the present simply ignore the mathematical complexities of polyelectrolytes, and instead make some intuitive remarks about the magnitude and signs of the electrostatic interactions between proteins in membranes, based essentially on Coulomb's law. (Recall that Coulomb's law states that the electrostatic force is directly proportional to the product of the charges, and inversely proportional to the square of the distance between them. It is also inversely proportional to the dielectric constant of the medium.). By Coulomb's law, proteins which bear like charges will repel each other in a membrane, whereas if there are proteins present which have opposite charges, they will attract each other. For a pair of identical proteins, the electrostatic interaction will be least at their isoelectric point (but not zero; see below), and the strength of their mutual repulsion will progressively increase as the pH is moved away from the isoelectric point. Since proteins often bear an appreciable number of ionizable groups, the strength of this repulsion may also become great enough at high or low pH to completely overshadow any of the other general effects which might favor protein aggregation. Although it is not evident from Coulomb's law by itself, it is clear from the Poisson–Boltzmann equation, as well as from the Debye–Hückel equation obtained from it, that the strength of the repulsive force between two such proteins is inversely related to the ionic strength. Adding salt, in other words, will decrease the electrostatic repulsion between proteins in a membrane.

*Permanent and Fluctuating Dipole Interactions.* The direct monopole Coulombic interaction between proteins which bear a net charge of zero is necessarily zero. Under zwitterionic conditions, dipole–dipole interactions will result in a net average *attraction* between the proteins. The assymmetrical electron distribution on proteins creates a permanent dipole moment, and thus virtually all proteins may be expected to display a net dipolar attraction. (The only exception to this rule is the case of multimeric complexes of sufficiently high symmetry, for which the net dipole moment could be zero due to internal compensation.)

Another type of electrostatic attraction between zwitterionic proteins arises from the fluctuation in charge distribution.[18] It can be rigorously shown that the interaction between the charge fluctuations on a pair of polyelectrolytes always results in a net attraction; this may also be described as the interaction between instantaneous dipoles.

The essence of the above paragraphs is this: when the pH is not at the isoelectric point of the proteins, there will be electrostatic repulsion between proteins of like charge, the strength of which increases as the pH is moved further and further away from the isoelectric point. If the pH is brought to the protein isoelectric point, on the other hand, there will be a net electrostatic *attraction* between them, because of the interactions between their permanent dipoles as well as between their fluctuating or instantaneous dipoles.

*Hydration and Ion Pair Formation.* A few more comments are in order which relate to the electrostatics of membranes. As we now understand membrane structure, the intrinsic proteins penetrate into the hydrophobic interior of the lipid bilayer, but in many cases an appreciable part of the protein remains exposed to the aqueous phase. One anticipates that the majority of the charged groups on the protein will be situated in a place where they can be hydrated, since the energy of a hydrated charged group is much lower than that of a naked charge. For a group to be hydrated, it must either reside on the portion of the protein which is exposed to the external aqueous phase or be part of an internal aqueous cavity or pore. In the absence of a complete hydration shell, the tendency for ion pair formation will be particularly strong. If this ion pairing requirement cannot be satisfied by interactions between different side chains on the same protein, or between the protein and the surrounding lipid polar groups, it may act as a powerful force for binding compatible proteins together in the membrane.

There are many ways in which the effective charge on membrane proteins can be modified. The general rule of thumb for polyelectrolytes is that the effective surface charge density, or, more precisely, the electrostatic potential at the surface, will be minimized by any means possible, since doing so also minimizes the electrostatic free energy of the system. First and foremost is the effect of pH, which directly affects the number of protons bound to the ionizable groups. The electrostatic potential also varies inversely with the ionic strength of the medium, with divalent counterions having a much stronger effect than monovalent ions.[19] Thus the presence of a small amount of divalent ions in the medium will have a major effect on reducing the effective surface charge density of membranes, leading perhaps to protein aggregation in the plane of the membrane, or, more noticeably, to the flocculation of a membranous vesicle suspension.

The interaction of extrinsic proteins with a membrane may often involve charge pairing reactions between the extrinsic protein and the intrinsic proteins or the lipids, which also yields a decrease of net effective surface charge. Charge pairing interactions are weakened by increased ionic strength, since the unpaired charged groups are stabilized by the presence of salt, and this is the reason why some extrinsic proteins (e.g., cytochrome *c*) can be solubilized from their membranes with salt.

## Nonbonded Dispersion Forces

The London–van der Waals nonbonded dispersion forces have long been recognized to be of importance for short-range (nearest-neighbor) interactions between nonpolar molecules, and also more recently for the role they play in determining the tertiary structure of proteins. The nonbonded energy of interaction between entire protein molecules has not been given much consideration, however, since for proteins in aqueous solution these forces are ordinarily overshadowed by the electrostatic forces. In the course of seeking to identify the various types of intermolecular interactions that may exist between proteins which are partially immersed in the hydrophobic portion of a lipid bilayer, we carried out some fairly crude calculations in order to see if the nonbonded interactions may in fact be of a significant magnitude.[15] The result was in the affirmative, but it will not be possible to evaluate the actual importance of this quantity in determining the protein distribution in membranes until all the other contributing factors have also been quantitatively analyzed.

The nonbonded dispersion energy between two atoms has a characteristic inverse sixth-power dependence on the distance of separation between them, and is directly proportional to the product of the polarizabilities of the atoms. It was Hamaker[20] who in 1937 discovered, in the course of studying the forces between colloidal particles, that the nonbonded energy summed over all the atoms in a pair of such particles can have an appreciable magnitude out to a considerable distance of separation. The method he used was to treat the particles as having a uniform atom density and to integrate the energy function over the two particle volumes, instead of carrying out a direct summation over all atom–atom pairs.

We have used the Hamaker method to compute the nonbonded energy for the interaction between pairs of proteins embedded in lipid.[15] Some simplifying assumptions had to be made, for which reason the results must be viewed as only order-of-magnitude estimates. Calculations were carried out for pairs of proteins having the shapes of spheres, cubes, and flat rectangular plates; the choice of geometry was dictated by the available integrated forms of the Hamaker equation. The calculations were performed for the artificial conditions of proteins entirely embedded in a uniform medium having the atom density of dipalmitoyl lecithin. That is, the calculation was for proteins not partially embedded in lipid bilayer and partially exposed to water but rather entirely immersed in a homogeneous lipoidal medium. Considerations due to Parsegian and Ninham[21] on the strength of nonbonded interactions acting through water would indicate that the results obtained with this homogeneous medium are underestimates of what would be obtained if a more realistic picture of proteins partially exposed to water were employed.

The mathematical methods and parameters employed in the calculations have been given elsewhere.[15] Only the principal results and the qualitative interpretations which may be derived from them are given here. Figures 2 and 3 give the results obtained. Figure 2 gives the net energy of interaction per protein, $\Delta W$, expressed as kcal/mole, as a

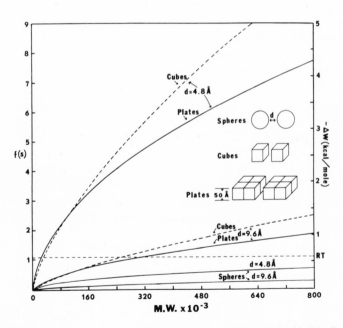

Fig. 2. Calculation of the nonbonded energy between proteins embedded in lipids. $\Delta w$ is the net energy of interaction per protein between a pair of proteins, and $f(s)$ is the geometric factor in the Hamaker equation, as defined by Vanderkooi and Bendler.[15] The results for spheres, cubes, and plates are given as a function of the molecular weight. The plates had a constant thickness of 50 Å and a square surface. In each case, $d$ is the distance of closest approach between the proteins and was held fixed at 4.8 Å or 9.6 Å. The thermal energy, $RT$, is indicated with a dashed line for comparison. From Vanderkooi and Bendler.[15]

*Garret Vanderkooi*

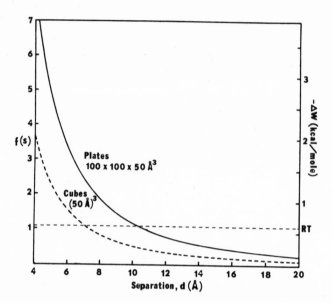

Fig. 3. Nonbonded energy between proteins embedded in lipid, expressed as a function of the distance of closest approach. Two cases are given: cubes of 50 Å$^3$ and plates having the size of a tetramer of these cubes. These volumes correspond to protein molecular weights of approximately 100,000 and 400,000, respectively. From Vanderkooi and Bendler.[15]

function of molecular weight, for three protein geometries, and at two assigned distances of closest approach of the interacting pair. The distances of closest approach employed were 4.8 and 9.6 Å, which correspond to the diameters of one and two aliphatic chains, respectively. The sphere and cube geometries are self-explanatory. Parallel, juxtaposed faces of the pair of interacting cubes were assumed. Square plates having a constant thickness of 50 Å were also studied, with the size of the square faces varying according to the molecular weight. The juxtaposed sides of the pair of plates were kept parallel to each other.

Several points are worthy of note. First, the nonbonded energy between proteins embedded in lipid is necessarily attractive; this statement is true independent of the particular parameters or geometries employed. The net attraction results mainly because of the difference in density between the proteins and lipids. Second, the energy of interaction between the rectangular solids (cubes and plates) is greater than the thermal energy, $RT$, over a large portion of the range presented. Taking $RT$ as 0.6 kcal/mole, $-\Delta W$ exceeds $RT$ for molecular weights in excess of 24,000–32,000 at 4.8 Å separation, or in excess of 240,000–320,000 at 9.6 Å separation. The value of $\Delta W$ is strongly dependent on shape, as well as on orientation (not shown); the energy of interaction between spheres is much less than between cubes or plates of the same volume, and does not exceed the thermal energy even for a molecular weight of 800,000. This shows that changes in protein shape, as through conformational changes, will have a marked effect on the nonbonded energy.

Figure 3 gives $\Delta W$ as a function of distance of separation for cubes and plates of constant volume. The cubes had a volume of $(50 \text{ Å})^3$, corresponding to a molecular weight of 100,000, whereas the plates had dimensions of $50 \times 100 \times 100$ Å, which is in effect a planar tetramer of the cubes. The energy exceeds $RT$ out to a distance of more than 10 Å for the tetramers, and to about 7 Å for the monomeric cubes. This shows that the nonbonded interaction between multimers is greater than that between monomers: the nonbonded energy between the monomers in Fig. 3 at a separation distance of 4.8 Å is $-1.38$ kcal/mole, while that between the tetramers at the same distance of separation is $-2.97$ kcal/mole. Thus if multimer formation is brought about by some other means, such as by partial chemical cross-linking of the intrinsic proteins, the nonbonded energy between pairs of multimers will be greater than between pairs of the monomers, and will increase the tendency for all of the proteins to become nonspecifically aggregated in the membrane.

### Summary on Protein-Related Interactions

The several protein-dependent physical effects which will influence the distribution of intrinsic proteins in the plane of a membrane may be summarized as follows:

a.  Entropy of mixing. The magnitude of this term, which favors dispersal of the proteins, is inversely proportional to the size of the proteins, for a given surface area ratio of intrinsic protein to lipid.

b.  Electrostatics. Coulombic repulsion is expected between proteins bearing like, uncompensated charge. Electrostatic attraction will be found between proteins having a net zero charge, on account of the permanent and fluctuating dipole interactions.

c.  Nonbonded energy. A net attraction between proteins in a membrane is predicted, which is sensitive to the size, shape, and distance of separation of the proteins.

d.  Cross-linking of intrinsic proteins. This has the effect of decreasing the entropy of mixing and increasing the nonbonded attractive force between particles, thereby favoring protein aggregation.

The principal factors favoring protein aggregation in a membrane are thus large protein or protein complex size and low uncompensated electrostatic charge. In the next section, lipid-dependent factors influencing the aggregation or dispersal of the proteins will be taken up.

## Lipid–Protein Interactions

The introduction of an intrinsic protein molecule into a protein-free lipid bilayer will inevitably cause some perturbation of the bilayer in the vicinity of the protein, with the perturbation decreasing as a function of the distance away from the protein, until a point is reached at which the properties of the bilayer are again indistinguishable from those of protein-free bilayer. Events which occur in the unperturbed lipid bilayer far from the protein can have no direct effect on the protein, since this lipid is too far away to be "seen" by the protein. The protein can only affect, and in turn be influenced by, the lipid which is in its immediate vicinity. This is the lipid which has been termed "boundary lipid"[22] or "halo lipid."[23] Changes which occur in the unperturbed bilayer, such as a thermal phase transition, will affect the protein by way of the intermediate of the halo or boundary lipid.

A clarification of terminology is in order. The term "boundary lipid" was originally used to describe the first layer of lipid around membrane proteins, which was shown to be highly immobilized on the ESR time scale.[22] In subsequent papers, [15,24] I have used this term to refer to the entire domain of lipid whose properties are perturbed by the protein, and this is the sense in which the term will be used here. The region of perturbation probably extends for more than one lipid layer away from the protein. The term "unperturbed bilayer," or simply "bilayer" when no ambiguity will result, will be used to refer to the lipid whose properties are essentially unaffected by the protein.

A distinction can and should be made between "boundary lipid" and the boundary "region." The boundary lipid is, as already defined, that lipid whose properties are affected by the presence of the proteins. The boundary region includes both lipid and peptide components, however, since the mobile amino acid side chains on the periphery of the proteins will be intermixed with the lipid chains. Because of this, changes in the characteristics of the boundary lipid will also necessarily affect the amino acid side chains which are in the boundary region, and hence the conformations of the proteins themselves.

### Surface Phases and Boundary Lipid

The boundary lipid may be looked upon as a one-dimensional "surface" phase of a two-dimensional solution, the surface being the boundary between unperturbed bilayer lipid and the protein or other object which may be inserted into the bilayer. This approach is by analogy to the description of two-dimensional surface phases of ordinary three-dimensional bulk solutions. In neither case is the dimensionality correct, however, since both the two- and one-dimensional surfaces have a finite thickness and volume. This thickness is much more important for the membrane problem than for the bulk phase case, since an appreciable fraction of the total lipid may be included in the lipid boundary, whereas the number of molecules in the surface phase of bulk solutions is usually vanishingly small in comparison with the number of molecules in the bulk phase.

The analogy between the one- and two-dimensional surface phases can be pressed a step further. Just as it is conventional to write the equilibrium conditions for the distribution of matter between surface and bulk phases in three dimensions, so also we can study the equilibrium between boundary and bilayer lipid, and can define partition coefficients for the relative solubility of other molecules in these two phases.

The amount of boundary lipid present should be nearly proportional to the combined lengths of exposed perimeters of the intrinsic proteins in the membrane. When the proteins are highly dispersed, the maximum amount of protein perimeter will be exposed and hence also the maximum amount of boundary lipid will be present. As the protein density is increased or if clustering or aggregation of the proteins occurs, the regions of influence of the proteins will begin to overlap, thereby decreasing the total amount of lipid which is perturbed by the proteins.

The significance of the preceding paragraph is that an important relationship exists between the boundary lipid–bilayer lipid equilibrium and the equilibrium of the dispersed and aggregated states of the proteins.[15,24] Factors which stabilize the bilayer as opposed to the boundary will tend to withdraw lipid molecules from the boundary. This can happen only if the total amount of boundary is decreased, which in turn means that the exposed protein perimeter must be decreased. A decrease in exposed protein perimeter implies aggregation of the proteins, or else actual removal of protein molecules from the membrane. The next logical step, therefore, is to inquire into the types of effects which may be expected to shift the boundary lipid–bilayer lipid equilibrium. Before doing so, however, it would be useful to give a brief review of some of the experimental observations which have been made on the physical and chemical characteristics of boundary lipid.

*Experimental Data on Boundary Lipid*

A brief summary will be given here of some of the presumed characteristics of boundary lipid, and the experimental observations upon which they are based. The membrane systems involving cytochrome $c$ oxidase and sarcoplasmic reticulum ATPase (SR-ATPase) have been most intensively studied, and hence much of the information quoted here is based on those two systems.

a.   The chemical composition of boundary lipid differs from that of the total lipid.

The oldest evidence on this point comes from the differential extractability of erythrocyte lipids with ether.[25,26] It was found that ether extracted all of the cholesterol but only one-fourth of the phospholipids from lyophilized erythrocyte membranes, which result implied that the remainder of the lipid was interacting with the proteins in some way.

Cytochrome oxidase membranes prepared from beef heart mitochondria by the method of Sun *et al.*[27] using Triton X-100 and Triton X-114 contain 25–30% phospholipid by weight, with half of the lipid being cardiolipin.[28] Since only 18% of the original mitochondrial membrane lipid fraction was cardiolipin, as determined by the same authors,[28] there is evidently a preferential binding of cardiolipin to the cytochrome oxidase proteins.

Warren *et al.*[29] have provided convincing evidence that cholesterol is excluded from the boundary region in SR-ATPase membranes. They showed that 25–30 phospholipid molecules per protein molecule must be present to give enzymic activity, and that these form a ring or annulus about the proteins. Up to 50 mole % of cholesterol may be present in the lipid fraction as a whole without appreciably affecting the enzymic activity, provided that there are sufficient phospholipid molecules present to fill the annulus. Enzymic activity falls precipitously, however, if the mole ratios are changed in such a way as to force cholesterol into the boundary layer. Thus under normal circumstances in which the enzyme is active, there must be a preferential binding of phospholipids to the proteins, and exclusion of cholesterol from the first lipid layer around the proteins.

b.   The enzymic activity of membrane-bound enzymes is dependent on the presence and composition of boundary lipid.

The work described in the preceding paragraph bears directly on the present point as well. Warren *et al.*[30] found that considerably higher SR-ATPase activities were obtained with unsaturated as opposed to saturated lecithins.

The presence of boundary lipid is required for cytochrome oxidase activity. Addition of phospholipid to a lipid-depleted cytochrome oxidase preparation gave a rapid increase in the enzymic activity up to a level of about 20% lipid by weight.[7] Further addition of lipid caused little change in the activity. Since the minimum amount of lipid required for full activity corresponds closely with the amount which interacts strongly with the proteins, as judged by the ESR spin label method,[22] it was inferred that the boundary lipid is required for activity but that additional bilayer lipid does not particularly affect the activity.

    c.   Boundary lipid does not take part in the lipid thermal phase transition.

    Träuble and Overath[23] studied the solubility of fluorescent dyes in *Escherichia coli* membranes as a function of temperature. They measured the differential solubility of the dyes between the liquid crystalline and gel states of the lipids both in the intact membranes and after extraction. Comparison of the results obtained in the two cases indicated that only about 80% of the lipids actually participated in the phase transition in the membrane. They concluded that the remaining 20% was in close proximity with the proteins.

    More direct evidence on this point comes from the calorimetric work of Papahadjopoulos *et al.*[31,32] who showed that $\Delta H$ for the thermal phase transition of model membranes consisting of myelin N-2 proteolipid protein plus lipid (lecithin or phosphatidylglycerol) decreased linearly as the proportion of protein was increased up to 50% by weight. The temperature of the phase transition did not change. These observations were interpreted to mean that a stoichiometric relationship exists between the amount of protein present and the amount of lipid whose properties are affected by the protein. Only the unaffected lipid participates in the phase transition and contributes to $\Delta H$. (It should be noted that this type of result was obtained only with proteins of the intrinsic type which make a major hydrophobic contact with the acyl chains of the lipids.)

    d.   The boundary lipid closest to the protein is highly immobilized on the ESR time scale.

    The work of Jost *et al.*[22,23] on cytochrome oxidase membranes demonstrated clearly that a certain fraction of the lipid in these membranes is highly immobilized, as shown by the ESR spin label method. The amount of immobilized lipid was proportional to the amount of protein present; the molar ratio of bound lipid phosphorus to protein was about 50, assuming a molecular weight of 200,000 per protein complex.[7]

    A similar result was obtained by Warren *et al.*[30] for SR-ATPase. They also found using ESR spin label methods that the 30 or so phospholipid molecules required for enzymic activity were in a state of restricted mobility. In another system, Hong and Hubbell[11] found that the "order parameter" of recombined rhodopsin membranes decreased in direct proportion to the amount of rhodopsin present in the membranes, indicating that the protein had the effect of decreasing the motional freedom of the lipid around it.

e.   The solubility of small molecules is different in the boundary region than in unperturbed bilayer lipid.

While this point can be rationally defended as being true on the basis of the characteristics already listed, it appears that unambiguous evidence in support of it is lacking. The solubility properties of the spin label, TEMPO, in membranes[34] may be interpreted to mean that it is less soluble in boundary lipid than in bilayer lipid. Also, the exclusion of cholesterol from the boundary region in SR-ATPase[29] may aptly be described in terms of a greater solubility of the cholesterol in the bilayer region. It may be that there are other molecules which will partition preferentially into the boundary region; these should be molecules which display one-dimensional surface activity.

f.   The permeability of the boundary region differs from that of unperturbed bilayer lipid.

It is well known that the addition of proteins of either the extrinsic or intrinsic type to lipid vesicles changes the permeability properties of the vesicles, usually in the direction of increasing the permeability. Papahadjopoulos *et al.*[31] have shown that the permeability of phos-phatidylserine and phosphatidylcholine vesicles to $^{22}Na$ increases in direct proportion to the amount of N-2 proteolipid protein added. This proportionality implies that the permeability of the boundary region differs from that of bilayer lipid, since there is no reason to believe in this case that the proteins themselves are porelike structures.

Clearly, extensive data on membranes now in hand can be related to or interpreted in terms of the properties of the boundary lipid or the boundary region. It is safe to say that the properties of at least one layer of lipid around each membrane protein are significantly and measurably perturbed. How far the perturbation extends beyond the first layer is an open question. Kleemann and McConnell[14] have given theoretical reasons for believing that the protein effect on the lipid may be of rather long range, especially if the temperature is close to that of the lipid thermal phase transition. (See also Addendum, p. 53.)

*Thermodynamics of Boundary Lipid*

From a theoretical point of view, it may be safely said that the introduction of an intrinsic protein into a lipid bilayer will necessarily affect the properties of the lipids in its immediate vicinity. A certain number of lipid–lipid interactions will be replaced by lipid–protein interactions. It is improbable that the enthalpy and entropy of a lipid molecule surrounded completely by other lipid molecules will be the same as those of a lipid molecule for which a portion of its interaction atmosphere is occupied by protein. The direction and magnitude of the changes in the enthalpy and entropy which occur when a lipid molecule is brought into the vicinity of a protein will depend on the state of the lipid as well as on the nature of the protein.

The experimental evidence in hand appears to indicate that the thermodynamic properties of boundary lipid are intermediate between those of the fluid (liquid crystalline) and gel states of bilayer lipid. The motional freedom of at least the first layer of boundary lipid has been shown experimentally (see above) to be less than that of fluid bilayer lipid. This implies that the entropy (randomness) of boundary lipid is less than that of fluid bilayer lipid. The entropy of first-layer boundary lipid is probably greater than that of gel bilayer lipid, however, since in the gel state the aliphatic chains are lined up in an orderly close-packed arrangement with limited motional freedom, whereas the lipid which is close to the surface of a protein is inevitably in a disordered state on account of the irregular contour of the protein surface. Thus the entropy of first-layer boundary lipid appears to be intermediate between the fluid and gel states. The enthalpy of boundary lipid must also be lower than that of fluid bilayer lipid; if it were not, there would be no explanation for the observed immobilization of and decrease in entropy of the boundary lipid relative to the fluid bilayer.

*Free Energy of Interaction.* Schellman[35] has studied the thermodynamics of ligand binding to proteins and other macromolecules. He defined the free energy of binding ($\Delta G_b$) as the difference in free energy between a system in which there are no binding interactions and one in which equilibrium binding has occurred. The ligand activity is taken to be the same before and after the interaction. $\Delta G_b$ is an excess quantity, which depends only on the change in free energy of those molecules involved in the interaction, and not on the free energy of any extra ligand molecules which may be present but are not involved in the interaction.

The free energy of interaction of boundary lipid with proteins will be defined in the same manner as the ligand-binding free energy of Schellman. The subscript $b$ on $\Delta G_b$ can then be taken to mean either "binding" or "boundary." $\Delta G_b$ may be expressed as a sum of the changes in partial molal free energies for all of the molecules involved in the interaction with a given reference protein. (For simplicity, a single protein species is assumed.) The reference state which exists before the interaction has occurred will be defined as one in which all of the lipid is present as unperturbed bilayer and the proteins are randomly dispersed in this bilayer but without interprotein interactions being present. When the interactions are turned on, the proteins undergo any necessary rearrangement to a possibly nonrandom distribution, and the lipids in the vicinity of the proteins become perturbed, to give the final equilibrium state which minimizes the total free energy of the system. The minimum of the total free energy will also be the minimum for $\Delta G_b$.

$\Delta G_b$, expressed on a per mole of protein basis, may be written as a sum of lipid and protein contributions in the following manner:

$$\Delta G_b = \int_r n(r)\Delta g(r)\, dr + \Delta G_{\text{protein}} \tag{3}$$

The lipid-dependent contribution has been expressed as an integral over all lipid molecules in the vicinity of the reference protein, which are perturbed by the protein. $n(r)$ is the mean number density of lipid molecules at a distance $r$ from the protein surface. $\Delta g(r)$ is the mean change in partial molal free energy for a molecule at $r$, for a system with interaction minus that without interaction. In other words, $\Delta g(r)$ is the mean difference in partial molal free energy between that for a boundary lipid molecule at a distance $r$ from the surface and one located in unperturbed bilayer. $\Delta G_{\text{protein}}$ includes all free energy changes arising from protein–protein interactions, as well as changes in the entropy of mixing which result from the possibly nonrandom distribution of proteins brought about by turning on the interactions. $\Delta G_{\text{protein}}$ also includes the contributions from any conformational change in the reference protein which may accompany the turning on of the interactions.

The major purpose of the present section is to examine the lipid-dependent contribution to $\Delta G_b$, since we are interested in seeing how changes in the lipid bilayer can affect the protein distribution by way of the boundary lipid. The quantity that must be examined in detail is therefore $\Delta g(r)$, which gives the free energy of interaction (per mole) for a lipid molecule located at $r$. For the purpose of the present discussion, let us assume that the protein–protein interactions in the membrane are weak, i.e., that $\Delta G_{\text{protein}} \approx 0$.

It must be understood that $n(r)$ and $\Delta g(r)$ are defined as mean values. The values which pertain to individual lipid molecules may vary as a function of position along the protein perimeter. For example, $\Delta g(r)$ for a first-layer lipid molecule which is hydrogen-bonded via a head-group interaction to the protein will be different from that for a first-layer molecule which is not so hydrogen-bonded. The mean number density, $n(r)$, will be decreased if a second protein approaches the reference protein to a distance less than $r$, thereby displacing some of the lipid molecules at that distance.

Figure 4 shows the deduced variation of $\Delta g(r)$ with $r$, for the cases of a protein in a fluid bilayer and in a gel bilayer. The curves were drawn so as to give the predicted result that the proteins in the fluid bilayer are

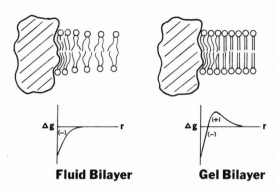

**Fluid Bilayer**          **Gel Bilayer**

Fig. 4. Presumed variation of the mean partial molal free energy of boundary lipid, $\Delta g(r)$, as a function of $r$, the distance from the protein surface, for the cases of fluid and gel bilayer. The effect of the protein on the lipid is depicted in the upper part of the figure; for fluid bilayer, the motional freedom of boundary lipid is less than and the density is greater than that of the unperturbed bilayer. For gel bilayer, the boundary lipid is less orderly than the bilayer.

randomly dispersed, whereas those in the gel bilayer are aggregated at equilibrium. This is in agreement with the observed protein distribution in several (but not all) membrane systems. Remember that we have assumed that $\Delta G_{\text{protein}}$ is negligible or of second-order importance in the membranes under consideration, with the properties of the boundary lipid being the determining factor in the protein distribution. With these reservations, it can be said that $\Delta g(r)$ must be negative or zero at all $r$ in the case of fluid bilayer. This will result in the maximum possible amount of boundary lipid being present, which is in turn the implication of a random protein distribution. Inevitably, $\Delta g(r)$ must go asymptotically to zero at larger $r$ in all cases, unless this is impossible on account of the close proximity of other proteins.

For the case of gel bilayer, in which the proteins are aggregated, it must be assumed that $\Delta g(r)$ is positive for some values of $r$. For $\Delta G_b$ to be minimized (with $\Delta G_{\text{protein}} = 0$), $n(r)$ must be as small as possible for all $r$ at which $\Delta g(r)$ is positive. Decreasing the value of $n(r)$ can be brought about only by protein aggregation. $\Delta g(r)$ is shown in Fig. 4 as having a negative value at small $r$, followed by a positive region, and finally decreasing to zero at large $r$. The negative value at small $r$ implies a strong favorable interaction of some or all of the first-layer boundary lipid molecules with the protein; polar headgroup–protein interactions may contribute to this binding. The assumption of a negative $\Delta g(r)$ for the first layer of boundary lipid serves to rationalize the observations that some lipid in many membranes is difficult to extract with mild solvents, and also that

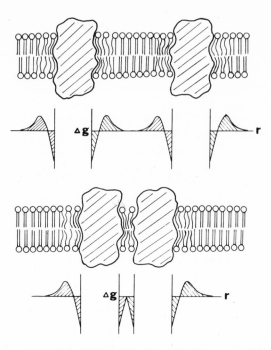

Fig. 5. Minimization of the free energy of boundary lipid by protein aggregation. In the upper part of the figure, the proteins are dispersed throughout a gel bilayer, causing the maximum amount of lipid to be perturbed. The total binding free energy is proportional to the sum of the shaded areas under the $\Delta g$ curves. By bringing the proteins together (lower part of figure), two of the four positive shaded areas have been eliminated, thus lowering the free energy. The lipid for which $\Delta g(r)$ is negative (if any) will remain between the pair of aggregated proteins.

some of the boundary lipid does not participate in the thermal phase transition, and is therefore in the same state both above and below the phase transition temperature. The positive $\Delta g(r)$ at larger $r$ must be assumed in order to explain the thermal phase transition-induced protein aggregation. How tightly the proteins become aggregated will depend on the detailed shape of the $\Delta g(r)$ curve, which will be a function of temperature. A counteracting force to aggregation, included in the $\Delta G_{protein}$ term but temporarily ignored, is the entropy of mixing, which will disfavor the partial loss of motional freedom that must accompany protein aggregation. Figure 5 shows in diagrammatic terms how the free energy of boundary lipid can be minimized by protein aggregation in a gel bilayer.

*Enthalpic and Entropic Contributions to $\Delta g(r)$.* By using a combination of physical argument and thermodynamic requirements, it is possible to tentatively resolve $\Delta g(r)$ into its enthalpic and entropic components, $\Delta h(r)$ and $\Delta s(r)$, respectively. These are related by the familiar equation $\Delta g(r) = \Delta h(r) - T\Delta s(r)$. In the case of proteins dispersed in fluid bilayer lipid, $\Delta g(r)$ is negative for all $r$ (Fig. 4). Since at least the first layer of boundary lipid has less motional freedom than fluid bilayer lipid (see above), the entropy of boundary lipid is less than that of the fluid bilayer, i.e., $T\Delta s$ is negative. Therefore, in order for $\Delta g$ to be negative, $\Delta h$ must be more negative than $T\Delta s$. This is intuitively reasonable; it is the favorable enthalpic interactions between the lipids and protein which compensate for the loss of entropy. Figure 6 illustrates the relationships among the various quantities.

For the case of protein in gel bilayer, it is probable that the entropy of all layers of boundary lipid, including the first, is greater than that of gel bilayer, since even though the boundary lipid is immobilized by the protein it is also necessarily in a somewhat disordered state on account of the irregular contours of the protein surface and the penetration of amino acid side chains into the first one or two lipid layers. By contrast, the unperturbed gel bilayer is in a crystalline array; hence the conclusion that boundary lipid has greater entropy than gel bilayer lipid. In Fig. 6, the entropy is depicted as decreasing monotonically and asymptotically from a positive value to the gel bilayer axis.

The enthalpy of boundary lipid relative to gel bilayer lipid must vary in the manner shown in Fig. 6 in order to satisfy point by point the equation $\Delta g(r) = \Delta h(r) - T\Delta s(r)$, if in fact the forms of the curves for $\Delta g$ and $\Delta s$ have been correctly deduced. Since $T\Delta s$ is presumed to be positive, and thus contributes negatively to $\Delta g$, then $\Delta h$ must be positive over part of its range in order to make $\Delta g$ positive.

*Anomalous Effects.* It is useful to see how various apparently anomalous observations on protein distribution can be interpreted in the

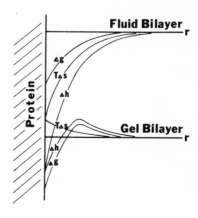

Fig. 6. Resolution of $\Delta g(r)$ into enthalpic ($\Delta h$) and entropic ($T\Delta s$) components, as a function of $r$, the distance from the protein surface. The two horizontal axes correspond to the asymptotic values of each of the thermodynamic functions in unperturbed fluid and gel bilayers, respectively.

present context. Haest *et al.* [36] have observed that no protein clustering accompanied the thermal phase transition in bacterial membranes enriched with branched-chain lipids. The chain branching prevents the gel lipid aliphatic chains from being as tightly packed as straight chain lipids, as shown by X-ray diffraction, [36] and thus the enthalpy change upon going from the fluid to the gel state will also be less than for straight-chain lipids. Thus there is less difference between the thermodynamic properties of the gel and fluid bilayer, and a scheme similar to that shown in Figs. 4 and 6 for the fluid bilayer may also apply to the gel state.

Kleemann and McConnell [14] found that sarcoplasmic reticulum ATPase proteins were strongly aggregated below the phase transition temperature in model membranes containing dimyristoyl phosphatidylcholine, but when cholesterol was added (20 mole % cholesterol), small clusters of proteins were observed, being randomly scattered throughout the surface. This finding is interpretable in light of the fact that addition of cholesterol is known to decrease the enthalpy change of the thermal phase transition, [37,38] thus making the gel and fluid states thermodynamically more similar.

*Effect of Small Molecules.* Small lipid-soluble molecules will partition in a potentially unequal manner between the boundary region and bilayer lipid. Whether the partition coefficient for this small molecule equilibrium is greater than unity (implying a greater solubility in the boundary than in the bulk) will depend on the nature of the small molecule as well as on the composition and state of the lipid. Small molecules which preferentially partition into the boundary region are the counterparts of surface active agents; these will lower the free energy of the boundary, thereby increasing the one-dimensional analogue of a spreading pressure, and give a tendency toward dispersal of the proteins. Conversely, small molecules which preferentially partition away from the boundary will increase the free energy of the boundary, and by the same reasoning tend to give protein aggregation. The problem is really not so simple, however, since the very presence of small molecules (e.g., alcohols or general anesthetics) in the boundary region will tend to fluidize this lipid, making it more bilayer-like. Thus the solubility in the boundary will in turn be a function of the small-molecule concentration in the boundary.

The emphasis in this chapter has been on the factors which may affect the protein distribution in a membrane. While small molecules may alter this distribution and thus influence the activity of some membrane-bound-enzyme systems, there are also at least two other general ways in which small molecules may have an effect on the enzymic activities. First, changes in the physical properties of the boundary lipid induced by small molecules may cause tertiary conformational changes of the proteins. Second, by changing the viscosity of the membrane as a whole, small molecules may cause changes in the activities of enzyme systems that depend upon diffusional motions of the proteins. These three possibilities must be borne in mind when one attempts to interpret enzyme inhibition or activation brought about by the seemingly innocuous molecules of the anesthetic type.

*Analogies between the Properties of Boundary Lipid and Protein-Bound Water.* Kuntz and Kauzmann[39] have reviewed properties of water associated with or bound to soluble proteins and polypeptides. Several remarkable analogies or similarities can be found between the properties of protein-bound water and the properties of boundary lipid as developed here. Proteins in aqueous solution do bind a certain amount of water. Roughly, a monolayer of water appears to be bound rather tightly, with water beyond the first layer being more weakly affected by the protein. A certain quantity of bound water is "unfreezable," as shown by calorimetry, IR, and dielectric measurements. The heat of fusion of the unfreezable water is very low, and consequently its freezing point is also very low. The unfreezable water has higher mobility than that of ice, as shown by the fact that it gives a relatively sharp proton NMR peak, but yet it has less mobility than liquid water. The enthalpies and entropies for transfer from bulk water to bound water are both negative, but not nearly so negative as for the process of going from liquid water to ice. Thus the motional and thermodynamic properties of bound water are intermediate between those of liquid water and ice. Each of these points has its counterpart in the membrane case, especially that at least one layer of boundary lipid does not participate in the thermal phase transition and that the properties of boundary lipid are intermediate between those of fluid and gel bilayer lipids.

## Addendum

A recent theoretical paper by Marčelja[40] came to my attention following the completion of the manuscript for this chapter. He has carried out statistical mechanical calculations on boundary lipid and concludes that two to three layers of lipid will be significantly perturbed by intrinsic membrane proteins.

ACKNOWLEDGMENTS

This work was supported in part by a grant from the National Science Foundation, BMS 75-14425.

## References

1. Wallach, D. F. H., and Zahler, P.H. (1966) *Proc. Natl. Acad. Sci. USA 56,* 1552–1559.
2. Glaser, M., Simpkins, H., Singer, S. J., Sheetz, M., and Chan, S. I. (1970) *Proc. Natl. Acad. Sci. USA 65,* 721–728.
3. Singer, S. J., and Nicolson, G. L. (1972) *Science 175,* 720–731.
4. Vanderkooi, G., and Green, D. E. (1970) *Proc. Natl. Acad. Sci. USA 66,* 615–621.
5. Vanderkooi, G. (1972) *Ann. N.Y. Acad. Sci. 195,* 6–15.
6. Bothorel, P., and Lussan, C. (1970) *C. R. Acad. Sci. Ser. D. Paris 271,* 680–683.
7. Vanderkooi, G. (1974) *Biochim. Biophys. Acta 344,* 307–345.
8. Strittmatter P., and Rogers, M. J. (1975) *Proc. Natl. Acad. Sci. USA 72,* 2658–2661.
9. Bennett, V., O'Keefe, E., and Cuatrecasas, P. (1975) *Proc. Natl. Acad. Sci. USA 72,* 33–37.
10. Pinto da Silva, P. (1972) *J. Cell Biol. 53,* 777–787.
11. Hong, K., and Hubbell, W. L. (1972) *Proc. Natl. Acad. Sci. USA 69,* 2617–2621.
12. Hackenbrock, C. R. (1977) in *Structure of Biological Membranes* (Abrahamsson, S., and Pascher, I., eds.), pp. 199–234, Plenum, New York.
13. Grant, C. W. M. and McConnell, H. M. (1974) *Proc. Natl. Acad. Sci. USA 71,*4653–4657.
14. Kleemann, W., and McConnell, H. M. (1976) *Biochim. Biophys. Acta 419,* 206–222.
15. Vanderkooi, G., and Bendler, J. T. (1977) in *Structure of Biological Membranes* (Abrahamsson, S., and Pascher, I., eds.), pp. 551–570, Plenum, New York.
16. Nicolson, G. L. (1973) *J. Cell Biol. 57,* 373–387.
17. Mille, M., and Vanderkooi, G. (1977) *J. Colloid Interface Sci. 61,* 455–474; 475–484.
18. Hill, T. L. (1960) *Introduction to Statistical Mechanics*, p. 361, Addison-Wesley, Reading, Mass.
19. Dolar, D., and Peterlin, A. (1969) *J. Chem. Phys. 50,* 3011–3015.
20. Hamaker, H. C. (1937) *Physica 4,* 1058–1072.
21. Parsegian, V. A., and Ninham, B. W. (1971) *J. Colloid Interface Sci. 27,* 332–341.
22. Jost, P., Griffith, O. H., Capaldi, R. A., and Vanderkooi, G. (1973) *Proc. Natl. Acad. Sci. USA 70,* 480–484.
23. Träuble, H., and Overath, P. (1973) *Biochim. Biophys. Acta 307,* 491–512.
24. Vanderkooi, G. (1975) *Int. J. Quantum Chem.: Quantum Biol. Symp. No. 2,* 209–219.
25. Parpart, A. K., and Ballentine, R. (1952) in *Modern Trends in Physiology and Biochemistry* (Barron, E. S. G., ed.), pp. 135–148, Academic Press, New York.
26. Roelofsen, B., Baadenhuysen, H., and van Deenen, L. L. M. (1966) *Nature 212,* 1379–1380.
27. Sun, F. F., Prezbindowski, K. S., Crane, F. L., and Jacobs, E. E. (1968) *Biochim. Biophys. Acta 153,* 804–818.
28. Awasthi, Y. C., Chuang, T. F., Keenan, T. W., and Crane, F. L. (1971) *Biochim. Biophys. Acta 226,* 42–52.
29. Warren, G. B., Houslay, M. D., Metcalfe, J. C., and Birdsall, N. J. M. (1975) *Nature 255,* 684–687.
30. Warren, G. B., Toon, P. A., Birdsall, N. J. M., Lee, A. G., and Metcalfe, J. C. (1974) *Biochemistry 13,* 5501–5507.
31. Papahadjopoulos, D., Vail, W. J., and Moscarello, M. (1975) *J. Membrane Biol. 22,* 143–164.
32. Papahadjopoulos, D., Moscarello, M., Eylar, E. H., and Isac, T. (1975) *Biochim, Biophys. Acta 401,* 317–335.
33. Jost, P., Griffith, O. H., Capaldi, R. A., and Vanderkooi, G. (1973) *Biochim. Biophys. Acta 311,* 141–152.

34. McConnell, H. M., Wright, K. L., and McFarland, B. G. (1972) *Biochem. Biophys. Res. Commun. 47*, 273–281.
35. Schellman, J. A. (1975) *Biopolymers 14*, 999–1018.
36. Haest, C. W. M., Verkleij, A. J., DeGier, J., Scheek, R., Ververgaert, P. H. J., and van Deenen, L. L. M. (1974) *Biochim. Biophys. Acta 356*, 17–26.
37. Ladbrooke, B. D., Williams, R. M., and Chapman, D. (1968) *Biochim. Biophys. Acta 150*, 333–340.
38. Hinz, H. and Sturtevant, J. M. (1972) *J. Biol. Chem. 247*, 3697–3700.
39. Kuntz, I. D., and Kauzmann, W. (1974) *Adv. Protein Chem. 28*, 239–345.
40. Marčelja, S. (1976) *Biochim. Biophys. Acta 455*, 1–7.

*Chapter 2*

# Structure of the Mitochondrion

## Robert A. Haworth and Douglas R. Hunter

### Introduction

It is fitting that a volume commemorating the scientific achievements of David Green should contain a chapter on mitochondrial structure, since from the earliest days he has been involved in its elucidation. This account is not meant to be a history of how notions of mitochondrial structure have progressed, nor is it comprehensive; it is rather a perspective on some recent advances in our understanding of mitochondrial configurations. First, the role of osmosis is evaluated, followed by a discussion of other physical factors and membrane structural properties which could be involved in the determination of the shape of cristae.

### Osmosis as Determinant of Mitochondrial Configuration

The role of osmosis as a determinant of mitochondrial configuration was early realized. Mitochondria swell when placed in a hypotonic medium. Quantitatively they behave like perfect osmometers.[1] In the electron microscope the matrix space is seen to be the compartment modulated osmotically by low molecular weight solutes; it expands until at sufficiently low tonicity the outer membrane is ruptured.[2] The expanded matrix space of deenergized mitochondria *in vivo* ("orthodox" configuration) can be contracted by including 0.25 M sucrose in the

***Robert A. Haworth and Douglas R. Hunter*** • Institute for Enzyme Research, University of Wisconsin, Madison, Wisconsin 53706.

incubation medium,[3,4] giving an appearance ("aggregated" configuration) similar to that of mitochondria isolated in a sucrose medium. Finally, active ion uptake induced by antibiotics causes an increase in mitochondrial volume commensurate with the osmotic effect of the accumulated ions[5]; in appearance the mitochondria show the orthodox configuration.[6] Blocking the energy source with inhibitors or uncouplers causes reequilibration of ion gradients and a return to the aggregated configuration.[5,6] All these observations find a simple and satisfactory explanation in terms of osmosis: in each case, a configurational change occurs when a change in concentration of an osmotically active solute takes place on one side of the membrane but not on the other side.

In addition to swelling due to active ion uptake, the swelling action of a number of agents such as fatty acid,[7] calcium,[8] and phosphate[9] has long been recognized. While the active uptake of calcium acetate or potassium (plus gramicidin) phosphate can be correlated with the osmotic effect of the accumulated ions,[10] calcium and phosphate both exert a swelling effect independent of this.[11, 9] The first comprehensive study on the regulation of mitochondrial configuration and function by low levels of calcium was carried out in Dr. Green's laboratory on adrenal cortex mitochondria.[12, 3, 13] We have extended this study in beef heart mitochondria, and have found that low levels of calcium will induce a reversible transition in the permeability of the mitochondrial inner membrane.[14] Mitochondria exposed to low levels of calcium rapidly accumulate it with no immediate ill-effect. After accumulation has ceased, the mitochondria still exhibit good respiratory control and are in the aggregated configuration (Figs. 1A and 2). After a lag period, however, one by one they undergo a rapid transition in the permeability of the inner membrane to small molecules such as sucrose (Fig. 3); they now exhibit no respiratory control (Fig. 2), and are in the orthodox configuration (Fig. 1C). Thus the configurational changes induced by these swelling agents can also be rationalized in terms of changes in osmotic support; when sucrose can permeate the inner membrane, it no longer exerts osmotic pressure to keep the mitochondria "aggregated," and so swelling occurs.

------------------------------------------------------------------------→

Fig. 1. Calcium-induced configurational transition. The medium contained 250 mM sucrose, 20 mM tris-Cl, pH 7.4 (30°C), 2 mg/ml heavy beef heart mitochondria, plus 100 nmoles calcium chloride/mg protein at time zero. Aliquots were fixed after different times of incubation by the addition of an equal volume of 2% glutaraldehyde in 250 mM sucrose and 50 mM potassium cacodylate, pH 7.4. (A) 15 sec, (B) 3½ min, and (C) 7 min.

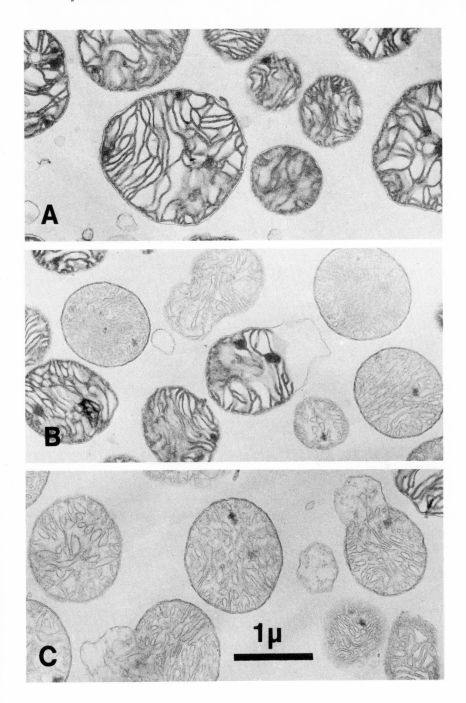

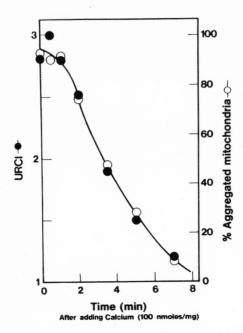

Fig. 2. Effect of calcium on configuration and coupling. This was the same experiment as that from which the electron micrographs in Fig. 1 were taken. Aliquots were removed at time intervals for the determination of the respiratory control index (URCI) with the substrate durohydroquinone and the uncoupler FCCP.

It has been claimed that the swelling action of calcium is linked to the formation of endogenous fatty acids.[15] This view was challenged when it was found that bovine serum albumin failed to inhibit calcium-induced swelling.[11, 9] We also found this, and moreover the swelling action of exogenous fatty acids was found to depend on the presence of calcium,[14] as does the swelling action of phosphate,[9] arsenate,[14] and thyroxine.[14] The effect of hydrogen peroxide[16] and diamide[17] may also be added to this list. It is clear now that levels of endogenous calcium in mitochondria isolated in the absence of chelators are sufficient to induce the transition in permeability, and the length of the lag period before the transition occurs can be shortened by the addition of various potentiators such as phosphate, arsenate, and fatty acid.

An important feature of the calcium-induced transition is its reversibility. Addition of magnesium or EGTA in the presence of an energy source restores impermeability to sucrose (Fig. 4) and good respiratory control.[14] These properties are restored very rapidly; no incubation is necessary. The mitochondria remain in the orthodox configuration, since the sucrose is trapped inside.[18] By contrast, orthodox adrenal cortex mitochondria were reported to contract again to the aggregated configuration in the presence of magnesium and an energy source, even in a sucrose medium.[13] This is the phenomenon of energized contrac-

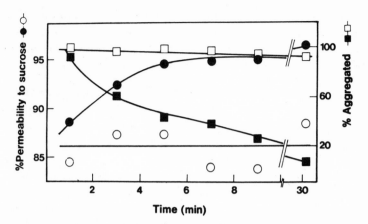

Fig. 3. Correlation between the calcium-induced configurational change and the change in permeability to sucrose. Permeability was measured from the extent to which [$^{14}$C]sucrose had permeated mitochondria pelleted at time intervals. See Hunter *et al.*[14] for details of methodology. ●, nmoles Ca$^{2+}$/mg; ○, zero Ca$^{2+}$; ■, 100 nmoles Ca$^{2+}$/mg; □, zero Ca$^{2+}$.

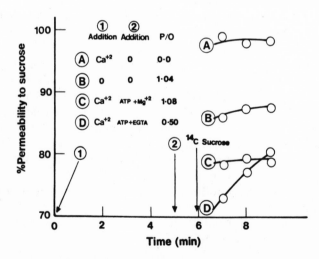

Fig. 4. Restoration of coupling and membrane impermeability. Mitochondria were turned orthodox by incubation in the presence of 150 nmoles calcium chloride/mg protein. After 5 min the additions listed were made. After a further 1-min incubation, [$^{14}$C]sucrose was added. Aliquots were removed at 1-min intervals thereafter, and the extent of sucrose permeation of the pelleted mitochondria was estimated as in Fig. 3. Aliquots were also removed at 10 min for measurement of P/O ratios in a medium lacking added magnesium, with durohydroquinone as substrate. Final concentrations of additions were ATP, 6 mM; MgCl, 6 mM; EGTA, 3 mM.

tion, which in salt media has received widespread attention.[15, 18–20] In these studies, swelling was induced by either calcium or agents now known to potentiate the swelling effect of endogenous calcium; in every case, energized contraction was induced by the addition of agents which would have reversed the membrane permeability change. An attractive osmotic mechanism for the process of active shrinkage is the active extrusion of ions.[21] This could work only where the membrane is sufficiently impermeable to ions for a gradient to be established. It may therefore be no coincidence that agents which induce active shrinkage are agents which restore impermeability.

These observations have some important sequellae. First, it is a mistake to view the effect of calcium as a degradative influence of no consequence, an artifact to be avoided. Some authors speak of "damage" or "stretching" of the membrane.[22, 18] It is clear, however, that the effect of calcium on the membrane is the cause of swelling rather than the result of it. Calcium must therefore be causing a specific structural change in the membrane to account for the dramatic change in its physical properties. The effect is reversible and under regulation.[16, 14] Mitochondria possess a specific uptake mechanism for calcium but have no known requirement for it. It is reasonable, therefore, to deduce that the effect of calcium on mitochondrial membrane permeability is a manifestation of a physiological role. Second, it is dangerous to draw conclusions about mitochondrial integrity, either *in vivo* or *in vitro*, from the configurational state of the mitochondria. Coupling is clearly a function of the membrane permeability rather than configuration. Third, care must be taken in interpreting configurational changes; many old data can be seen in a new light once the role of the calcium-induced transition is appreciated. Three examples will be given. First, configurational changes induced in rat liver mitochondria by active accumulation of calcium phosphate have been wrongly interpreted as being due to the osmotic effect of the calcium phosphate[23]; in retrospect, this is a clear instance of the calcium-induced transition. Second, high molecular weight solutes have been reported to cause configurational transitions in mitochondria along with inhibition of oxidation of some substrates.[24] The configurational changes are equivalent to the aggregated-to-orthodox transition, and yet, in spite of claims to the contrary,[25] the osmotic effect of solutes which cannot permeate the outer membrane should theoretically be to decrease both the intracristal and matrix volumes. The answer to the paradox is that each high molecular weight solute contains either calcium or some impurity which potentiates the calcium-induced transition. The effect of each of these solutes on configuration and function can be shown to be referable to the induction of this transition.[26] Finally, it could well be that calcium-dependent permeability changes are operative in the metabolically linked configurational changes documented by Hackenbrock.[27] He observed changes in configuration of mouse liver mitochondria from aggregated ("condensed" in his terminology) to orthodox on incubation in state I or state IV. The configuration was reversible on initiation of state III respiration in the latter but not the former case. The situation is complicated by the presence of phosphate in the medium, which will potentiate the transition, and also of magnesium and ADP, which will tend to prevent the transition and reverse it, once it has occurred.[14] Hackenbrock has argued convincingly that the change to the orthodox configuration could not be the result of active ion uptake: measurements of ion movements showed no net uptake and the configurational change occurred in the presence of uncouplers.[28, 29] Other features of the configurational change which link it to the calcium-induced transition are the efflux of endogenous ions and nucleotides,[29] influx of sucrose,[30] and inhibition by inhibitors of respiration.[28]

This is not to say that mitochondria do not undergo conformational changes in direct response to energization. Such changes were carefully documented in Dr. Green's laboratory and led to the postulate that such changes might be a basis for energy transduction.[31–33] These conformational changes were found to correspond to various degrees of expansion of the matrix space; they could be mimicked in a specific sequence by either active or passive swelling of mitochondria in salt media.[34, 35] It is reasonable, therefore, to conclude that energization of mitochondria produces configurational changes by virtue of a small degree of active ion uptake.

As a summary of the role of osmosis in determining mitochondrial configuration, we can now adequately rationalize all known matrix volume changes in terms of an osmotic mechanism, whether it be active ion uptake, active ion extrusion, or passive influx via a calcium-induced permeability change.

### Other Factors That Determine Mitochondrial Conformation

To say this, however, is not to have said everything about mitochondrial configuration. Different configurations are distinctive not only by their matrix volume, which osmosis determines, but also by the characteristic shape of the cristae. For any given matrix volume (except fully expanded), there is in principle an infinite variety of ways in which the membrane could invaginate into cristae. Mitochondria in different tissues *in vivo* do, in fact, exhibit almost an infinite variety of configurations.[36] For example, cristae are seen with triangular[37] or square cross-sections,[38] and these are distributed in a variety of orderly arrays.[39] What factors determine the particular configuration a crista

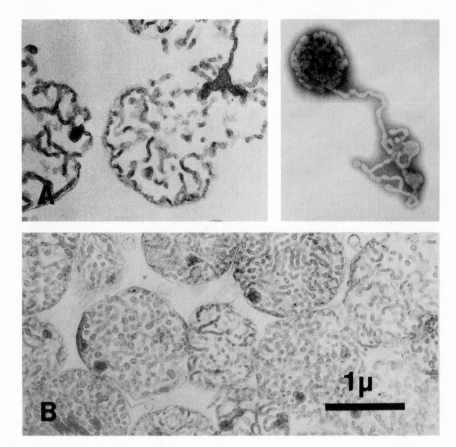

Fig. 5. Nonosmotic configurational transformations, as seen in thin section. The incubation medium contained 250 mM sucrose and 10 mM tris-Cl, pH 7.8. Before treatment, the appearance of the mitochondria was as in Fig. 1A. A: Mitochondria after 10 min incubation at room temperature in the presence of 1% silicotungstate, pH 7.4. Inset shows the appearance by negative staining. B: Mitochondria after 30 min incubation at 0°C in the presence of 0.04 mg lysolecithin/mg protein. By courtesy of H. Komai.

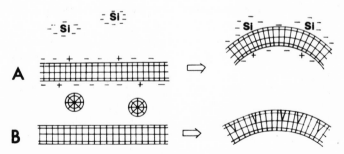

Fig. 6. Interpretive diagrams of the change in membrane shape shown in Fig. 5. A: For silicotungstate. B: For lysolecithin.

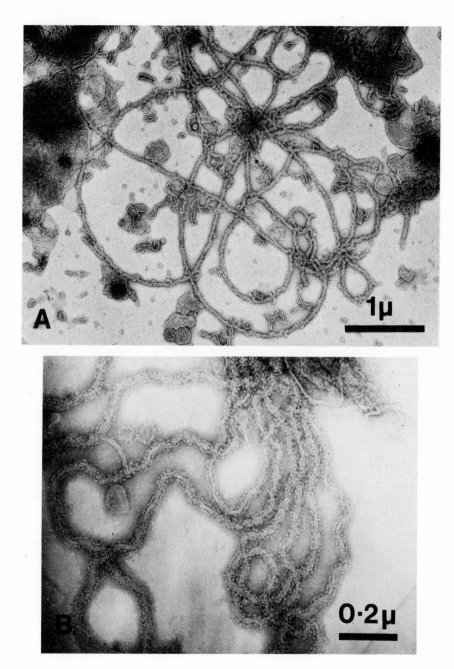

Fig. 7. Tubular cristae of swollen mitochondria as visualized by negative staining.

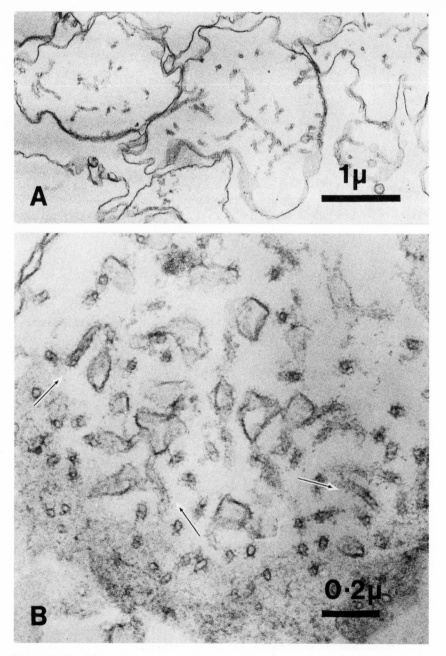

Fig. 8. Tubular cristae of swollen mitochondria as visualized in thin section. Arrows show a row of headpieces.

will assume? This question is difficult to answer in detail. We can, however, identify two factors which are likely to play significant roles in determining cristal shape.

First, there is the role of the surface pressure of the two halves of the bilayer lipid. A major difference between a flat crista and a tubular crista is that the latter requires compression of one half of the bilayer and expansion of the other half. This condition should be favored by asymmetrical surface charge densities such as pH gradients or ion binding might impose. An example of such a configurational change induced by silicotungstate is shown in Fig. 5A. Lysolecithin has a similar effect, possibly by a more mechanical mechanism (Figs. 5B and 6). In each case, there is no significant increase in matrix volume, as judged by the density of staining of the matrix, but the cristae rearrange from a configuration with a small degree of membrane surface curvature to a tubular configuration having a high degree of convexity on the intracristal surface of the membrane. It should be noted that this extent of convexity can be achieved only by the membrane going tubular, unless it is permitted to vesiculate.

The second factor which may well determine the shape of cristae is the internal structure of the membrane. Osmotically swollen mitochondria with tubular cristae typically show rows of tripartite repeating units (TRU: headpiece, stalk, basepiece) running longitudinally down the tubes.[40] We have recently refined the swelling process to the point where the tubes are irreducibly narrow and contain just two rows of TRU, over lengths of up to 20 $\mu$m.[41] These are shown in Fig. 7 by negative staining and in Fig. 8 in thin section. Figure 9 is a diagrammatic representation of the structure of the tubes. But to what extent are these ribbons of TRU products of the tube-forming process rather than determinants of it? Are the ribbons artifacts of preparative methods for electron microscopy? And are they still functional?

Sonication of such swollen mitochondria chops up the tubes into short lengths. Differential centrifugation allows a fraction to be isolated which contains mostly tubular ETP (Fig. 10A). This fraction exhibited a P/O ratio of 1.9 with NADH as substrate; it is therefore not substantially damaged. In fact, ETP made by an identical method from unswollen mitochondria showed a P/O of only 1.3 (Fig. 10B). It may be, therefore, that the high surface curvature of the tubular configuration favors tight coupling. The fact that the tubes did not revert to spherical vesicles on sonication supports the notion that the TRU ribbons are determining the tubular configuration. Finally, since conditions of specimen preparations are quite different for the two methods of visualization (Figs. 7 and 8), we can be confident that these appearances accurately reflect a real structure and are not artifactual. Taken with the data on functional

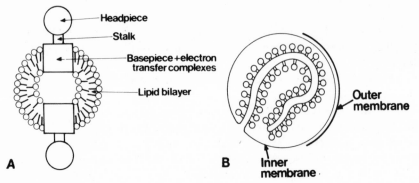

Fig. 9. Interpretive diagram of the structure and topology of the tubes shown in Figs. 7 and 8. A: Cross-section of a tube. B: Relationship of tube to rest of inner membrane.

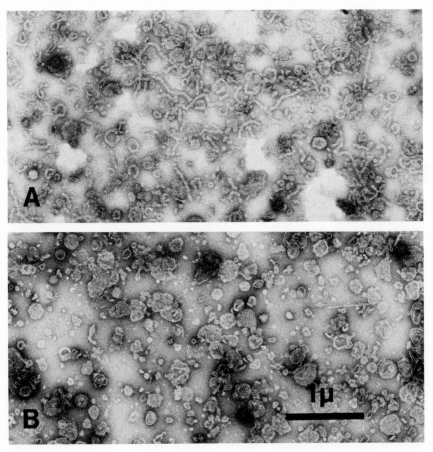

Fig. 10. A: ETP derived from swollen mitochondria with tubular cristae by sonication and differential centrifugation. The fragments retain their tubular configuration. B: ETP derived from unswollen mitochondria by the same methodology.

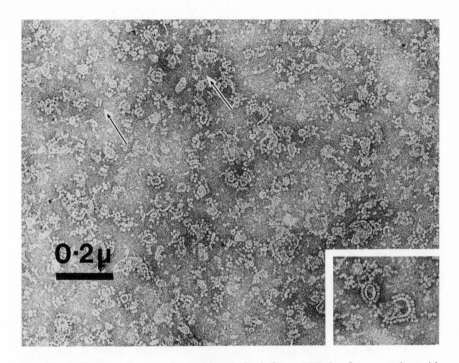

Fig. 11. "ATP-Pi exchangease" particles derived from ETP by fragmentation with lysolecithin and differential centrifugation as seen by negative staining. Arrows and inset indicate particles which clearly show ribbon structure.

integrity, we can, moreover, conclude that the "headpiece-out" conformation of the TRU is a viable conformation. This conformation may be an artifact in the sense that it may be produced during the swelling process,[42] but the TRU still works.

Independent evidence for the existence of ribbons in the normal membrane comes from the observation that they can be lifted out intact from ETP by the action of lysolecithin. Figure 11 is a picture of particles which catalyze ATP-Pi exchange.[43] The particles were isolated from ETP by treatment with lysolecithin and differential centrifugation. The ribbon structure is clearly seen (inset and arrows). These particles are 90% deficient in electron transfer complexes[43]; this allows us to identify the basepiece of the TRU as the backbone of the ribbon. The electron transfer complexes presumably interact with the TRU ribbon in the place of the membrane; we have elsewhere presented an argument for a particular arrangement of the complexes.[41]

If ribbons are present in the normal membrane, how could they operate to modulate membrane shape? It is clear from the enormous length of the tubes (Fig. 7) that the long ribbons must form by

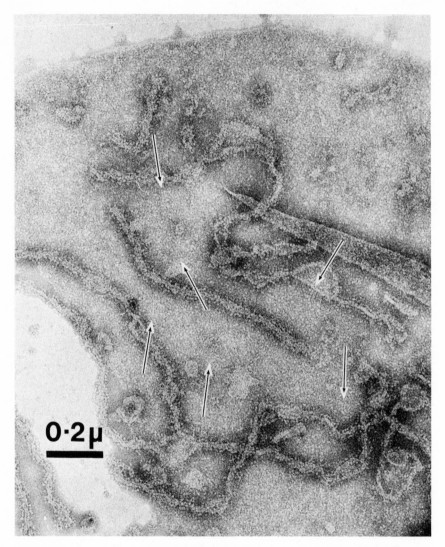

Fig. 12. Surface features of a flat part of the inner membrane, as seen by negative staining. Ribbon structure can be seen best by viewing along the arrows.

end-to-end association of shorter ribbons. Ribbon structures are visible in the normal flat membrane[44] (Fig. 12, arrows). We view these as being in a dynamic steady state of end-to-end association and dissociation, a flexibility derived from their being interspersed by bilayer lipid. Formation of tubes could occur by invagination of a single pair of ribbons to form a short tube (Fig. 13, arrow A); the process is then continued and promoted indefinitely by end-to-end association of further ribbons from the normal flat membane (Fig. 13, arrow B). This figure also suggests that the "headpiece-out" conformation of the TRU might be

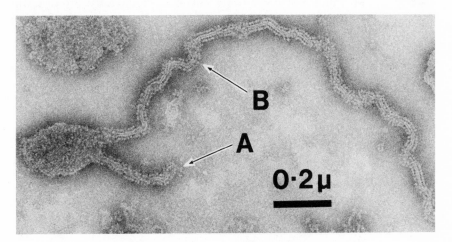

Fig. 13. Formation of tubes from the flat membrane, seen in fragments of swollen mitochondria by negative staining. Arrow A, a short tube; arrow B, a longer tube.

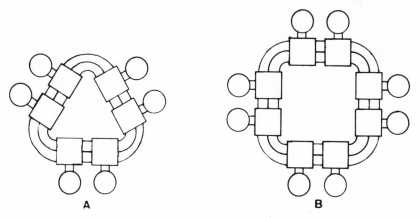

Fig. 14. A proposal for the cross-sectional structure of (A) a triangular and (B) a square tubular crista.

related to the high curvature of the membrane: this conformation is invariably seen in the tubes, while it is often absent in the vesicles.

Returning to the question of regularity in cross-section of cristae, we can begin to see how ribbon structure might promote this regularity. If, as we have found for two-ribbon tubes, the ribbons in multiribbon tubes are longitudinally oriented, then the cristae could fold preferentially at the more flexible interribbon bilayer lipid (Fig. 14).

In conclusion, it is worth contrasting the above observations on ribbon structure with the current popular concept of membrane structure as a lipid continuum containing individual freely diffusing protein units.[45] The appearance of the tubes and their behavior on sonication are both unexpected on the basis of this fluid mosaic model. The evidence for fluidity in the membrane is, however, undeniable,[48] and we do in fact need to invoke a degree of fluidity in the membrane to allow for rearrangement of ribbons. For a long time, David Green has championed the viewpoint of the membrane as a continuum of repeating lipoprotein units, each unit a self-contained complex of electron-transferring and phosphorylating enzymes.[46, 47] Ribbon structure succeeds in combining the concept of a protein continuum with that of a lipid continuum, thereby accounting for all observed physical properties of the membrane.

## References

1. Cleland, K. W. (1952) *Nature 170*, 497.
2. Stoner, C. D., and Sirak, H. D. (1969) *J. Cell Biol. 43*, 521–538.
3. Allmann, D. W., Munroe, J., Wakabayashi, Y., Harris, R. A., and Green, D. E. (1970) *J. Bioenergetics 1*, 87–107.
4. Williams, C. H., Vail, W. J., Harris, R. A., Caldwell, M., and Green, D. E. (1970) *J. Bioenergetics 1*, 147–180.
5. Chappell, J. B., and Crofts, A. R. (1965) *Biochem. J. 95*, 393.
6. Packer, L., Wigglesworth, J. M., Fortes, P. A. G., and Pressman, B. C. (1968) *J. Cell Biol. 39*, 382–391.
7. Lehninger, A. L., and Remmert, L. F. (1959) *J. Biol. Chem. 234*, 2459.
8. Slater, E., and Cleland, K. W. (1953) *Biochem. J. 53*, 557.
9. Azzi, A., and Azzone, G. F. (1966) *Biochim. Biophys. Acta 113*, 438–444.
10. Chappell, J. B., and Crofts, A. R. (1966) in *Regulation of Metabolic Processes in Mitochondria* (Tager, J. M., *et al.*, eds.), pp. 293–316, Elsevier, New York.
11. Chappell, J. B. and Crofts, A. R. (1965) *Biochem. J. 95*, 378–386.
12. Allman, D. W., Wakabayashi, T., Korman, E. F., and Green, D. E. (1970) *J. Bioenergetics 1*, 73–86.
13. Allmann, D. W., Munroe, J., Wakabayashi, Y., and Green, D. E. (1970) *J. Bioenergetics 1*, 331–353.
14. Hunter, D. R., Haworth, R. A., and Southard, J. H. (1976) *J. Biol. Chem. 251*, 5069–5077.

15. Wojtczak, L., and Lehninger, A. L. (1961) *Biochim. Biophys. Acta 51*, 442–456.
16. Hunter, D. R. and Haworth, R. A., in preparation.
17. Siliprandi, D., Toninello, A., Zoccarato, F., Rugoto, M., and Siliprandi, N. (1975) *Biochem. Biophys. Res. Commun. 66*, 956–961.
18. Massari, S., and Azzone, G. F. (1972) *Biochim. Biophys. Acta 283*, 23–29.
19. Vignais, P. V., Vignais, P. M., and Lehninger, A. L. (1964) *J. Biol. Chem. 239*, 2002.
20. Crofts, A. R. and Chappell, J. B. (1965) *Biochem. J. 95*, 387–392.
21. Brierley, G. P. and Stoner, C. D. (1970) *Biochemistry 9*, 708–713.
22. Azzone, G. F., and Permonte, G. (1969) *The Energy Level and Metabolic Control in Mitochondria* (Papa, S., *et al.*, eds.), Adriatica Editrice, Bari.
23. Hackenbrock, C., and Caplan, A. I. (1969) *J. Cell Biol. 42*, 221–234.
24. Harris, E. J., Tate, C., Manger, J. R., and Bangham, J. A. (1971) *J. Bioenergetics 2*, 221–232.
25. Bakeeva, L. E., Chentsov, Y. S., Jasaitis, A. A. and Skulachev, V. P. (1972) *Biochim. Biophys. Acta 275*, 319–332.
26. Haworth, R. A. and Hunter, D. R., in preparation.
27. Hackenbrock, C. R. (1966) *J. Cell Biol. 30*, 269–297.
28. Hackenbrock, C. R. (1968) *J. Cell Biol. 37*, 345–369.
29. Hackenbrock, C. R., Rehn, T. G., Gamble, J. L., Weinbach, E. C. and Lemasters, J. J. (1971) in *Energy Transduction in Respiration and Photosynthesis* (Quagliariello, E., Papa, S., and Rossi, C., eds.), pp. 285–305, *Adriatica Editrice, Bari*.
30. Hackenbrock, C. R. and Gamble, J. L. (1971) in *Probes of Structure and Function of Macromolecules and Membranes* (Chance, B., Lee, C. P., and Blasie, J. K., eds.), pp. 339–344, Academic Press, New York.
31. Penniston, J. T., Harris, R. A., Asai, J. and Green, D. E. (1968) *Proc. Natl. Acad. Sci. USA 59*, 624–631.
32. Harris, R. A., Penniston, J. T., Asai, J. and Green, D. E. (1968) *Proc. Natl. Acad. Sci. USA 59*, 830–837.
33. Green, D. E., Asai, J., Harris, R. A. and Penniston, J. T. (1968) *Arch. Biochem. Biophys. 125*, 684–705.
34. Asai, J., Blondin, G. A., Vail, W. J. and Green, D. E. (1969) *Arch. Biochem. Biophys. 132*, 524–543.
35. Korman, E. F., Blondin, G. A., Vail, W. J. and Green, D. E. (1970) *J. Bioenergetics 1*, 379–386.
36. Munn, E. A. (1944) *The Structure of Mitochondria*, Chaps. 1 and 2, Academic Press, New York.
37. Blinzinger, K., Newcastle, N. B. and Hager, H. (1965) *J. Cell Biol. 25*, 293–303.
38. Fain-Maurel, M. A. (1968) *C. R. Hebd. Seanc. Acad. Sci. Paris 267*, 1614–1616.
39. Korman, E. F., Harris, R. A., Williams, C. H., Wakabayashi, Y., Green, D. E. and Valdivia, E. (1970) *J. Bioenergetics 1*, 387–404.
40. Fernandez-Moran, H., Oda, T., Blair, P. V. and Green, D.E. (1964) *J. Cell Biol. 22*, 63–100.
41. Haworth, R. A., Komai, H., Green, D. E., and Vail, W. J. (1977) *J. Bioenergetics 9*, 151–170.
42. Sjostrand, F. S. (1968) in *Regulatory Functions of Biological Membranes*, BBA Library, Vol. 11 (Jarnfelt, J. ed.), pp. 1–20, Elsevier, New York.
43. Sadler, M. H., Hunter, D. R. and Haworth, R. A. (1974) *Biochem. Biophys. Res. Commun. 59*, 804–812.
44. Stiles, J. W., Wilson, J. T. and Crane F. L. (1968) *Biochim. Biophys. Acta 162*, 631–634.
45. Singer, S. J. and Nicholson, G. L. (1972) *Science 175*, 720–731.

46. Green, D. E. and Perdue, J. F. (1966) *Proc. Natl. Acad. Sci. USA 55*, 1295–1302.
47. Green, D. E., Korman, E. F., Vanderkooi, G. and Wakabayashi, T. (1970) in *Autonomy and Biogenesis of Mitochondria and Chloroplasts*, pp. 1–17, North-Holland, Amsterdam.
48. Case, G. D., Vanderkooi, J. M. and Scarpa, A. (1974) *Arch. Biochem. Biophys. 162*, 174.

## Chapter 3

# Permeability of Membrane as a Factor Determining the Rate of Mitochondrial Respiration: Role of Ultrastructure

*Jerzy Popinigis*

## Introduction

Despite the great progress made by biochemistry in recent years, there are still problems presented by phenomena the significance of which is unknown. One of these is the role played by transitions in the configurational state of mitochondria. We know that mitochondria may oscillate between two distinct ultrastructural forms, but why is far from being understood. At present, we may say that our rather good knowledge of the mechanisms working at the "molecular level" of mitochondria decreases when we encounter more complex systems and is rather poor at the level of "supramolecular organization."

It is known from the history of sciences that progress in our knowledge occurs in two stages: stage I is the accumulation of experimental data and usually takes years, and stage II is that in which rapid expansion of information allows the drawing of a general conclusion. It is my contention that the ultrastructure–metabolism interrelationship problem represents an approach to the end of stage I, although data are still incomplete.

*Jerzy Popinigis* • Laboratory of Bioenergetics, Department of Physiology and Biochemistry, "Jedrzej Sniadecki" Academy of Physical Education, 80-336 Gdańsk, Poland.

The following phenomena seem to be well documented:

a. The existence of two morphologically distinct forms of the same mitochondria confirmed by thin section, negative staining, and freeze-etching and -cleaving. [1-16]

b. The occurrence of metabolically induced ultrastructural transformations in isolated mitochondria [1-15] as well as *in situ.* [8,9,14]

c. The existence of a group of substances possessing the ability to exert a direct effect on mitochondrial ultrastructure, or to modify the occurrence of ultrastructural transitions, such as sucrose, [5-8,13,15,16] ADP, [17,18] fatty acids, [19-22] *Bordetella* endotoxin, [23] Ca, [6,24,25] halothane, [26] lipopolysaccharide from *Salmonella typhimurium,* [27] basic proteins, [28] ACTH, [29,30] insulin, [31] and uncouplers. [11,32]

On the other hand, if we assume that the mitochondrion in its two distinctive configurations is to some extent like an allosteric enzyme in its two "ground" conformational states, we may understand better which information is still lacking. Two problems seem most important:

a. Comparison of chemical, physical, and biological properties of the same mitochondria in relation to their two "ground" configurational states.

b. Determination of which metabolic functions of mitochondria are changed via ultrastructural transition.

Undoubtedly, this what we can see by electron microscopy; that is, at the so-called gross morphology level the states of the individual components within the system are expressed. Since the organization of membrane systems in "orthodox" O and "aggregated" A mitochondria (for nomenclature, see Korman *et al.* [33]) is quite different, we have to consider that an individual component of the system, e.g., an enzyme, has different molecular and functional correlations within the system in O and A mitochondria, respectively. At present we know that O and A mitochondria differ from each other in their electron microscopic appearance, water content, and ion content. [9,34] But still lacking are such important observations as the surface area characteristic of membrane surfaces exposed to an external medium and sign, charge density, protein/lipid ratio, and permeability properties. Since the transition in the configurational state changes membrane geometry, this should be followed by transition in membrane permeability. [35,36] Change in membrane geometry may induce redistribution of phospholipids. [37] All these together with the redistribution of water may have a profound effect on the metabolic properties of mitochondria. That this is indeed the case is indicated by observations that regulatory and catalytic properties of some enzymes undergo transitions together with transitions in the configurational state of mitochondria.

At present, the best-documented control is for the malic enzyme in adrenal cortex mitochondria. It was found that ultrastructural transformation of mitochondria from mode A to mode O precedes and conditions activation of this enzyme.[38] Another enzyme for which there is no doubt that its activity is affected by the configurational state of mitochondria is ATPase.[39,40] Recently, Salmenperä[30] documented the role of configurational state O for the activity of enzymes involved in steroidogenesis. Still relevant are the pioneer works of Allmann *et al.*,[5-7] who were the first to point out the role of ultrastructure in efficient oxidative phosphorylation. But generally such information is still scanty.

All above mentioned data clearly indicate the importance of problem (b) above. Moreover, for many workers such a question may be to some extent confusing because at present the role of ultrastructure is generally misinterpreted and ultrastructural transition itself is thought to be only an "end-product" of some metabolic events. Therefore, the proposal that mitochondrial function is controlled by the following causal sequence

Intramitochondrial aqueous environment
Configuration of the *de facto* unit
Conformation of the supramolecule
Work performance

as originally proposed by Green and Ji,[41] is still awaiting general acceptance. I think that even the recent findings in Green's laboratory indicating that ultrastructural transition induces disappearance of one set of coupling functions and appearance of another set of these functions[42] did not attract as much attention as they should have. Therefore, it is my contention that in order to facilitate discussion about the metabolic role of ultrastructural transitions and to increase general interest in this problem it is sometimes well to describe experimental data that are to some extent hard to explain without going into ultrastructural interpretation. The purpose of this chapter is to show some of my experimental data which are within such a scope and in addition to present some of my suggestions for discussion of their importance. But before describing details of my experiments, I should like to summarize briefly the results that in my opinion are the most interesting.

First, I should like to show that basic protein, protamine, which is well known as an inhibitor of the electron transport chain, does not necessarily inhibit mitochondrial respiration. This protein's effects on

respiration of mitochondria can be modified by influencing the configurational state, and thus, depending on the conditions, are inhibitory,
without effect, or stimulatory.

As a second interesting result I should like to show that butylmalonate, which is known as an inhibitor of succinate entry and oxidation in
mitochondria, may exert an opposite, stimulatory effect on that oxidation when added after protamine. Ultrastructural data will be presented
which permit a simple explanation of this phenomena.

Finally, it will be shown that the general anesthetic halothane exerts
its effect on succinate oxidation in a manner dependent on the initial
"ground" configurational state of mitochondria, and when added in the
same concentration may show either a stimulatory or an inhibitory effect
on respiration.

### Review of Experiments

The main purpose of my investigation was to answer the question to
what extent three phenomena—the state of mitochondrial configuration, the permeability of mitochondrial membrane to anions, and
respiratory activity measured with succinate as substrate—are interre-
lated and dependent on each other.

To determine this interdependence, the following "model" agents
were used:

a. Protamine—a basic protein with molecular weight about 8000,
   containing arginine as a main amino acid. Is supposed to interact
   with negative charges at the surface of inner membrane. Inhibits
   activity of cytochrome $c$ oxidase.[43,44] Is supposed to be "a specific
   inhibitor of electron transport chain between cytochromes $c$ and
   $a$ at the outer surface of the inner mitochondrial membrane."[45]
   With colodion membrane[46] and lipid black film,[47] was found to
   induce the phenomenon of anion-selective permeability. May be
   considered to be a polycation simulating action of arginine-rich
   histone. High affinity of interaction with phospholipids, especially with cardiolipin.

b. Gramicidin—an antibiotic possessing ionophoretic activity. Increases permeability of mitochondrial membrane to alkali cations
   and ammonia ion.[48–50] May enhance accumulation of anion
   inside the mitochondrion in a cation-dependent manner.[48,49,51]
   Is known to induce ultrastructural transition A → O under
   conditions of alkali cation accumulation or in direction O → A in
   the presence of $NH_4^+$.[50]

    c. Valinomycin—another ionophore increasing permeability of the inner mitochondrial membrane to potassium ion.[48] Under conditions of active accumulation of this ion, transport and accumulation of anions are also facilitated.[48,49,51] May therefore rearrange ultrastructure of mitochondria, changing configurational mode A to mode O.[48]

    d. Benezene-1,2,3-tricarboxylate—an agent which inhibits citrate entry into the mitochondrion.[52, 53] Is thought to inhibit the malate–citrate exchange reaction through its direct effect on the translocator catalyzing this reaction.

    e. Butylmalonate—a specific inhibitor of succinate and malate entry into the mitochondria.[54,55] Has a direct inhibitory effect on the phosphate–dicarboxylate exchange reaction.

    f. Sucrose—a bisaccharide not permeable through the inner mitochondrial membrane, and therefore an inducer of the O → A transition in mitochondrial structure.[5–8,13,15,16] There is an inverse relation between the concentration of sucrose in the incubating medium and the volume of the matrix space of the mitochondria.[56] When used in higher concentration, it suppresses respiratory activity of mitochondria,[57,58] probably through a direct inhibitory effect on anion substrate transport into the mitochondria.[58,59]

    g. Halothane—1,1,1-trifluoro-2-bromo-2-chloroethane. A general anesthetic known as an inducer of A → O transition in mitochondrial structure.[26]

The reviewed experiments were performed using mitochondria isolated from rat liver (RLM), swine liver (SLM), or beef heart (HBHM). Most of the experiments presented here have already been published before in individual papers.[60–70]

At the time I started my experiments on the effects of protamine on mitochondrial respiration, this polycation was believed to be a typical inhibitor of the respiratory chain acting directly at the level of cytochrome $c$ oxidase.[43,44] Recently, Konstantinov *et al.*[45] established more precisely that protamine acts between cytochromes $c$ and $a$ at the outer surface of the inner membrane. Reexamining the effects of this polycation on succinate oxidation in RLM, we found that the inhibitory effect of protamine on that respiration was released upon addition of DNP, citrate, and ammonia plus gramicidin, but not by gramicidin plus potassium[60–62] (Fig. 1). We also found that in submitochondrial particles, obtained by sonic disruption of RLM, succinate oxidation was not affected by protamine[60] (Fig. 1). Since gramicidin plus potassium not only was unable to overcome protamine inhibition but also, when added

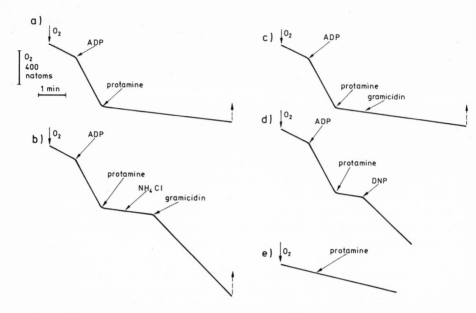

Fig. 1. Effect of protamine on succinate oxidation in RLM (expts. a–d) or sonic particles (expt. e). Respiration was measured with a Clark oxygen electrode in 3.5 ml of pH 7.3 medium containing 15 mM KCl, 50 mM tris chloride, 5 mM $MgSO_4$, 5 mM potassium phosphate, and 4 $\mu$g rotenone. Substrate: 10 mM potassium succinate. RLM in an amount corresponding to 8 mg of protein and sonic particles corresponding to 1 mg of protein were used. Additions: 300 $\mu$g protamine, 1 mM ADP, 10 mM $NH_4Cl$, 3 $\mu$g gramicidin, and 0.1 mM DNP. The dashed vertical arrows indicate stages of the experiment at which mitochondria were removed for electron microscopy. For ultrastructure, see Fig. 2.

directly after protamine, induced the phenomenon of irreversible respiratory inhibition, we performed electron microscopic examination fixing mitochondria for electron microscopy at different stages of metabolic experiments. These[61,62] experiments gave us three pieces of information (see Fig. 2):

a.  In the presence of protamines the ultrastructure of mitochondria was changed; inhibition of respiration by protamine was accompanied by extreme expansion of the matrix space, leading to disappearance of intracristal spaces.

b.  When the ability of mitochondria to respire was restored by the addition of $NH_4Cl$ plus gramicidin[61] (see Fig. 2) or by citrate,[62] the above mentioned ultrastructural transitions were found to be partially reversed.

c.  Common addition of protamine plus gramicidin in a medium containing potassium resulted in damage of mitochondria.

On the basis of these results, the following mechanism of protamine action have been proposed:

Effect of protamine on proton conductivity of membrane
Inhibition of respiration
Ultrastructural transition

We explained the ability of citrate or $NH_4+$ plus gramicidin, as well as DNP, to overcome protamine-inhibited respiration by assuming that these agents may restore proton movement across the mitochondrial membrane. In a similar manner we interpreted the ability of citrate or $NH_4+$ plus gramicidin to counteract ultrastructural effects of protamine. Even damage of mitochondria, which was observed upon addition of protamine plus gramicidin, we explained as related to enhanced $\Delta pH$.[61,62]

The unusual potency of protamine to rearrange mitochondrial ultrastructure was confirmed later with SLM and HBHM.[63] However, almost at the same time we obtained some interesting experimental data which aroused doubts as to our interpretation of protamine action. First,

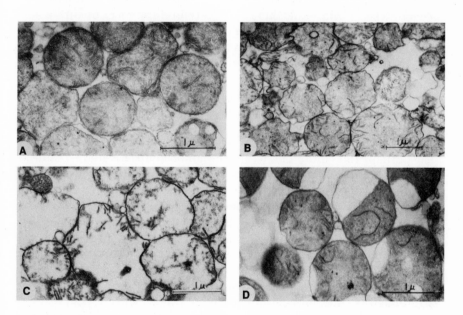

Fig. 2. Ultrastructure of RLM. A: Succinate, 7 min at state 4. B: Succinate, ADP, protamine. C: Succinate, ADP, protamine, gramicidin. D: Succinate, ADP, protamine, $NH_4Cl$, gramicidin. For details, see Fig. 1 and Popinigis *et al.*[61]

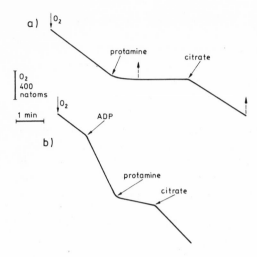

Fig. 3. Ability of citrate to overcome the inhibitory effect of protamine on succinate oxidation in HBHM. The medium (5 ml) contained 15 mM KCl, 50 mM tris chloride, 5 mM potassium phosphate, 5 mM $MgCl_2$, and 3 $\mu$g rotenone. HBHM: 6 mg. Substrate: 20 mM tris succinate. Where indicated, protamine (50 $\mu$g/mg protein) and 20 mM tris citrate were added. Final concentration of ADP in expt. B was 0.4 mM. The dashed vertical arrows indicate stages of the experiment at which mitochondria were fixed for electron microscopy. For morphology, see Fig. 7.

in the next experiments it was observed that citrate is able to overcome protamine inhibition on succinate oxidation and also in the case of HBHM (see Fig. 3). This result was confusing since in ox heart mitochondria the translocator for tricarboxylic anion is absent.[71,72] Second, when using SLM, there were certain differences in citrate and DNP action on succinate oxidation in the presence of protamine[64] (see Fig. 4). Using low (2 mM) and high (20 mM) succinate concentrations, it was found that the ability of citrate to increase oxygen consumption in the presence of protamine does not depend on substrate concentration. However, DNP reversed the inhibitory effect of protamine only with the high succinate concentration. As it is known that the entry of citrate and that of succinate are competitive in relation to each other,[73] lack of correlation between the concentration of both these anions indicated that perhaps the citrate entry into the mitochondria is not a required condition for a reversal of the protamine-induced inhibition. For these reasons, in later experiments it was decided to check whether citrate would be more efficient in reversing protamine-induced inhibition when its penetration into the mitochondria is facilitated or when it is restricted.

In order to facilitate citrate transport, experiments were carried out in the presence of valinomycin and the results were compared with those performed in the presence of substances reducing the penetrability of citrate across the mitochondrial membrane: benzene-1,2,3-tricarboxylate and sucrose. Moreover, to settle the question of whether the reversing effect of citrate is associated with its ability to remove protons, these experiments were performed by means of apparatus permitting simultaneous measurement of oxygen consumption and proton movement. The results of these experiments[65] are shown in Fig. 5. It was found that when the permeability of mitochondrial membrane to citrate was increased by valinomycin, this anion lost its properties of reversing protamine inhibition on succinate oxidation. On the other hand, when permeability was decreased by adding benzene-1,2,3-tricarboxylate or sucrose, the power of citrate to stimulate protamine-inhibited mitochondrial oxygen consumption increased. It was interesting to note that higher the sucrose concentration in the incubation medium the more evident the ability of citrate to stimulate oxygen consumption. It was also observed that the reversing action of citrate on protamine-inhibited respiration is not accompanied by net proton transport to the mitochondria. The maximum reversing effect was shown by citrate when there were no changes in concentration of

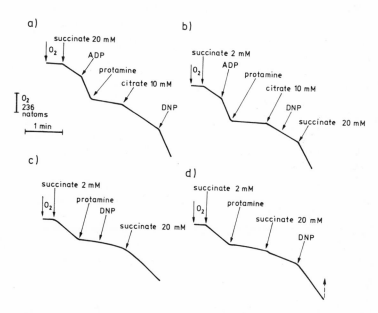

Fig. 4. Effects of DNP on protamine-inhibited succinate oxidation in SLM: dependence on substrate concentration.[64] Respiration was measured in medium containing 15 mM KCl, 50 mM tris chloride, 2 mM MgCl₂, and 4 μg rotenone. In expts. a and b, medium was supplemented with 5 mM tris phosphate. Final volume was 5 ml, pH 7.3. SLM was added in an amount corresponding to 12 mg of protein. Other additions were as indicated in the figure: 25 μg protamine per milligram of protein, 1 mM ADP, 0.1 mM DNP.

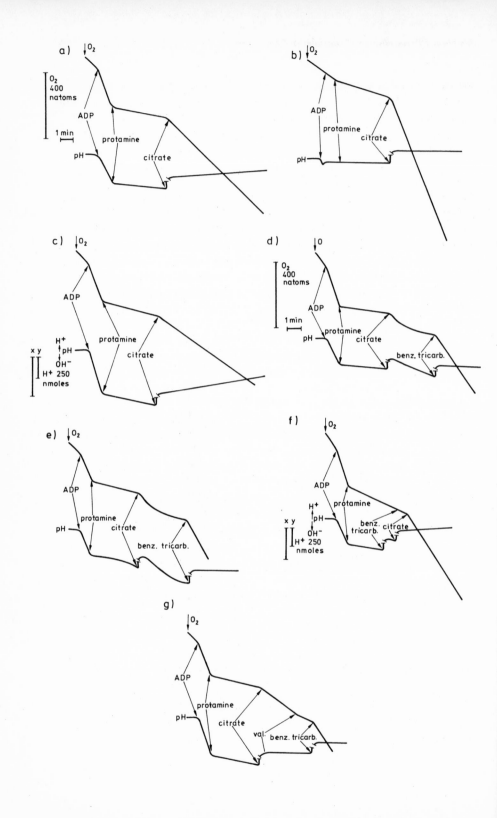

hydrogen ion in the medium, i.e., in conditions when the rate of proton transport into the mitochondria equaled their rate of exit from mitochondria.

Obviously, the results obtained were extremely interesting because they showed that citrate's ability to reverse protamine action need not be conditioned by the transport of this ion into the mitochondria. On the other hand, by showing that the metabolic effect of the anion may be inversely proportional to its ability to penetrate the mitochondrial membrane, we indicated the difficulty of solving the problem of how the anion, not penetrating the mitochondrial membrane, can exert its effect on mitochondrial metabolism. From the cellular studies being carried out at present, we know that the metabolic processes going on inside the cell are affected by several agents unable to penetrate the cell membrane. However, the mechanism of this phenomenon is still obscure. Two explanations are most attractive: the first assumes that stimulus from the surface of membrane may be transduced by changes in the physical state of lipids; the second postulates the role of contractile protein and cooperation of calcium. Naturally, similar mechanisms may be postulated for citrate when it does overcome the action of protamine. This anion, for example, may be considered as an agent removing magnesium from the matrix or mobilizing magnesium in the intracristal space of mitochondria by the formation of chelates. This may be followed by change in Mg/Ca ratio in membranes, leading to changes in the state of phospholipids and/or the state of contractile protein. But methods available did not permit me to perform such studies. Therefore, I concluded that I would have to consider another possible explanation of protamine action without going into investigations "at the molecular level." To do that, I might need to take into account Green's suggestions that ultrastructural transition may not only be the result of metabolic events but also condition the occurrence of transitions in metabolic states of the mitochondria. I decided to check by experiment the assumption that inhibition of respiration by protamine is the consequence of transition in mitochondrial structure. The following sequence of "cause and effects" was the most probable:

Fig. 5. Effect of citrate on protamine-inhibited succinate oxidation in RLM.[65] RLM (4 mg of protein) was added to 2.5 ml of pH 7.2 media containing 10 mM KCl, 3 mM MgSO$_4$, 2.5 mM potassium phosphate, 5 mM tris succinate, 3 $\mu$g rotenone, and (in expt. a) 250 mM sucrose, (b) 400 mM sucrose, (c) 100 mM sucrose, (d) 100 mM sucrose + 75 mM KCl, (e) 100 mM sucrose + 150 mM KCl, (f) 100 mM sucrose + 75 mM KCl, (g) 100 mM sucrose. Additions: 1 mM ADP, 500 $\mu$g protamine, 10 mM tris citrate, 10 mM tris benzene-1,2,3-tricarboxylate, 1 $\mu$g valinomycin. Marks $x$ and $y$ indicate changes in pH upon addition of 250 nmoles HCl in the absence ($x$) or after the addition ($y$) of citrate.

Primary effect of protamine on ultrastructure of mitochondria
Attainment of an ultrastructural form unfavorable for respiratory
chain activity
Inhibition of respiration

The advantages of this assumption were as follows:

a. It created extremely interesting possibilities for an experimental
   solution of the problem.
b. It explained opposite effects of gramicidin plus ammonia
   and gramicidin plus potassium on protamine-inhibited respiration.

In studies made in the laboratory of Chappell[50] it was found that
gramicidin evokes a dual effect on mitochondrial ultrastructure. In the
presence of ammonia in the incubation medium this antibiotic was
found to contract mitochondria, i.e., to induce an ultrastructural transition in direction O → A, opposite to that induced by protamine. On the
other hand, in the presence of potassium, gramicidin induced swelling
of mitochondria as a consequence of potassium accumulation. Since in
this latter case the ultrastructural transition occurred in the same
direction as in the case of protamine, i.e., A → O, this could not result in
the release of inhibition of respiration. Instability of the membrane in
such conditions could be explained by assuming that protamine increases conductivity of the membrane to anions[46,47] and of gramicidin
to cations. Experimental damage of the membrane upon common
addition of agents increasing the conductivity of the membrane to
anions and cations had already been described in detail by Mueller and
Rudin.[74]
    Generalizing the fact that the ability of gramicidin plus ammonia to
overcome the inhibitory effect of protamine on respiration is warranted
by the ability of this system by itself to induce the O → A transition in the
ultrastructure of mitochondria, I could presume that other systems
might act by the same mechanism. This indicated the necessity for
continuing ultrastructural studies. Some of them are described here.
    Figure 6 shows sequential rearrangements in the structure of SLM
when respiration of these mitochondria was first inhibited by protamine
and then restored by DNP. It may be seen that under the influence of
protamine typical for respiring mitochondira, configuration O underwent a change.[63] As a consequence of extreme expansion of matrix
space, intracristal spaces as well as spaces between inner boundary
membrane and outer membrane ceased to be visible. This indicates that
mitochondria reached their most extreme form of configuration O,
which presumably conditions inhibition of respiration. DNP did not

significantly affect the appearance of whole mitochondria but exerted a very marked effect on the state of intermembrane spaces. If in mitochondria treated with protamine the inner boundary membrane and outer membrane appeared to form a hybrid typical of the phenomenon of fusion, the protamine-and DNP-treated mitochondria showed reappearance of these spaces. In almost all mitochondria there are places where both an outer and an inner membrane are clearly seen again.

The possibility that inhibition of respiration may occur as a consequence of an excessive shift in mitochondrial ultrastructure in direction O was later confirmed with HBHM. For example, when protamine was added to "twisted" HBHM respiring in the presence of succinate and phosphate and then fixed for electron microscopy, we found the morphology presented in Fig. 7. Protamine exerted a more profound effect on the ultrastructure of HBHM but the direction of the transition was the same as observed earlier with RLM or SLM. Moreover, in HBHM intracristal spaces were filled with dark-staining material and completely closed.[63,66]

A visible rearrangement of mitochondrial morphology occurred upon addition of citrate. In almost all mitochondria the intracristal spaces are free again and clearly visible. Citrate-induced transitions in morphology of HBHM are better seen when we compare the appearance of protamine-treated and protamine plus citrate treated HBHM under high magnification.

As these experiments clearly supported the possibility that the metabolic effect of protamine, i.e., inhibition of respiration, is counteracted directly at the ultrastructural level, I could imagine that if under experimental conditions one might succeed in eliminating the protamine excessive shift of the ultrastructure of mitochondria in direction O, then this protein not only ought not to inhibit the respiratory activity of mitochondria but also should even stimulate respiration under conditions in which respiratory activity is restricted by the rate of anion substrate penetration into the mitochondria.

From a review of the biochemical literature in order to find a substance which could prevent a protamine-induced excessive morphological shifting in direction O, and at the same time restrict the penetration of substrate into the mitochondria, it was found that sucrose is the ideal agent for this purpose. This bisaccharide in common use for isolating mitochondria is known to penetrate the outer membrane easily, but it does not go through the inner membrane. Experiments showed that the higher the sucrose concentration in the medium suspending the mitochondria, the more contracted the matrix space.[56] In electron microscopy studies it was found that there is a direct relation between

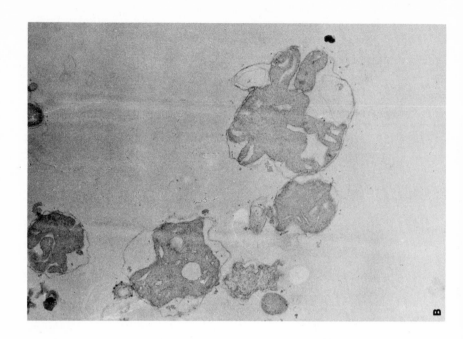

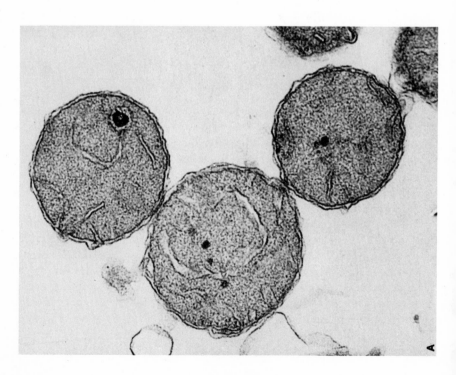

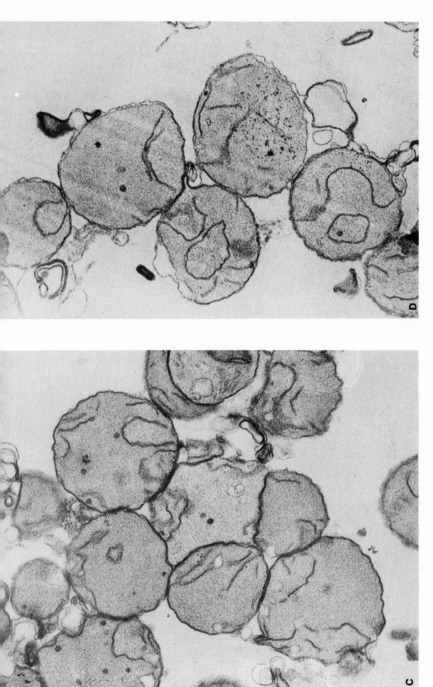

Fig. 6. Ultrastructure of SLM. A: Control, state 4 respiration. ×60,000 (reduced 33% for reproduction). B: Succinate followed by butylmalonate. ×45,000 (reduced 33% for reproduction). For experimental conditions, see Fig. 8 (expt. e). C: Succinate oxidation inhibited by protamine. ×45,000 (reduced 23% for reproduction). For experimental conditions, see Popinigis *et al.*[63] D: Succinate oxidation first inhibited by protamine and then restored by DNP. ×54,000 (reduced 33% for reproduction).

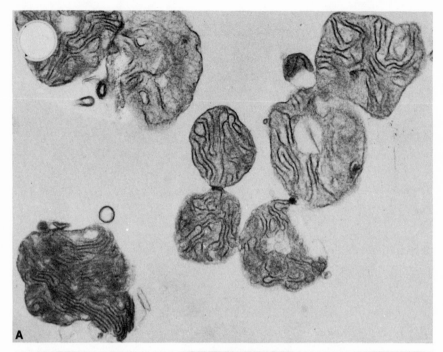

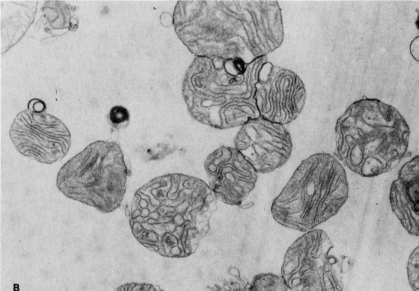

Fig. 7. Ultrastructure of HBHM. A: Succinate oxidation inhibited by the addition of protamine. ×45,000 (reduced 25% for reproduction). For experimental conditions, see Popinigis *et al.*[63] B: Succinate oxidation first inhibited by the addition of protamine and then restored by citrate. ×30,000 (reduced 33% for reproduction). For experimental

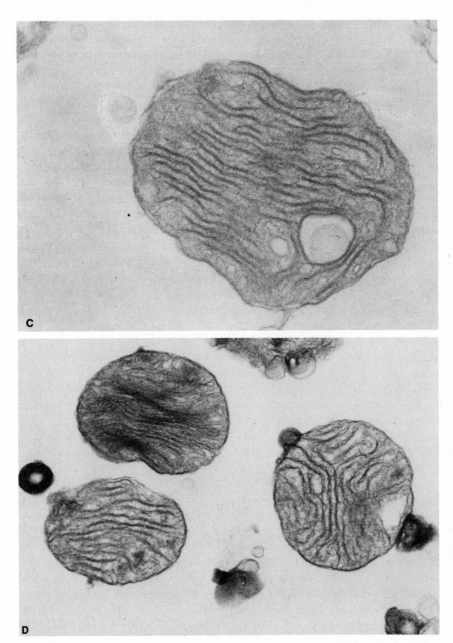

conditions, see Fig. 3. C: Succinate oxidation inhibited by the addition of protamine. ×84,000 (reduced 30% for reproduction). For experimental conditions, see Fig. 3. D: Succinate oxidation first inhibited by the addition of protamine and then restored by citrate. ×60,000 (reduced 33% for reproduction).

the sucrose concentration used and the mitochondrial configuration. Mitochondria suspended in a medium of low sucrose concentration (60–70 mM) showed configuration O, but the same mitochondria suspended in isoosmotic or higher concentrations of sucrose featured configuration A.[16,25] This indicated that sucrose induces ultrastructural transition in a direction opposite to that caused by protamine. Moreover, mitochondria suspended in media containing high sucrose concentrations have a slower respiratory rate,[57,58] probably because of the ability of sucrose to suppress activity of translocators,[59] i.e., proteins which presumably condition the entry of anions into the mitochondria.

With this information as a basis, we investigated the effect of protamine on succinate oxidation in media of varying sucrose content. The results of these experiments are given in Fig. 8. They show a direct correlation between configurational state of mitochondria and the effect of protamine on respiration. The reactions of the same mitochondria suspended in media of low and high sucrose concentrations were completely different.[67,70] If in mitochondria incubated in media with low concentrations of sucrose protamine caused an inhibitory effect, then in the same mitochondria incubated in medium with high sucrose concentration (0.5 M) the rate of respiration remained unaffected or even (in the presence of citrate) was stimulated by the protamine. The presence of citrate could be avoided by the use of higher concentrations of succinate (not shown). The most essential fact was that these varying effects, i.e., inhibition, no effect, or stimulation, could be produced by using the same protamine concentration, only modifying the configurational state of mitochondria by the use of various media.

Despite the fact that our assumption was fully confirmed by the results obtained, it seemed advisable to have more documentary evidence for the ultrastructural interpretation of the phenomena. It was necessary to show that the effect of sucrose was not conditioned merely by its presence in the medium as one of the components of the medium but that its activity was conditioned by its effects on mitochondrial structure. For this reason, the experiment with the effect of protamine on succinate oxidation in medium containing high concentrations of sucrose plus citrate was repeated in the presence of halothane.[67,69] The action of the latter compound, belonging to the group of general anesthetics, will be discussed later. It does possess the ability to cause changes in the configurational state of mitochondria. Suspended in sucrose, mitochondria showing configurational state A changed their ultrastructure under the influence of halothane to mode O.[26] It was found experimentally (Fig. 8) that the introduction of halothane into incubation medium containing high concentrations of sucrose and citrate completely changes the mitochondrial response to protamine.

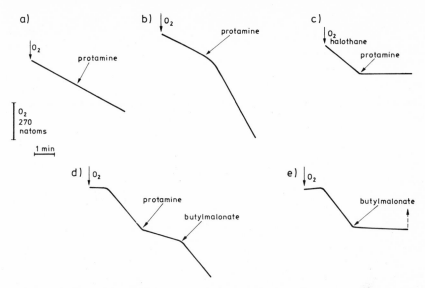

Fig. 8. Three effects of protamine on succinate oxidation in RLM. In all experiments the same amount of RLM (5 mg protein) was used. Also, protamine was added in the same concentration (corresponding 50 $\mu$g/mg mitochondrial protein). Only compositions of the media were changed. Medium (3 ml) consisted of the following. Expt. a: 500 mM sucrose, 0.5 mM potassium phosphate, 2 mM potassium succinate, and 4 $\mu$g of rotenone, pH 7.0. Expt. b: The above medium supplemented with 1.5 mM tris citrate. Expt. c: Composition the same as in expt. b, but supplemented with 10 mM halothane. Expt. d: 220 mM sucrose, 5 mM $MgCl_2$, 10 mM KCl, 20 mM tris chloride, 1 mM potassium phosphate, 0.5 mM EGTA, and 20 mM tris succinate; where indicated, 30 mM tris butylmalonate was added; pH of the medium was 7.3. In expt. e the inhibitory effect of butylmalonate on succinate oxidation in SLM was investigated. Medium (5 ml) consisted of 15 mM KCl, 50 mM tris chloride, 5 mM potassium phosphate, 5 mM $MgCl_2$, and 5 $\mu$g rotenone. Substrate: 5 mM tris succinate. SLM: 8 mg protein. Where indicated, 35 mM tris butylmalonate was added, pH 7.3. The dashed vertical arrow indicates the stage of experiment in which sample was fixed for electron microscopy.

Instead of the stimulation of respiration typical for this medium, an opposite, inhibitory effect was observed, such as protamine caused in media with a low sucrose content. This result explicitly proved it was not the presence of sucrose but the osmotic influence exerted by it on the inner membrane that was the required condition for protamine to be able to stimulate respiration. The result also had another advantage; it showed that when adding two compounds in succession, if the first agent is capable of causing transitions in the mitochondrial ultrastructure, the action of the second agent may be atypical.

There remained the question of whether and to what extent the ability of protamine to stimulate respiration might be related to its ability to increase transport of anions. If the ability of protamine to stimulate respiration in media containing high concentrations of sucrose and citrate could be explained in this way, did the same effect condition the inhibition appearing in hypo- or isoosmotic medium?

It was decided to solve the problem by means of butylmalonate. This compound is known to be a relatively specific inhibitor of succinate transport into the mitochondria.[54,55] This effect conditions the ability of butylmalonate to inhibit succinate oxidation. An attempt to reverse the respiratory inhibition caused by one agent by the use of another agent which is also inhibiting respiration would seem to be a paradoxical experiment. If, however, the activity of these agents is based on opposite mechanisms, the metabolic effect obtained as a final result of their joint use ought to be different from that of each separately. That was expected in this case. Assuming that the inhibitory effect of protamine on respiration is conditioned by lack of a permeability barrier to anions, by using a high concentration of succinate in the medium it could be supposed that respiratory inhibition would be conditioned by an excessive accumulation of substrate inside the mitochondrion. This does not mean that we expected the substrate in some specific way to inhibit its own oxidation. On the contrary, it was thought that protamine's increasing the permeability of the inner mitochondrial membrane to anions, including succinate, would induce a new Donnan equilibration of anions on both sides of the inner membrane. In such conditions succinate anion, which in the absence of protamine played an additional role as a "osmoticum" for the inner membrane losing its osmotic properties, would allow the inner membrane to expand. When the degree of matrix expansion became excessive, this might result in inhibition of respiration. Butylmalonate was thought to be an agent which in restoring the impermeability of the inner membrane to succinate should at the same time restore osmotic forces caused by this anion on the inner mitochondrial membrane. Therefore, such action should induce reduction of the volume of the matrix space, leading to restoration of mitochondrial respiration. Experiments fully confirmed these suppositions.[70] The oxygen electrode trace shown in Fig. 8 (expt. d) indicates that, indeed, when respiration of mitochondria incubated in the presence of high concentrations of succinate was inhibited by protamine, subsequent introduction of butylmalonate restored oxygen consumption.

In order to find whether butylmalonate, like other, earlier systems overcoming protamine inhibition directly at the level of ultrastructure, can induce an O → A ultrastructural transition, we performed one simple experiment. We inhibited succinate oxidation by the addition of butylmalonate and fixed the sample for electron microscopy. The morphology of butylmalonate-treated mitochondria is shown in Fig. 6. It is evident that butylmalonate inhibiting succinate oxidation in SLM induces an ultrastructural transition in a direction opposite to that induced by protamine. Therefore, if we assumed that an excessive shift in mitochondrial ultrastructure to direction O conditioned the inhibitory effect of protamine on respiration, an ultrastructural transition in the opposite direction caused by butylmalonate seems to condition inhibition of respiration by this second agent.

So we found that we had to consider the possibility that the permeability of the inner mitochondrial membrane to anions may undergo changes depending on the configurational state of mitochondria. This means that transition in the configurational state of mitochondria may be the primary event conditioning the occurrence of transitions in the permeability of the inner membrane to anions. We knew that there are two main factors controlling anion transport through the membrane. One is permeability of membrane to cations, the second is fixed negative charge. But from a structural point of view, anion conductance of the membrane depends on the degree of its "stretch." Azzone and Piemonte,[75] studying anion transport in mitochondria, noted that as mitochondria increase the volume of their matrix spaces the anion conductance of the inner membrane subsequently increases, also. In this context, we could not exclude that all abovementioned effects of protamine are related to its ability to induce a transition in configurational state in direction $A \rightarrow O$ rather than these effects being dependent on incorporation of additive positive charges at the surface of the inner membrane. Moreover, this indicated that other substances possessing the same ability as protamine to induce $A \rightarrow O$ transition in mitochondrial structure may exert protaminelike metabolic effects. We chose halothane for investigation because this general anesthetic has a potency like that of protamine to induce transition in the configurational state of mitochondria in direction $A \rightarrow O$[26] but differs from protamine in other physical and chemical properties. On the other hand, for us, most important was the fact that this agent is unable to neutralize negative charges or to incorporate positive charges at the membrane surface.

Experiments with halothane were performed using two different media. One was slightly hypotonic, allowing mitochondria to adopt configurational state O; the second was hypertonic with sucrose in order to induce the most extreme form of configurational state A. It may be seen on Fig. 9 that halothane when introduced in the same 4 mM concentration to both media exerted opposite effects on the rate of respiration. In hypotonic medium (Fig. 9a) the observed effect was inhibitory, whereas in the hypertonic medium (Fig. 9b) the rate of oxygen consumption was stimulated. Further increase in halothane

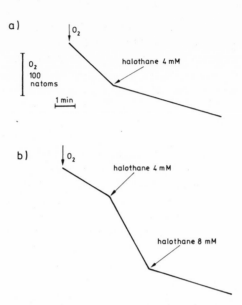

Fig. 9. Two effects of halothane on succinate oxidation in RLM: dependence on the "ground" configurational state of mitochondria. RLM (2 mg protein) was added to 3 ml of pH 7.0 media consisting of (expt. a) 60 mM sucrose, 30 mM tris chloride, 0.5 mM potassium phosphate, and 10 mM tris succinate, or (expt. b) 500 mM sucrose, 0.5 mM potassium phosphate, and 10 mM tris succinate. Both media also contained rotenone in a concentration of 4 μg/ml. Halothane was introduced where indicated.

concentration also caused inhibition in this medium. Since in both experiments the amount of mitochondria added was the same, data of these experiments clearly indicate that the so-called ground configurational state of mitochondria determines the permeability properties of the membrane and the response of mitochondrial respiration to added effectors.[69]

### Concluding Comments

In studying any metabolic problem we may consider it as solved when all the mechanisms involved are defined at the so-called molecular level. In this chapter, in order to show how important to present-day biochemistry is the suggestion of Green and Ji about the role of suprastructure in the regulation of metabolic processes, I have purposely omitted discussion of the events occurring at the molecular level. I focused my attention on only one problem—the role of ultrastructure in regulation of respiration of mitochondria.

The results presented indicate that a significant interrelationship exists for three phenomena: the configurational state of mitochondria, the anion permeability of the inner membrane, and the respiratory activity of mitochondria. My experimental data, together with the observations of other authors,[32, 76-78] allow me to present for discussion the following correlation:

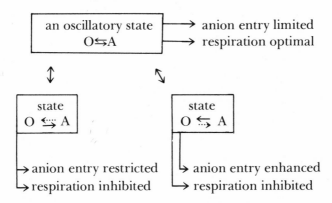

*Regulation of Respiration*

As it is known that mitochondrial respiration may support either active contraction or active swelling of mitochondria,[76-78] optimal conditions for respiration seem to exist when mitochondria are allowed to freely change their configurational state, i.e., during an oscillatory state. This oscillatory state seems to be obligatory for energy dissipation; i.e., only when this condition is fulfilled may the energy formed by the respiratory chain be dissipated. In this sense, uncoupling *per se* is conditioned by the ability of mitochondria to undergo quick ultrastructural transitions in which both phases of the oscillatory state or exclusively one phase (i.e., contraction or swelling) is energy driven. To fulfill the condition for the oscillatory state in this latter case, the occurrence of passive ultrastructural movements in an opposite direction is required. When respiration supports active swelling, then to achieve optimal conditions for respiration the occurrence of a passive contraction phase is required. And, *vice versa*, when the contraction phase is energy dependent, optimal conditions for respiration are provided by the presence of a passive swelling system in the incubation medium. It is worth mentioning that such an oscillatory state may take place within any configurational state and not affect the morphology of mitochondria. Disequilibrium between two phases of the oscillatory state always leads to inhibition of respiration. Therefore, from the structural point of view (see diagram presented above) inhibition of respiration may result from excessive shift in configurational state of mitochondria in either direction O or in direction A. When the changes have not yet become irreversible, the ultrastructurally conditioned inhibition of respiration may be overcome. When inhibition of respiration depends on excessive shift in mitochondrial structure (O ⇆ A) in direction O, a reversing effect would be exerted by those substances which induce contraction of mitochondria and which therefore would restore the occurrence of reversible O ⇆ A transitions. Similarly, when inhibition of respiration is conditioned by excessive shifting of the configurational state of mitochondria in direction A (O ⇆ A), the rate of respiration would increase upon addition of agents allowing mitochondria to swell, as this permits regaining an O ⇆ A oscillatory state. This explains why protamine and halothane increased the rate of succinate oxidation in mitochondria which had a contracted matrix and at the same time exerted an inhibitory effect on respiration of mitochondria with an expanded matrix.

*Problem of Permeability*

Changes in degree of permeability of a membrane to anions seem to follow ultrastructural transitions; i.e., anion conductance of the membrane increases together with increase in membrane stretch. Therefore, transition in the morphology of mitochondria from initial configurational state A to state O would condition the induction of anion conductivity of the membrane, i.e., transition from an anion-impermeable state of the membrane to another state in which the membrane is easily penetrable by anions. An ultrastructural transformation in the opposite direction, i.e., from configurational state O to state A, would reduce the anion conductivity of the membrane. Experiments which we are carrying out at the present time indicate that the ability to induce transition in configurational state O → A is obligatory for other transport inhibitors also to stop anion entry into the mitochondria. At present, we do not have enough experimental data to explain such a relationship between configurational state of mitochondria and anion conductivity of the membrane; we think that the most important factor is change in membrane geometry.

It is important to mention that changes in membrane permeability may not only occur as a consequence of ultrastructural transition but also be primary events conditioning the occurrence of such transitions. An example of such dependence is represented by the effects of halothane shown in Fig. 9 (expt. b). First, isolated mitochondria were suspended in medium containing 0.5 M sucrose, therefore creating conditions in which under the influence of high osmotic pressure mitochondria adopted the O $\leftrightsquigarrow$ A configuration. As is indicated in the preceding diagram, in such conditions both succinate penetration into the mitochondria and the rate of oxygen consumption of these mitochondria are restricted. Under the influence of 4 mM halothane, sucrose is allowed to pass the inner membrane and therefore the effective osmotic pressure decreases, allowing mitochondria to adopt the O $\leftrightarrows$ A configuration. Now we have a situation in which the rate of respiration is no longer restricted by succinate penetration and the rate of respiration is maximal because of energy dissipation by the O $\leftrightarrows$ A oscillatory state. But what may happen when we further increase the permeability of the membrane to sucrose, increasing the halothane concentration? Further decrease in effective osmotic pressure will induce an inhibitory state of respiration again, but in this case the inhibition is dependent on shifting the mitochondrial structure to configurational state O $\leftrightsquigarrow$ A. Now this inhibition may be overcome by substances restoring the O $\rightleftarrows$ A oscillatory state.

It is my contention that using the Green and Ji model we may explain other problems which at present are difficult to understand.

ACKNOWLEDGMENTS

I am greatly indebted to Professors M. Żydowo, L. Żelewski, S. Angielski, and E. Boj from the Medical School, Gdańsk, and Professors D. E. Green and C. H. Williams, University of Wisconsin, Madison, for their permission to carry out my experiments in their laboratories; to Professors Y. Takahashi, T. Wakabayashi, T. Wrzołkowa, and Dr. A. Roszkiewicz for taking electron microscopic pictures for my experiments; to all my colleagues and the coauthors of my papers for cooperation and discussions; to Professor W. Rzeczycki for his interest and encouragement; to Professors S. Ji and W. J. Vail for stimulating correspondence; and to Dr. Michał Woźniak for his advice.

## References

1. Hackenbrock, C. R. (1966) *J. Cell Biol. 30*, 269.
2. Hackenbrock, C. R. (1968) *J. Cell Biol. 37*, 345.
3. Green, D. E., Asai, J., Harris, R. A., and Penniston, J. T. (1968) *Arch. Biochem. Biophys. 125*, 684.
4. Harris R. A., Asbell, M. A., Asai, J., Jolly, W. W., and Green, D. E. (1969) *Arch. Biochem. Biophys. 132*, 545.
5. Allmann, D. W., Wakabayshi, T., Korman, E. F., and Green, D. E. (1970) *J. Bioenergetics 1*, 73.

6. Allmann, D. W., Munroe, J., Wakabayshi, T., Harris, R. A., and Green, D. E. (1970) *J. Bioenergetics 1*, 87.

7. Allmann, D. W., Munroe, J., Wakabayashi, T., and Green, D. E. (1970) *J. Bioenergetics 1*, 331.

8. Williams C. H., Vail, W. J., Harris, R. A., Caldwell, M., Green, D. E., and Valdivia, E. (1970) *J. Bioenergetics 1*, 147.

9. Hackenbrock, C. R., Rehn, T. G., Gamble, J. L., Weinbach, E. C., and Lemasters, J. J. (1971) in *Energy Transduction in Respiration and Photosynthesis* (Quagliariello, E., Papa, S., and Rossi, C. S., eds.), p. 285, Adriatica Editrice, Bari.

10. Muscatello, U., Guarriera-Bobyleva, V., and Buffa, P. (1972) *J. Ultrastruct. Res. 40*, 215, 235.

11. Muscatello, U., Guarriera-Bobyleva, V., Pasquali-Ronchetti, I., and Ballotti-Ricci, A. M. (1975) *J. Ultrastruct. Res. 52*, 2.

12. Vail, W. J., and Riley, R. K. (1971) *Nature 231*, 525.

13. Wrigglesworth, J. M., Packer, L., and Branton, D. (1970) *Biochim. Biophys. Acta 205*, 125.

14. Vail, W. J., and Riley, R. K. (1972) *Proc. W. Va. Acad. Sci. 44*, 103.

15. Green, D. E., and Harris, R. A. (1970) in *The Physiology and Biochemistry of Muscle as a Food* (Briskey, J., Cassens, R. G., and March, B. B., eds.), p. 239, University Wisconsin Press, Madison.

16. Packer, L. (1973) *Biokhimyia 38*, 1288.

17. Weber, N. E., Blair, P. V. and Martin, B. (1970) *Biochem. Biophys. Res. Commun. 41*, 821.

18. Stoner, C. D., and Sirak, H. D. (1973) *J. Cell Biol. 56*, 51.

19. Lehninger, A. L., and Remmert, L. F. (1959) *J. Biol. Chem. 234*, 2459.

20. Avi-Dor, Y. (1960) *Biochim. Biophys. Acta 39*, 53.

21. Zborowski, J., and Wojtczak, L. (1963) *Biochim. Biophys. Acta 70*, 596.

22. Wojtczak, L., Bogucka, K., Sarzała, M. G., and Załuska, H. (1969) in *Mitochondria, Structure and Function* (Ernster, L., and Drahota, Z., eds.) p. 79, Academic Press, New York.

23. Harris, R. A., Harris, D. L., and Green, D. E. (1968) *Arch. Biochem. Biophys. 128*, 219.

24. Harris, R. A., Asbell, M. A., Asai, J., Jolly, W. W., and Green, D. E. (1969) *Arch. Biochem. Biophys. 132*, 545.

25. Hunter, D. R., Haworth, R. A. and Southard, J. H. (1976) *J. Biol. Chem. 251*, 5069.

26. Taylor, C. A., Williams, C. H., Wakabayashi, T., Valdivia, E., and Green, D. E. (1972) in *Cellular Biology and Toxicity of Anesthetics* (Fink, B. R., ed.), p. 117, Williams and Wilkins, Baltimore.

27. Greer, G. G., Epps, N. A., and Vail, W. J. (1973) *J. Infect. Dis. 127*, 551.

28. Johnson, C. L., Goldstein, M. A., and Schwartz, A. (1973) *Arch Biochem. Biophys. 157*, 597.

29. Kimmel, G. L., Peron, F. G., Haksar, A., Bedigian, E., Robidoux W. F., and Lin, M. T. (1974) *J. Cell Biol. 62*, 152.

30. Salmenperä, M. (1976) *J. Ultrastruct. Res. 56*, 277.

31. Wagle, S. R., Ingebretsen, W. R., and Sampson, L. (1973) *Biochem. Biophys. Res. Commun. 53*, 937.

32. Blair, P. V. and Munn, E. A. (1972) *Biochem. Biophys. Res. Commun. 49*, 727.

33. Korman, E. F., Addink, A. D. F., Wakabayashi, T., and Green, D. E. (1970) *J. Bioenergetics 1*, 9.

34. Hackenbrock, C. R., and Gamble, J. L. (1971) in *Probes of Structure and Function of Macromolecules and Membranes*, (Chance, B., Lee, C. P., and Kent Blasie, J., eds.), p.339, Academic Press, New York.

35. Richardson, I. W., Ličko, V., and Bartoli, E. (1973) *J. Membrane Biol. 11*, 293.

36. Chan, S. I., Sheetz, M. P., Seiter, C. H. A. Feigenson, G. W., Hsu, M., Lau, A., and Yau, A. (1973) *Ann. N.Y. Acad. Sci. 222*, 499.

37. Israelachvili, J. N. (1973) *Biochim. Biophys. Acta 323*, 659.
38. Pfeiffer, D. R., Kuo, T. H., and Tchen, T. T. (1976) *Arch Biochem. Biophys. 176*, 556.
39. Cereijo-Santalo, R. (1972) *Arch Biochem. Biophys. 150*, 542.
40. Nicholls, D. G., and Lindberg, O. (1972) *FEBS Lett. 25*, 61.
41. Green, D. E., and Ji, S. (1973) in *Membrane Structure and Mechanism of Biological Energy Transduction* (Avery, J., ed.), p. 159, Plenum, New York.
42. Southard, J. H., and Green, D. E. (1974) *Biochem. Biophys. Res. Commun. 61*, 1310.
43. Smith, L., and Conrad, H. (1958) *Fed. Proc. 17*, 313.
44. Person, P., and Fine, A. S. (1961) *Arch. Biochem. Biophys. 94*, 392.
45. Konstantinov, A. A., Maslov, S. P., Severina, I. I., and Skulachev, V. P. (1975) *Biokhimyia 40*, 401.
46. Solner, K. (1945) *J. Phys. Chem. 49*, 47, 171.
47. Mueller, P., and Rudin, D. O. (1968) *Nature 217*, 713.
48. Pressman, B. C. (1969) in *Mitochondria, Structure and Function* (Ernster, L., and Drahota, Z., eds.), p. 315, Academic Press, New York.
49. Rottenberg, H., and Solomon, A. K. (1969) *Biochim. Biophys. Acta 139*, 48.
50. Chappell, J. B., and Crofts, A. R. (1965) *Biochim. J. 95*, 393.
51. Harrris, E. J., Van Dam, K., and Pressman, B. C. (1967) *Nature 213*, 1126.
52. Robinson, B. H., Williams, G. R., Halperin, M. L., and Leznoff, C. C. (1971) *Eur. J. Biochem. 20*, 65.
53. Robinson, B. H. (1971) *FEBS Lett. 16*, 267.
54. Robinson, B. H., and Chappell, J. B. (1967) *Biochem. Biophys. Res. Commun. 28*, 249.
55. Meijer, A. J., and Tager, J. M. (1969) *Biochim. Biophys. Acta 189*, 136.
56. Tarr, J. S., and Gamble, J. L. (1966) *Am. J. Physiol. 211*, 1187.
57. Slater, E. C., and Cleland, K. W. (1967) *Biochem. J. 53*, 557.
58. Johnson, D., and Lardy, H. (1958) *Nature 181*, 701.
59. Chappell, J. B., and Haarhoff, K. N. (1967) in *Biochemistry of Mitochondria* (Slater, E. C., Kaniuga, Z., and Wojtczak, L., eds.), p. 75, Academic Press, New York.
60. Popinigis, J., Rzeczycki, W., and Swierczyński, J. (1970) *FEBS Lett. 8*, 149.
61. Popinigis, J., Swierczyński, J., and Wrzołkowa T. (1970) *FEBS Lett. 9*, 309.
62. Popinigis, J., Wrzołkowa, T., and Swierczyński, (1971) *Biochim. Biophys. Acta 245*, 70.
63. Popinigis, J., Takahashi, Y., Wakabayashi, T., Hull, R. M., and Williams, C. H. (1971) *FEBS Lett. 19*, 221.
64. Popinigis, J., Smoly, J. M., Brucker, R. F., and Williams, C. H. (1972) *Abstr. Commun. Fed. Eur. Biochem. Soc.*, 634.
65. Popinigis, J., Wozniak, M., Nowicka, C., and Gmaj, P. (1973) *FEBS Lett. 32*, 339.
66. Popinigis, J., and Takahashi, Y. (1973) *Physiol. Chem. Phys. 5*, 477.
67. Popinigis, J. (1974) *FEBS Lett. 41*, 46.
68. Popinigis, J., and Wozniak, M. (1975) *Ann. Med. Sect. Pol. Acad. Sci. 20*, 113.
69. Popinigis, J. (1976) in *Bioenergetics and Mitochondria* (Russanov, E., and Balevska, P., eds.), p. 53, Publishing House of the Bulgarian Academy of Sciences, Sofia.
70. Popinigis, J. (1976) in *Bioenergetics and Mitochondria* (Russanov, E., and Balevska, P., eds.), p. 151., Publishing House of the Bulgarian Academy of Sciences, Sofia.
71. England, P. J., and Robinson, B. H. (1969) *Biochem. J. 112*, 8P.
72. Sluse, F. E., Meijer, A. J. and Tager, J. M. (1971) *FEBS Lett. 18*, 149.
73. Quagliariello, E., and Palmieri, F. (1971) in *Energy Transduction in Respiration and Photosynthesis* (Quagliariello, E., Papa, S., and Rossi, C. S., eds.), p. 205, Adriatica Editrice, Bari.
74. Mueller, P., and Rudin, D. O. (1967) *Nature 213*, 603.
75. Azzone, G. F., and Piemonte, G. (1969) in *The Energy Level and Metabolic Control in Mitochondria* (Papa, S., Tager, J. M., and Slater, E. C. eds.), p. 115, Adriatica Editrice, Bari.

76. Chappell, J. B., and Crofts, A. R. (1965) *Biochem. J.* *95*, 378.
77. Crofts, A. R., and Chappell, J. B. (1965) *Biochem. J.* *95*, 387.
78. Packer, L., and Wrigglesworth, J. M. (1969) in *The Energy Level and Metabolic Control in Mitochondria* (Papa, S., Tager, J. M., Quagliariello, E., and Slater, E. C., eds.), p. 125, Adriatica Editrice, Bari.

## Chapter 4

# Chemistry of Intrinsic Membrane Protein Complexes: Studies on Complex III and Cytochrome c Oxidase

## Roderick A. Capaldi

### Membrane Structure and Intrinsic Membrane Proteins

The picture of membranes that has emerged recently is one with the lipids in the form of a bilayer with proteins sitting on or penetrating into or completely through the lipid continuum. [1,2] Proteins which enter the bilayer are then called "intrinsic membrane proteins" to distinguish them from proteins which are loosely associated with the surface of the membrane, the "extrinsic membrane proteins." [1,3]

Intrinsic membrane proteins occupy a unique environment within the cell, partly buried in a hydrocarbon medium and partly exposed to the aqueous phase, and because of this they might be expected to have properties which distinguish them from extrinsic proteins or nonmembrane proteins in general. The key requirement is that intrinsic proteins provide an apolar face for association with fatty acid tails while maintaining a polar surface exposed to water; i.e., they must be amphipathic. In almost all cases, the amphipathic character of intrinsic proteins is reflected in their primary structural characteristics. Globular proteins such as purple membrane protein and rhodopsin, which are deeply buried in the lipid bilayer, contain an unusually large percentage of hydrophobic amino acids; i.e., they have a low polarity [see ref. (4)].

*Roderick A. Capaldi* • Institute of Molecular Biology, Department of Biology, University of Oregon, Eugene, Oregon 97403.

Other proteins, such as cytochrome $b_5$,[5] cytochrome $b_5$ reductase,[6] and glycophorin,[7] are intrinsic to the membrane, but the bulk of the protein is not deeply buried within the bilayer. Instead, they have segments of chain which are unusually hydrophobic and which serve to anchor them in the bilayer. Yet other proteins, such as the major lipoprotein in the outer membrane of *Escherichia coli,* do not have particularly hydrophobic segments, but the polypeptide chain, when folded into an $\alpha$ helix, gives a hydrophobic face and a hydrophilic face to the protein.[8] It has been proposed that such proteins aggregate in membranes to give a structure with a polar interior (which may act as a pore through the bilayer) and a continuous hydrophobic exterior for interaction with lipids[9] (Fig. 1).

Clearly, for single polypeptide chains from which the above examples have been drawn, there are many structural arrangements which provide the amphipathic structure needed to stabilize intrinsic membrane proteins in the lipid bilayer.

## Multicomponent Complexes and the Mitochondrial Electron Transport Chain

Many proteins in membranes do not exist as single polypeptide chains but are multisubunit complexes, often composed of several distinct proteins integrated into a functional unit. The clearest examples are the electron transfer complexes in the mitochondrial inner membrane which span this membrane and serve to transfer electrons in a series of discrete steps from NADH or succinate to molecular oxygen while conserving the energy generated in these reactions for ion transport or ATP synthesis.[10] Each electron transfer complex contains functionally distinct proteins, i.e., the cytochromes and/or nonheme iron proteins involved in the individual steps of electron transfer. Our goal is to understand the structural organization of these electron transfer complexes and in particular the structure of ubiquinone cytochrome $c$ reductase (also called reduced coenzyme Q cytochrome $c$ reductase or complex III) and cytochrome $c$ oxidase (cytochrome $aa_3$ or complex IV).

Complex III and cytochrome $c$ oxidase can be isolated by detergent solubilization of mitochondria followed by ammonium sulfate fractionation[10] or immunoprecipitation.[11] There is good evidence that the nondenaturing detergents used act primarily to break protein–lipid and lipid–lipid interactions[12] in a membrane. It is not unreasonable, then, to conclude that the association of polypeptides, hemes, and other electron acceptors isolated together in stoichiometric amounts in these electron transfer complexes represents stable structural units as well as functional segments of the mitochondrial inner membrane.

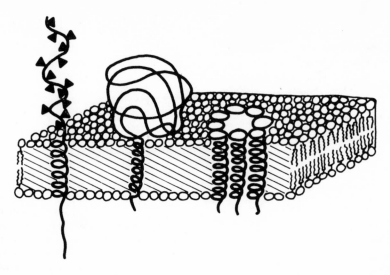

Fig. 1. Schematic of a membrane showing a lipid bilayer and three types of intrinsic membrane proteins. The one on the left is a glycoprotein such as glycophorin. The center protein has a small segment of polypeptide chain within the bilayer, e.g., cytochrome $b_5$. The right-hand protein is an aggregate of several helical polypeptides forming a pore through the lipid as proposed for the lipoprotein of *E. coli.*

## Ubiquinone Cytochrome Reductase (Complex III)

Complex III, as isolated, functions to transfer electrons from reduced ubiquinone to cytochrome $c$ while conserving the energy of this reaction for coupled events. The purified complex contains components required for both electron transport and energy-conserving functions. The purest preparations contain $b$ heme, $c_1$ heme, and nonheme iron centers in the molar ratio 2:1:1.[13] The simplest interpretation of the many studies of the redox properties of the various electron acceptors is that there are two spectrally distinct $b$ cytochromes ($b_K$ and $b_T$) of high potential and two electron acceptors, cytochrome $c_1$ and nonheme iron protein, of low potential.[14] Proposals that there are other electron transfer components in complex III are generally unsupported.

## Polypeptide Composition

Estimation of the number and molecular weights of polypeptides in complex III is made difficult by limitations in the commonly used SDS-polyacrylamide gel electrophoresis procedures. As shown in Fig. 2,

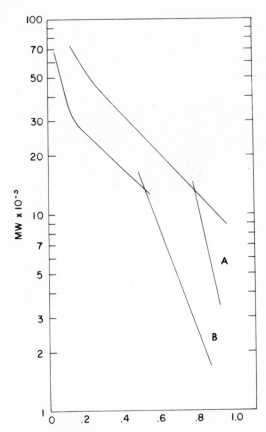

Fig. 2. Standard curves constructed for 10% Weber–Osborn gels (A) and 12.5% Swank–Munkres gels (B). The linear relationship between log molecular weight and migration does not hold for relatively large proteins in any gel system. As shown by the break in the plots, polypeptides of molecular weight below 12,000 are not so well resolved as larger components in either gel system. Of the two procedures, the Swank–Munkres system does a much better job of separating lower molecular weight proteins.

most gel procedures do not separate low molecular weight polypeptides very well. Complex III and cytochrome $c$ oxidase each contain several subunits of low molecular weight which are very difficult to resolve. We have obtained the best separation of polypeptides in complex III on 9% or 10% gels (total acrylamide), made with high levels of cross-linker (1:10 bisacrylamide–acrylamide),[15] by electrophoresing in the Swank–Munkres buffer system, which contains both SDS (0.1%) and 8 M urea.[15] The typical polypeptide profile obtained is shown in Fig. 3. Eight major bands are resolved, with molecular weights ranging from close to 50,000 to below 5000.

The identity of the various polypeptides has been attempted by fractionating complex III into cytochrome $c_1$ rich and cytochrome $b$ rich fractions or by purifying cytochrome $c_1$, cytochrome $b$, or nonheme iron protein by published procedures.[15] Polypeptide IV is the cytochrome $c_1$ hemoprotein, while polypeptide III has been tentatively identified as the cytochrome $b$ apoprotein. According to Yu et al.,[17] it can be isolated with heme attached. The identities of other polypeptides are listed in Table I.

An added problem in analyzing the polypeptide composition of complex III is that the migration of some polypeptides is different under different gel conditions.[18] This is clearly shown by the two-dimensional gel in Fig. 4, in which complex III has been electrophoresed in the horizontal dimension in the Weber–Osborn gel system[9] and in the vertical dimension in the Swank–Munkres gel system. Polypeptides not only run with different molecular weights (as listed in Table 1) but also change places in the profile. It can be seen in Fig. 3 that VII migrates behind VI in Weber–Osborn gels but ahead of it in Swank–Munkres gels. Effects like this make comparison of results from different laboratories difficult, unless identical gel procedures have been used.

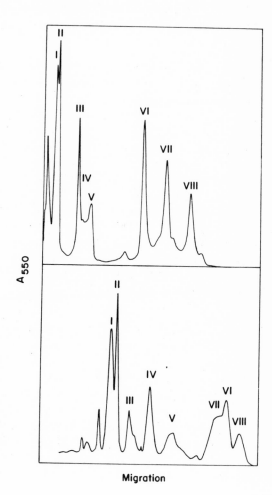

Fig. 3. Densitometric traces of SDS-polyacrylamide gels of complex III. The upper trace shows a 9% gel of enzyme run in the Swank–Munkres gel system. Components are numbered in the reverse of the order in which they migrate on this gel. The bottom trace shows enzyme run on a 10% Weber–Osborn gel.

Table I. Differing Molecular Weight Determinations for the Polypeptides in Complex III

| | | Estimated molecular weights | | | |
|---|---|---|---|---|---|
| Polypeptide | Tentative assignment | On Swank–Munkres gels | Tris acetate gels | Weber–Osborn gels | Gdn HCl columns |
| I | Core protein I | 45,500 | 50,000 | 46,600 | |
| II | Core protein II | 44,500 | 46,000 | 41,500 | |
| III | Cytochrome $b$ | 28,600 | 31,500 | 38,000 | |
| IV | Cytochrome $c_1$ | 26,700 | 29,000 | 30,000 | |
| V | Nonheme Fe protein | 24,600 | 25,000 | 23,300 | |
| VI | Cyt $c_1$ assoc. | 15,000 | 14,000 | 12,700 | 12,700 |
| VII | Cyt $b$ assoc. | 9,000 | 12,500 | 14,300 | 9,800 |
| VIII | Antimycin binding (?) | 4,800 | 9,000 | 5,600 | |

Yet another anomaly is that the migration of polypeptide III depends on the conditions used to dissociate the complex for electrophoresis. If samples are incubated for electrophoresis on Weber–Osborn gels at room temperature, component III runs as a broad band with a molecular weight centered around 45,000. If the enzyme is dissociated in SDS by warming to 37°C for 2 hr, III migrates around 35,000 while if the enzyme is heated to 100°C before electrophoresis, a major portion of this component migrates at 30,000. It is not surprising then that the identification of III with cytochrome $b$ has been missed in several studies (see refs. 20, 21). The most likely explanation for the strange behavior of III is that cytochrome $b$ resists denaturation in SDS but is unfolded by heating. On Swank–Munkres gels, III runs at around 30,000 whether the sample is heated before electrophoresis or not. Presumably having 8 M urea in the dissociation medium aids in the unfolding of the protein by SDS at lower temperatures.

## Primary Structural Characteristics of the Components

We have devised methods to purify each of the eight polypeptides seen. Components III, IV, V, and VI have been obtained by gel filtration in SDS or Gdn HCl, while I, II, VII, and VIII were obtained by preparative gel electrophoresis.[11,15] The amino acid compositions of several of the components are compared in Table II.

Polypeptide III is hydrophobic in character, with a polarity below 40%; the remaining components and the complex as a whole have polarities similar to those of water-soluble proteins.

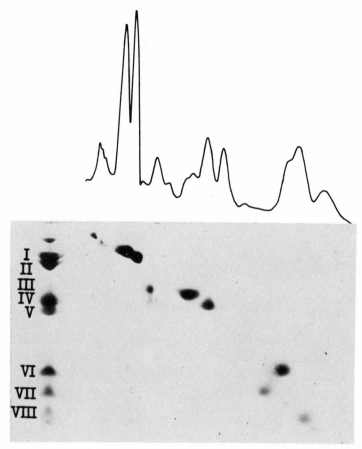

Fig. 4. Two-dimensional SDS-polyacrylamide gel electrophoresis showing the relationship of the bands in complex III as separated on Weber–Osborn gels and Swank–Munkres gels. Complex III was dissociated in 3% SDS and 10% $\beta$-mercaptoethanol by incubating at room temperature for 2 hr, and 100 $\mu$g of this sample was applied to a tube gel and electrophoresed in the Weber–Osborn buffer system (trace at top). Polypeptides separated in this gel were electrophoresed into a slab gel run in the Swank–Munkres system and identified against a standard run along the left-hand side of the slab. From Capaldi *et al.*[18]

## Functions of the Different Polypeptides

The role of cytochrome $b$ and cytochrome $c_1$ in electron transfer activity is clear. The importance of the nonheme iron protein and other components to electron transport and/or the energy-conserving function of complex III is less well understood.

Based on reconstitution studies it has been claimed that the nonheme iron protein is not needed for electron transfer between ubiquinone and cytochrome $c$.[22] However, this study is open to criticism, for several reasons (see ref. 13). In considering the role of polypeptides in the electron transfer function of complex III, we have followed up the preliminary observation of Baum *et al.* [23] that low levels of trypsin rapidly destroy activity while cleaving only one or two polypeptides. Complex III in both the oxidized and reduced forms was cleaved with trypsin, 1/50 at 4°C, and the loss of activity and the cleavage of individual polypeptides were followed. As shown in Fig. 5, the loss of activity in the oxidized form of the enzyme exactly parallels the cleavage of polypeptide V. Similarly, in the reduced state this was the only component cleaved in parallel with loss of electron transfer activity (result not shown). Apparently, then, the nonheme iron protein is important to the structure of the complex if not involved directly in electron transport. Labeling studies with [$^{14}$C]iodoacetamide have similarly suggested a structural role for core protein I. Under conditions in which the reagent causes loss of electron transfer activity, core protein I is the only protein to be labeled.[24]

## Molecular Weight of Complex III

As described already, the best preparations of complex III contain 4 nmoles $c_1$ heme per milligram of protein, from which a minimum

*Table II. Amino Acid Compositions of Components of Complex III*

|  | Complex III | Polypeptides | | |
|---|---|---|---|---|
|  |  | III | IV | VI |
| Lys | 5.1 | 4.7 | 5.5 | 5.8 |
| His | 3.1 | 2.8 | 2.6 | 0.5 |
| Arg | 5.7 | 3.5 | 5.7 | 6.0 |
| Asp | 8.2 | 7.8 | 8.9 | 9.4 |
| Thr | 5.4 | 6.8 | 3.9 | 3.8 |
| Ser | 7.3 | 6.1 | 6.6 | 6.8 |
| Glu | 10.3 | 4.9 | 10.8 | 17.5 |
| Pro | 6.1 | 7.2 | 9.5 | 6.0 |
| Gly | 7.6 | 7.5 | 8.3 | 6.5 |
| Ala | 10.1 | 7.7 | 8.7 | 8.3 |
| Val | 6.3 | 5.8 | 6.5 | 6.4 |
| Met | 2.0 | 2.5 | 2.8 | 3.3 |
| Ile | 4.0 | 8.5 | 3.0 | 3.6 |
| Leu | 10.9 | 15.0 | 10.8 | 11.1 |
| Tyr | 3.8 | 2.2 | 2.7 | 2.4 |
| Phe | 4.2 | 6.9 | 3.7 | 2.6 |

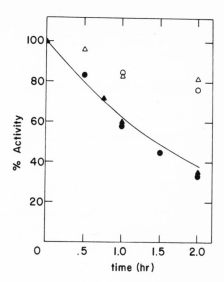

Fig. 5. Cleavage of complex III with trypsin (1/50, w/w) at room temperature. The loss of activity of the oxidized form of the complex with time is plotted. Superimposed on this graph is the extent (as a percentage of Coomassie blue stain peak) to which polypeptides V ($\bullet$,$\blacktriangle$) and VI ($\bigcirc$,$\triangle$) were cleaved by proteolytic digestion in two different experiments. Other polypeptides were not cleaved noticeably during the time of the reaction.

molecular weight of 250,000 can be calculated for the complex. Light-scattering measurements[25] and sedimentation studies[13] give a molecular weight of 260,000–300,000 for complex III dispersed in bile salts. Our recent cross-linking experiments with DTBP and DSP give a molecular weight of 310,000 for the fully cross-linked complex,[26] indicating that complex III is a monomer in Triton X-100 as well as in bile salts.

The stoichiometry of the different polypeptides in complex III has not been determined directly. There are two $b$ cytochromes for each $c_1$ heme. Weiss and his associates have shown that in *Neurospora* all the $b$ heme is associated with polypeptides of molecular weight around 30,000 (equivalent to polypeptide III in beef heart).[27] An aggregate containing two copies of III and one copy each of the other polypeptides would have an aggregate molecular weight of 245,000, which is close to the minimum molecular weight calculated from the $c_1$ heme concentration.

### Cytochrome c Oxidase

Cytochrome $c$ oxidase, as isolated, functions to transfer electrons from reduced cytochrome $c$ to molecular oxygen while conserving the energy of this reaction for coupled events. Purified cytochrome $c$ oxidase contains two spectrally distinct hemes ($a$ and $a_3$) and copper in a 1:1 molar ratio with heme.[28] It is generally assumed that the two hemes and two copper atoms act as four electron acceptors in the reaction $4H^+ +$

$4e^- + O_2 \rightarrow 2H_2O$, the heme $a$ and one Cu atom being high-potential centers and heme $a_3$ and a second Cu being low-potential centers.[29]

## Polypeptide Composition

The limitations in SDS-polyacrylamide gel electrophoresis discussed for complex III are an even greater problem in examining the polypeptide composition of cytochrome $c$ oxidase, and there are added difficulties which result from the extremely hydrophobic character of some of the component polypeptides.

Cytochrome $c$ oxidase contains three subunits with molecular weights below 15,000 which are not well separated in the commonly used gel systems. In addition, all preparations of enzyme contain small amounts of impurities which contribute to the gel scan in the low molecular weight range and confuse the issue of how many subunits are present.

We have obtained the best resolution of subunits of cytochrome $c$ oxidase on a 12.5% gel made with high levels of cross-linker by electrophoresing in the Swank–Munkres buffer system.[30] By increasing the length of the gel, a very clear separation of subunits from impurities can be obtained (Fig. 6). Under these conditions, seven different subunits were identified. These are listed with their molecular weights in Table III.

Components labeled $a, b, c$ and small bands around subunit VII are impurities of the preparation. The impurity labeled $c$ can be removed completely without loss of electron transfer activity by incubating the complex with trypsin 1/50 for 2 hr at 4°C (Fig. 7). Impurities $a, b$ and

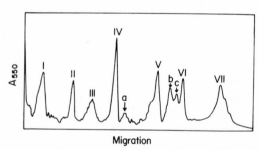

Fig. 6. Densitometric trace of an SDS-polyacrylamide gel of cytochrome $c$ oxidase. Cytochrome $c$ oxidase (100 μg) was dissociated in 3% SDS, 10 mM β-mercaptoethanol, 8 M urea and run on a 12.5% polyacrylamide gel, 20 cm long, in the Swank–Munkres gel system. Bands labeled $a$, $b$, and $c$ are impurities.

*Table III. Molecular Weights of Subunits of Cytochrome c Oxidase*

|  | Mol. wt.[a] |
|:---:|:---:|
| I | 35,400 |
| II | 24,100 |
| III | 21,000 |
| IV | 16,800 |
| V | 12,400 |
| VI | 8,200 |
| VII | 4,400 |

[a] Averaged from SDS gel and Gdn HCl column procedures.

other minor contaminants are depleted by gel filtration in Tween 80 or Triton X-100 again without loss of activity, and in cleaning up cytochrome $c$ oxidase for cross-linking or labeling studies we use trypsin treatment followed by column chromatography in Triton X-100 as a final purification step.

As with polypeptides in complex III, subunits of cytochrome $c$ oxidase behave very differently under different conditions of gel electrophoresis. Figure 8 shows a two-dimensional gel in which a sample of enzyme has been run using the Weber–Osborn gel system in the horizontal dimension and the Swank–Munkres system in the vertical direction. Certain polypeptides have very different molecular weights in the two gel systems as listed in Table III. In the Weber–Osborn gel, subunit III runs ahead of II and subunit VI ahead of V. Subunit III always runs as a broad band. It runs so close to II on Weber–Osborn gels and in tris acetate gels that this component was not resolved in several studies of the subunit structure of cytochrome $c$ oxidase.[23,31–33] The fact that subunits run in a different order under different conditions of gel electrophoresis makes comparison of results from different laboratories difficult.

The polypeptide profile of cytochrome $c$ oxidase also depends on the conditions used to dissociate the complex, regardless of the gel procedure used. When samples of enzyme are heated to 100°C prior to electrophoresis, some of I and III become aggregated and do not enter the gel. If samples which have been lipid-depleted by organic solvents, by detergent exchange, or by repeated cycles of ammonium sulfate fractionation are heated to 100°C prior to electrophoresis, I, II, and III may disappear from the gel scan completely. In these cases, the aggregate of I, II, and III produced is so large that it does not appear at the top of the gel but must remain suspended in the electrophoresis buffer.

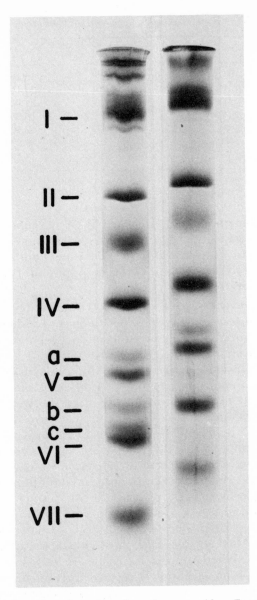

Fig. 7. Effect of trypsin on the polypeptides of cytochrome $c$ oxidase. Enzyme was incubated with trypsin (1/50, w/w) for 2 hr at room temperature. The reaction was stopped with trypsin inhibitor and the mixture was eluted through a Sepharose 4B column in Triton X-100 to remove trypsin, trypsin inhibitor, and any small, cleaved fragments of subunits. The left-hand gel is a standard of unreacted enzyme. On the right is treated enzyme. Note that impurity $b$ is diminished and $c$ is completely removed by the proteolytic digestion, and this occurs without loss of enzymic activity.

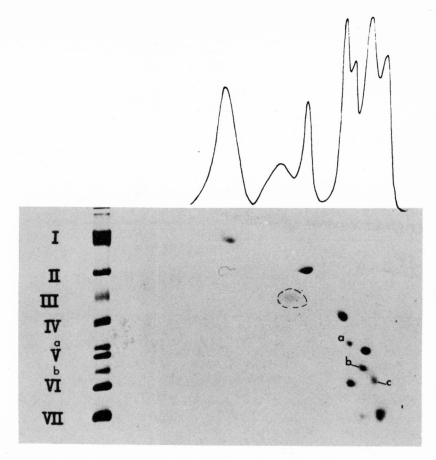

Fig. 8. Two-dimensional SDS-polyacrylamide gel electrophoresis of cytochrome *c* oxidase showing the relationship of bands observed in Weber–Osborn gels with those seen in the Swank–Munkres gel system. From Capaldi *et al.* [18]

## Primary Structural Characteristics of Components

We have purified the seven subunits of cytochrome *c* oxidase by gel filtration. Subunit I was obtained by gel filtration in SDS, II and III by gel filtration in both SDS and 8 M urea, and IV–VII by column chromatography in 6 M Gdn HCl. The amino acid compositions of all of the components have been determined (refs. 30,34; see also 35). As a second approach, each subunit has now been obtained by preparative gel electrophoresis and its amino acid composition has been examined (Table IV).

*Table IV. Amino Acid Compositions of Subunits of Cytochrome c Oxidase*[a]

| | Cytochrome oxidase complex | Polypeptides | | | | | | |
|---|---|---|---|---|---|---|---|---|
| | | I | II | III | IV | V | VI | VII |
| Lys | 5.1 | 3.3 | 3.8 | 4.0 | 9.3 | 6.9 | 7.5 | 8.5 |
| His | 3.7 | 2.8 | 2.8 | 3.1 | 2.9 | 2.5 | 3.0 | 2.7 |
| Arg | 3.7 | 2.1 | 3.0 | 2.7 | 3.3 | 5.7 | 5.7 | 3.7 |
| Asp | 7.8 | 7.8 | 7.5 | 7.2 | 9.3 | 9.9 | 8.7 | 7.7 |
| Thr | 6.7 | 7.5 | 7.6 | 8.4 | 4.4 | 5.7 | 4.5 | 5.5 |
| Ser | 6.9 | 7.6 | 10.8 | 8.1 | 7.0 | 5.7 | 8.4 | 8.5 |
| Glu | 8.3 | 4.9 | 10.7 | 9.7 | 11.7 | 11.4 | 10.8 | 8.2 |
| Pro | 5.9 | 5.5 | 5.8 | 5.7 | 6.1 | 7.4 | 3.7 | 6.5 |
| Gly | 8.7 | 11.2 | 7.6 | 9.8 | 8.2 | 8.5 | 12.2 | 10.5 |
| Ala | 8.3 | 8.3 | 5.9 | 8.6 | 8.2 | 9.2 | 10.5 | 9.3 |
| Val | 6.5 | 6.4 | 4.9 | 4.9 | 6.3 | 5.6 | 4.8 | 5.4 |
| Met | 3.4 | 4.1 | 4.2 | 2.9 | 3.0 | 1.1 | 1.8 | 2.6 |
| Ile | 5.9 | 6.2 | 4.6 | 4.8 | 3.8 | 4.7 | 3.3 | 2.6 |
| Leu | 9.0 | 11.4 | 12.8 | 9.8 | 8.0 | 9.5 | 7.2 | 9.7 |
| Tyr | 4.2 | 3.7 | 4.4 | 3.3 | 4.3 | 2.9 | 2.3 | 3.3 |
| Phe | 5.9 | 7.2 | 3.6 | 7.0 | 4.2 | 3.4 | 5.6 | 5.2 |

[a] For subunits isolated by gel filtration and preparative gel electrophoresis.

Of the seven subunits, polypeptide I is the most hydrophobic (polarity 35.7%), II and III are both relatively hydrophobic, and IV–VII are hydrophilic in amino acid composition.

### Functions of the Different Polypeptides

The roles of the different subunits of cytochrome $c$ oxidase in electron transfer or coupling functions have proved difficult to assess because conditions which to date have been used to dissociate the subunits also release the heme and copper moieties. There have been several reports that subunits I, II, and III can be removed without dissociating the heme and copper moieties and without loss of electron transfer activity, but none of these studies is very convincing. In two studies, repeated detergent solubilization and ammonium sulfate precipitation steps were used.[36,37] These conditions delipidate the enzyme, and it is now known that the large subunits of cytochrome $c$ oxidase once aggregated by lipid depletion are not dissociated by SDS and do not enter the gel even if present. Another study claimed to remove subunits I, II, III, IV, and VII from cytochrome $c$ oxidase by trypsin cleavage of the complex without loss of activity.[38] However, in our hands proteolytic digestion removes impurities but does not cleave any of the subunits of the complex. More recently, Phan and Mahler[39] claim to have removed I, II, and III from cytochrome $c$ oxidase by hydrophobic chromatography. Their "four-subunit" enzyme (IV–VII) retains full activity and contains both heme and copper moieties.[39]

## Molecular Weight of Cytochrome c Oxidase and Stoichiometry of Subunits

The cleanest preparations of cytochrome $c$ oxidase contain 12–14 nmoles of heme $a$ per milligram of protein. From this it can be calculated that the minimum molecular weight of a two-heme (two-copper) complex is 140,000–165,000. In the course of detergent-binding studies we have measured the molecular weight of the enzyme complex suspended in Triton X-100 and find that cytochrome $c$ oxidase exists as a dimer or four-heme complex (molecular weight around 325,000) under these conditions.[40] Molecular weight determinations in Tween 80 also give a value consistent with a dimer or four-heme complex.[28,40] This dimer retains high activity when assayed in Tween 80 or other amphiphiles containing a flexible hydrocarbon chain.[40]

We have also examined the molecular weight of the enzyme after lipid depletion with bile salts (deoxycholate and cholate) and found that this causes a significant dissociation of the dimer into monomers or two-heme complexes. Enzyme delipidated with bile salt was not fully reactivated with activating detergents.[40] Evidence that cytochrome $c$ oxidase exists as a dimer in Triton X-100 or Tween 80 dispersion has also been obtained from cross-linking experiments.[41]

Recently, we have obtained evidence that the dimer may be the structurally stable species in the membrane. Cytochrome $c$ oxidase isolated by modification of the method of Sun *et al.*[42] forms a crystalline lattice which is ammenable to electron diffraction studies of the type developed by Henderson and Unwin[43] in their work on purple membrane protein. Electron diffraction studies of cytochrome $c$ oxidase have shown that the lattice in this preparation is formed from two opposing membranes in a vesicle. The lattice has a space group P $2_1 2_1 2$ and pairs of molecules are related by a crystallographic two-fold axis perpendicular to the plane of the membrane.[44] This is a good indication that cytochrome $c$ oxidase is associated into a dimer within the lipid bilayer.

## Summary and Conclusions

Complex III and cytochrome $c$ oxidase provide two examples of intrinsic membrane proteins which are multicomponent complexes, and each appears to play both a structural and a functional role in the mitochondrial inner membrane. The results available to date suggest that complex III is present as a monomer and cytochrome $c$ oxidase as a dimer in the inner membrane, and this is consistent with their known stoichiometries.[45] Whether these complexes diffuse separately in the membrane or are bridged by cytochrome $c$ into a permanent aggregate remains to be determined.

Complex III contains eight different polypeptides and cytochrome $c$ oxidase seven different components. There is no convincing evidence that either complex can be subdivided to give functional units with less than the number of subunits just listed. Complex III and cytochrome $c$ oxidase do not have any polypeptides in common. Both contain hydrophobic and hydrophilic subunits. An attractive possibility is that the hydrophobic subunits form the core of the complex within the bilayer with hydrophilic polypeptides on the surface exposed parts. To test this, we are conducting labeling studies with membrane-impermeable probes which we hope will also give the orientation of complex III and cytochrome $c$ oxidase in the mitochondrial inner membrane. The results from these studies taken together with the data from cross-linking experiments, which give the near-neighbor relations of polypeptides in each complex, should give us a picture of the organization of the terminal part of the electron transport chain from which the path of electrons in the membrane can be traced.

## References

1. Capaldi, R. A., and Green, D. E. (1972) *FEBS Lett. 25*, 251.
2. Singer, S. J., and Nicolson, G. L. (1972) *Science 175*, 720.
3. Vanderkooi, G. (1972) *Ann. N.Y. Acad Sci. 195*, 6.
4. Capaldi, R. A., and Vanderkooi, G. (1972) *Proc. Natl. Acad Sci. USA 69*, 930.
5. Spatz, L., and Strittmatter, P. (1971) *Proc. Natl. Acad Sci. USA 68*, 1042.
6. Spatz, L. and Strittmatter, P. (1972) *J. Biol. Chem. 248*, 793.
7. Tomita, M., and Marchesi, V. T. (1975) *Proc. Natl. Acad Sci. USA 72*, 2964.
8. Braun, V. (1975) *Biochim. Biophys. Acta 415*, 335.
9. Inouye, M. (1974) *Proc. Natl. Acad Sci. USA 71*, 2396.
10. Hatefi, Y., Haavik, A. G., and Griffiths, D. E. (1962) *J. Biol. Chem. 237*, 1681.
11. Bell, R. L. (1977) Ph.D. thesis, University of Oregon.
12. Helenius, A., and Simons, K. (1975) *Biochim. Biophys. Acta 415*, 29.
13. Rieske, J. S. (1976) *Biochim. Biophys. Acta 456*, 195.
14. Erecinska, M., Wilson, D. F., and Miyata, Y. (1976) *Arch. Biochem. Biophys. 177*, 133.
15. Bell, R. L., and Capaldi, R. A. (1976) *Biochemistry 15*, 996.
16. Swank, R. T., and Munkres, R. D. (1971) *Anal. Biochem. 39*, 462.
17. Yu, C. A., Yu, L., and King, T. E. (1975) *Biochem. Biophys. Res. Commun. 66*, 1194.
18. Capaldi, R. A., Bell, R. L., and Branchek, T. (1977) *Biochem. Biophys. Res. Commun. 74*, 425.

19. Weber, K., and Osborn, M. (1969) *J. Biol. Chem. 244*, 4406.
20. Gellerfors, P., and Nelson, B. S. (1970) *Eur. J. Biochem. 52*, 433.
21. Das Grupta, V. D., and Rieske, J. S. (1973) *Biochem. Biophys. Res. Commun. 54*, 1247.
22. Yamashita, S., and Racker, E. (1969) *J. Biol. Chem. 244*, 1220.
23. Baum, H., Silman, H., Rieske, J. S., and Lipton, S. H. (1967) *J. Biol. Chem. 242*, 4876.
24. Gellerfors, P., Lunden, M., and Nelson, B. D. (1976) *Eur J. Biochem. 67*, 463.
25. Tzagoloff, A., Yang, P. C., Wharton, D. C., and Rieske, I. S. (1965) *Biochim. Biophys. Acta 96*, 1.
26. Smith, R. J., and Capaldi, R. A. (1977) *Biochemistry 16*, 2629.
27. Weiss, H. (1976) *Biochim. Biophys. Acta 456*, 291.
28. Kuboyama, M., Yong, F. C., and King, T. E. (1972) *J. Biol. Chem. 247*, 6375.
29. Malmstrom, B. G. (1974) *Q. Rev. Biophys. 6*, 389.
30. Downer, N. W., Robinson, N. C., and Capaldi, R. A. (1976) *Biochemistry 15*, 2930.
31. Eytan, G. D., Carroll, R. C., Schatz, G., and Racker, E. (1975) *J. Biol. Chem. 250*, 8598.
32. Kornblatt, J. A., Baraff, G., and Williams, G. R. (1973) *Can. J. Biochem. 51*, 1417.
33. Rubin, J. S., and Tzagoloff, A. (1973) *J. Biol. Chem. 248*, 4269.
34. Briggs, M., Kamp, P. K., Robinson, N. C., and Capaldi, R. A. (1975) *Biochemistry 14*, 5123.
35. Steffens, G., and Buse, G. (1976) *Hoppe Seyler's Z. Physiol Chem. 357*, 1125.
36. Komai, H., and Capaldi, R. A. (1973) *FEBS Lett. 30*, 272.
37. Hare, J. F., and Crane, F. L. (1974) *Subcell. Biochem. 3*, 1.
38. Yamamoto, T., and Orii, Y. (1974) *J. Biochem. 75*, 1081.
39. Phan, S. H., and Mahler, H. (1976) *J. Biol. Chem. 251*, 257.
40. Robinson, N. C., and Capaldi, R. A. (1977) *Biochemistry 16*, 375.
41. Briggs, M. M., and Capaldi, R. A. (1977) *Biochemsitry 16*, 73.
42. Sun, F. F., Prezbindowski, K. S., Crane, F. L., and Jacobs, E. E. (1968) *Biochim. Biophys. Acta. 153*, 804.
43. Henderson, R., and Unwin, N. (1975) *Nature 257*, 28.
44. Henderson, R., Capaldi, R. A., and Leigh, J. (1977) *J. Mol. Biol. 112*, 631.
45. Slater, E. C. (1972) in *Mitochondria/Biomembranes* (VanderBergh, S. G., Borst, P., VanDeenen, L. M., Riermersma, J. C., Slater, E. C., and Tager, J. M., eds.), North-Holland, Amsterdam.

*Chapter 5*

# Reversible Mitochondrial ATPase as a Model for the Resolution of a Membrane Complex

## A. E. Senior

### Introduction

Quite early David Green coined the term "repeating unit" to express his concept of the self-contained nature of the ATPase enzyme and the electron transfer complexes in the mitochondrial inner membrane. Like many of David's concepts it underwent many contortions over the years, but in essence it stood the test of time. In the case of the ATPase it provided a basis for rationalization of the multiplicity of "factors" ("coupling," "energy transfer," etc.) as components of the overall ATPase complex.

I was fortunate enough to arrive in Madison at the time when the remarkably complicated nature of the ATPase complex was emerging. Alex Tzagoloff and David MacLennan had just shown that "headpiece" and "stalk" (two parts of Green's "repeating unit") could be identified with specific proteins or sets of proteins and were bound to "membrane factor" proteins in the "basepiece." The next logical step was of course to fully characterize the individual proteins, define their structural relationships to each other, and, by application of all the modern, sophisticated techniques of protein chemistry, figure out how ATP is synthesized. Easier said than done!

*A. E. Senior* • University of Rochester Medical Center, Rochester, New York 14642.

Much experimental progress has been made, however, at the Enzyme Institute and in many other laboratories around the world, and it has revealed a fundamental and universal feature of living cells. The reversible mitochondrial ATPase has been characterized, and we now have a thorough albeit incomplete picture of its structure. An almost identical enzyme complex has been found in thylakoid membranes of chloroplasts and in the cell membrane of many species of prokaryote. In mitochondria and chloroplasts the enzyme functions to synthesize ATP; in prokaryotes the enzyme may catalyze oxidative phosphorylation, photophosphorylation, or (a major function) the ATP-driven accumulation of nutrients and ions. Clearly the perfection of this enzyme was a giant step in evolution of living things, allowing large-scale, efficient generation and utilization of energy. Hence understanding this enzyme is of topical interest.

In this chapter I shall summarize current views on the structure of the enzyme and then I shall try to show by use of examples how our ability to resolve and reconstitute the enzyme has aided comprehension of its biosynthesis, assembly, and mechanism of action. For a complete description of the enzyme, I refer the reader to recently published reviews,[1-5] and to a forthcoming "update" review.[6] In this chapter I have referenced only new papers not referred to in my earlier review.[1]

### Structure of the ATPase Complex: Resolution Techniques

Figure 1 is a diagram of the beef heart mitochondrial ATPase complex. Evidence for this structure has been drawn from two main techniques—electron microscopy and resolution. In electron micrographs of negatively or positively stained mitochondrial inner membrane such structures are seen, and their disappearance during resolution may be correlated with removal of certain protein components. Further understanding of the number and nature of the individual components has been derived from the sequential resolution and reconstitution of the complex as detailed below. The diagram shown in Fig. 1 is probably not fully correct, but it accurately reflects present knowledge and provides a basis for future experimentation. Points of uncertainty will be discussed below.

### $F_1$

$F_1$ corresponds to Green's "headpiece" or Racker's "inner membrane spheres." It is the spherical (~9 nm diameter) part of the complex which abuts into the matrix. This component is removed by sonication of the mitochondrial membranes in the presence of EDTA at low ionic

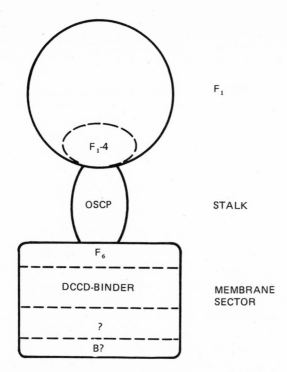

Fig. 1. Mitochondrial ATPase complex. The localization of OSCP in the stalk, of $F_6$ on the matrix side of the inner membrane, and of B on the exterior surface of the inner membrane is tentative. Cross-linking studies have recently revealed spatial relationships between subunits of $F_1$ (to be reviewed elsewhere; see ref. 6), which are not shown in detail here.

strength, often at elevated pH (~9). This component carries the site at which ATP is synthesized and hydrolyzed, and the purified soluble enzyme is a stable active oligomycin- and DCCD*-insensitive ATPase. There are six types of subunit in the $F_1$; five are very tightly bound to each other, and the sixth is a loosely bound regulatory protein called the "$F_1$ inhibitor." So $F_1$ is itself a very complicated molecule, with a total molecular weight of around 360,000.

*OSCP*

OSCP (oligomycin-sensitivity-conferring protein) binds $F_1$ to the membrane and is probably the stalk or one part of the stalk, although it is not possible to make this assignment definitively with current technol-

* Abbreviations: DCCD, dicyclohexylcarbodiimide; NBD-Cl, 7-chloro-4-nitrobenzo-2-oxa-1,3-diazole; NEM, *N*-ethylmaleimide.

ogy. OSCP is removed from $F_1$-depleted inner membrane by extraction with ammonia in the presence of EDTA and salt. The purified sparingly soluble protein has molecular weight around 18,000, is basic, and is quite temperature sensitive. It has no intrinsic enzyme activity and is assayed by its ability to bind $F_1$ back to depleted membrane preparations, with concomitant emergence of either membrane-bound oligomycin or DCCD-sensitive ATPase activity or energy coupling, e.g., ATP-Pi exchange, ATP-driven reversed electron transfer from succinate to NAD, or oxidative phosphorylation. $F_1$ and OSCP form a soluble complex together.

## $F_6$

$F_6$ is a protein which is extracted from $F_1$- and OSCP-depleted membranes by thiocyanate. The protein has also been successfully purified recently from a detergent-solubilized preparation of the whole ATPase complex, and its molecular weight was found to be about 8000.[7] It is extremely heat stable, and is assayed by its ability to bind $F_1$ back to depleted membrane preparations in the presence of saturating amounts of OSCP, with concomitant emergence of membrane-bound oligomycin- or DCCD-sensitive ATPase activity and energy coupling. $F_6$ is probably the same protein as $F_{c2}$; they have similar properties. Racker's group has equated the two in recent publications. In Fig. 1, I have put $F_6$ in the membrane sector. However, this point is ambiguous; one could just as easily put it in the stalk on current evidence.

### B Factors

Several B-type factors have been described recently; all are sensitive to —SH reagents and are prepared from acetone powders or lyophilized mitochondria. Their reported molecular weights vary from around 12,000 to around 45,000, and their interrelationship is unclear.[8-10] They are assayed by their ability to enhance energy coupling in submitochondrial particles which are partly depleted of ATPase components or (more recently) by their ability to enhance ATP-Pi exchange in liposomes into which the whole ATPase complex is incorporated.[11] They appear to enhance energy transduction in uncoupled particles like small amounts of oligomycin by blocking "proton leaks" in the membrane, and for this reason I have put them in the membrane sector in Fig. 1. This is a very tentative assignment (but is supported also by recent *in situ* labeling by an inhibitory —SH reagent[12]).

## DCCD-Binding Protein

DCCD-binding protein is a membrane sector protein which has been purified from both the mitochondrial and bacterial ATPase complexes after extraction into organic solvents. It is now thought that this protein forms part of a proton channel through the membrane sector of the enzyme.

The membrane sector of the complex is thought to contain about four proteins, which vary in size from 10,000 to 55,000 (ref. 13; see also refs. 14,15). One of these is the DCCD-binding protein, and the B factor (or factors) and $F_6$ may account for one or more of the other proteins, but this part of the complex is not yet fully characterized and much more work needs to be done here on both the protein and lipid components.

Preparations of the whole ATPase complex have been obtained from yeast and heart mitochondria by detergent solubilization. The preparations from yeast have been obtained in monodisperse form and contain about ten subunits. Electron microscopy of the dispersed preparations confirms the structure shown in Fig. 1. The best-characterized preparations from beef heart have 10–14 bands on SDS gels,[13, 14] some of which can be shown to be contaminants derived from electron transfer components. The yeast and beef heart complexes may be reconstituted into liposomes *à la* Racker, giving ATP-Pi exchange activities and/or H + pumping,[11, 14, 16] and these preparations represent the simplest energy coupling systems of mitochondrial origin devised to date. Resolution of the whole ATPase complex, combined with electron microscopic examination of the resolved preparations, has amply confirmed conclusions drawn earlier from resolution of whole inner membrane preparations. There is as yet no consensus regarding the stoichiometry of the subunits in the complex, and I shall not therefore discuss it further here.

By and large, there are only a few differences between the mitochondrial ATPase complex and the chloroplast and bacterial enzymes. These are, briefly, as follows: (a) the $F_1$-inhibitor protein appears to be one of the five tightly bound subunits of the chloroplast and bacterial enzymes; (b) trypsin treatment of chloroplast and bacterial $F_1$ readily removes the three smaller subunits, leaving a "two-subunit" "ATPase-active" enzyme which cannot bind back to the membrane, whereas trypsin treatment of mitochondrial $F_1$ does not; (c) although DCCD and NBD-Cl are universal inhibitors, other inhibitors are more specific, e.g., oligomycin is inactive in chloroplasts and bacteria.

This short summary of the structure of the complex is in danger of turning into a review, and so I will now go on to describe recent work on the complex which has used resolution techniques.

### Applications of Resolution Techniques in Understanding the ATPase Complex

There are three questions (at least) which when answered would usefully extend our knowledge. They are (a) how is the complex assembled? (b) how is the complex held together? (c) how does the complex function as an electrogenic $H^+$ pump and, conversely, how is ATP synthesized?

#### Biosynthesis and Assembly of the Complex

Many features of the biosynthesis and assembly of the complex have become clear, very largely as a result of Alex Tzagoloff's work which (at his suggestion) I will briefly summarize here. Tzagoloff has worked with yeast and has shown that yeast has $F_1$, OSCP, and membrane sector which may be resolved and reconstituted.[17] Yeast ATPase is DCCD sensitive, and there is evidence for an $F_1$-inhibitor protein.[18, 19] Tzagoloff showed that all the subunits of $F_1$ and OSCP are synthesized on cytoplasmic ribosomes using mRNA coded by nuclear DNA, but that membrane sector subunits are synthesized in mitochondria on mitochondrial ribosomes, and are coded for by mitochondrial DNA. Both $F_1$ and OSCP could be found in cytosol of yeast cells in which mitochondrial protein synthesis was inhibited, and they were competent in reconstitution experiments using normal "membrane sector" preparations made by extraction of $F_1$ and OSCP from yeast submitochondrial particles. Using these techniques, Tzagoloff is now moving on to try to map mitochondrial DNA using yeast mutants and to understand how the cytoplasmic and mitochondrial systems are integrated.[20] Other topical questions are how are $F_1$ and OSCP physically intercalated through the inner membranes of mitochondria during assembly? And why is it that the mitochondrion synthesizes any of its own protein at all? One answer to this last question partly suggested by Raff and Mahler[21] is that the proteins which are synthesized are very hydrophobic (in order for them to act as electrical insulators and specific proton conductors?) such that transcellular transport would be impossible for them—they must be laid down *in situ* and the rest of the membrane is then assembled around them.

In bacteria and yeasts, mutant organisms may have alterations in specific subunits of the ATPase complex. Resolution allows one to readily localize the defect. Thus in *Escherichia coli uncA* mutants are defective in the $F_1$ sector of the ATPase, whereas *uncB* mutants are defective in the membrane sector. Hybridization experiments (e.g., *uncA* membrane sector with wild-type $F_1$) are useful in confirming this type of diagnosis. These mutants together with the yeast mutants are likely to provide considerable insight into mechanism of action of the ATPase complex.

## *How Is the Complex Held Together?*

Vadineanu et al.[22] have recently made some quantitative measurements of $F_1$, OSCP, and $F_6(F_{c2})$ binding in submitochondrial particles depleted of these components. The total amount of $F_1$ that could be bound to particles was around 0.55 nmole/mg protein, about stoichiometric with cytochrome $c_1$, as may also be calculated by aurovertin titration[23] and by NBD-Cl reaction.[24] $F_1$ bound with $K_d$ of around 0.1 $\mu$M to depleted particles in the presence of OSCP and EDTA (no $Mg^{2+}$). The $K_d$ for OSCP binding to the membrane appears from the data to be around 0.02–0.04 $\mu$M, and the $K_d$ for $F_6$ binding to the membrane higher (less tight) than this. In the presence of Mg, however, $F_1$ binding was very much tighter ($K_d$ ~2–4 nM). The role of $Mg^{2+}$ in binding the $F_1$ seems paramount. Uncoupled submitochondrial particles made in the presence of EDTA (e.g., "A" particles, "ETP") are almost certainly uncoupled because they have lost some $F_1$ and/or OSCP by dissociation and are leaky to protons. The "structural" role of chemically modified $F_1$ previously described by Racker seems to be a leak-plugging effect. Vadineanu et al.[22] showed that the oligomycin titer efficacious in restoring respiratory control in uncoupled particles was exactly equal to the number of $F_1$ molecules removed during preparation. This lends weight to the view that oligomycin added to submitochondrial particles which are uncoupled because they are partly depleted of OSCP and $F_1$ binds preferentially to the ATPase complexes which are partly depleted,[22, 25] thus first plugging leaks and restoring coupling at low concentrations before inhibiting at higher levels. Montecucco and Azzi[26] showed similarly that the membrane sector subunit which binds DCCD becomes very much more accessible to the hydrophilic ascorbic acid after removal of $F_1$ and OSCP. Thus this approach has emphasized the key role of $Mg^{2+}$ in the integration of the complex and the importance of protecting the membrane sector subunits from direct exposure to the hydrophilic environment.

Attempts have been made to identify specific amino acids which are involved in binding on the various components of the complex and have so far proved rather negative. I have shown that sulfhydryl groups are not involved in the binding of OSCP and $F_1$; Fessenden-Raden found $F_6$ to be unaffected by sulfhydryl reagents. Some reagents, e.g., iodine and dinitrofluorobenzene, are known to inactivate the ATPase activity of $F_1$ without affecting membrane-binding properties. Despite disappointing results so far, however, this line of attack seems potentially profitable.

Some progress has been made in deciding which of the $F_1$ subunits is involved in binding using the bacterial $F_1$. Bacterial $F_1$ consisting of only four subunits (lacking the $F_1$-4 or $\delta$ subunit) cannot rebind to membranes[27-29] and the binding ability is completely restored by addition of purified $F_1$-4 or $\delta$ subunit.[30] So it certainly seems that this subunit is the important one in rebinding of solubilized $F_1$, at least in bacteria. On this aspect of the problem, the mitochondriacs have lagged behind and we don't yet know whether the $F_1$-4 subunit plays such a key role in binding in the mitochondrion.

### *The ATPase Complex as an Electrogenic $H^+$ Pump and ATP Synthesis Machine*

Potentially one great advantage of resolving the $F_1$, OSCP, B, and $F_6$ away from the membrane is that, once soluble, these proteins are then more amenable to structural and functional characterization by protein chemistry and analytical techniques. Most of this effort has been aimed at $F_1$, which carries the ATP hydrolysis and synthesis site(s).

*Use of Photoaffinity Labeling Reagents.* 2,4-Dinitrophenol increases mitochondrial $F_1$-ATPase activity and binds to $F_1$ with relatively high affinity. The question has been repeatedly raised over the years as to whether this binding has anything to do with uncoupling. Stockdale and Selwyn[31] had shown that there was no correlation between activation of $F_1$-ATPase activity and uncoupling capability for a series of substituted phenols. Using the compound 4-fluoro, 3-nitrophenylazide (FNPA) (Fig. 2), we have been able to answer this question decisively.[32] We found that FNPA does activate $F_1$, and does compete with [$^{14}$C]dinitrophenol for a binding site ($K_d$ about 90 $\mu$M for dinitrophenol) on $F_1$. Like DNP, FNP activates by decreasing $K_m$ATP. When incubated with FNPA in the concentration of 0–200 $\mu$M and exposed to flashes of light, $F_1$ is rapidly *inactivated*, presumably by covalent insertion of FNPA into the

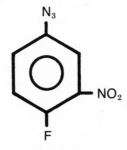

Fig. 2. FNPA (4-fluoro, 3-nitrophenylazide) structure.

binding site. Control experiments showed that the inactivation was not reversed when the enzyme was subjected to gel filtration on Sephadex G25, and that photolysis of FNPA before addition of $F_1$ or photolysis of $F_1$ alone produced no inactivation. DNP binding to the inactivated enzyme was reduced. The 50% inactivated enzyme retained about 40% DNP binding. Most important, FNPA was not an uncoupler in either ETP$_H$ (where ATP-Pi exchange was measured) or in coupled rat liver mitochondria (where $O_2$ uptake was followed). There are two conclusions to be drawn. First, the increase of $F_1$-ATPase activity caused by dinitrophenol seems to be completely separate from uncoupling activity, and, second, there is a relatively high-affinity binding site for dinitrophenol on beef heart $F_1$ to which FNPA (an uncharged species) may also bind. This site may be the site at which a regulatory ligand binds *in vivo*.[33]

An interesting recent paper described the use of an azido analogue of ATP.[34] The azido-ATP derivative inhibited $F_1$ on exposure to light, and was a substrate. About 0.8 mole of ATP analogue bound to 1 mole of $F_1$ after inactivation and was not liberated after SDS treatment. This seems a very promising approach to discovering what amino acid side chains form the site for ATP binding.

*Chemical Modification of $F_1$ Aimed at Characterizing the Active Site(s).* Early work using tetranitromethane showed that a small number of tyrosine residues (two or three) were involved in ATP binding and probably hydrolysis. Subsequently, the reagent NBD-Cl was found to be a very much more specific inhibitor. NBD-Cl binds to one tyrosine in $F_1$ with inactivation of ATPase[35, 36]; the tyrosine is on the $F_1$-2 or $\beta$ subunit. The inhibition seems to suggest on the surface that a tyrosine is at the active site of ATP hydrolysis. The situation is actually more complicated. NBD-Cl does not inhibit the ATP-induced quenching of aurovertin fluorescence which is normally seen in $F_1$; thus binding of ATP itself is unaffected.[37] Like DCCD, NBD-Cl is inhibitory in mitochondria, chloroplasts, and bacteria, inhibiting both ATP synthesis and hydrolysis.

Evidence is accumulating that an arginine residue(s) forms part of the nucleotide-binding site in many enzymes which bind ATP. Butanedione and phenylglyoxal, two reagents with high specificity for modification of reactive arginine residues in enzymes, inhibit ATPase activity of $F_1$,[38] and one or two essential arginine residues were implicated. There was no indication of an essential —SH group(s) in $F_1$ in early experiments or in more thorough tests.

*Conformational Changes in the ATPase Complex.* It has been possible to show that rearrangements of conformation occur on $F_1$ during energy transduction and that they may be induced either by respiratory chain activity or directly by imposition of proton gradients. In mitochondria the evidence is of three kinds. Aurovertin, a fluorescent antibiotic, binds to membrane-bound $F_1$, and on induction of energy transduction a quenching of fluorescence is seen.[23, 39] As I alluded to before, the $F_1$-inhibitor protein is a regulatory subunit of the $F_1$, and its binding to membrane $F_1$ is very much loosened (affinity lowered) in induction of respiration in submitochondrial particles. Uncouplers counteracted this change in affinity, which is presumably a reflection of the conformational change on the $F_1$. A third kind of evidence for induced conformational changes in $F_1$ comes from evidence regarding the "exchangeability" of "tightly bound" nucleotides (discussed in the next section).

In chloroplasts, Ryrie and Jagendorf have found that tritium is incorporated extensively into $F_1$ when chloroplasts are exposed briefly to light or a "pH jump" in the presence of tritiated water, and this tritium is effectively sequestered. In a very similar fashion, the "tightly bound" nucleotides of chloroplast membranes (which are known to be associated with the $F_1$; see below) are normally not "exchangeable," but they become transiently freely exchangeable on exposure of the membranes to light or a "pH jump," and after the transient exchange radioactive nucleotide bound on the $F_1$ is effectively sequestered. Yet another line of evidence involved the use of NEM. It was found that chloroplasts exposed to light reacted with NEM. Purification of $F_1$ resolved from the reacted membranes showed that the NEM reacted specifically with a group in the $F_1$-3 ($\gamma$) subunit of the enzyme. This reaction occurred only in the light conditions, and presumably reflects the transient exposure of a particular buried group.[40] Thus there is a lot of evidence that parts of $F_1$ which are normally protected from the hydrophilic environment become transiently exposed to it on the membrane during energy transduction, i.e., the $F_1$ changes its conformation.

*"Tightly Bound" Nucleotides.* It has been found that the ATPase complex in bacteria, chloroplasts, and mitochondria contains so-called tightly bound nucleotides (ATP and ADP).[41–44] They are bound to the $F_1$ part of the complex, and they are bound so tightly that they are not dislodged during resolution and purification of the soluble $F_1$ or by extensive washing of membrane preparations. They cannot be studied by equilibrium binding methods because they do not exchange fast enough. However, on induction of oxidoreduction, or in "pH jump" experiments they transiently become "exchangeable."[42, 44-47] This transient change in binding affinity is abolished by uncouplers, and several factors which influence phosphorylation have parallel affects on the induced transient exchange of nucleotides. Recently, Richard Leimgruber and I have been able to show that when all the "tightly bound" nucleotide was removed from soluble mitochondrial $F_1$, the enzyme completely lost its ability to reconstitute either reversed electron transfer

Table I. Effects of Removal of "Tightly Bound" Nucleotides from Soluble $F_1$

| Property | Trypsin treatment[a] | High ionic strength treatment[b] |
|---|---|---|
| Content of "tightly bound" ADP | ADP → zero | ADP → zero |
| ATPase activity | Unchanged | Unchanged |
| Competitive inhibition by ADP of ATPase activity (normal $K_i$ ~85 $\mu$M) | $K_i$ = 95 $\mu$M, competitive | $K_i$ = 102 $\mu$M, competitive |
| $F_1$-inhibitor effect on $F_1$ | Lowered affinity, maximal inhibition slightly lowered | Slightly lowered affinity, maximal inhibition slightly lowered |
| Rebinding to depleted membrane | Unchanged | Unchanged |
| Oligomycin sensitivity of rebound ATPase activity of $F_1$ | Slightly less sensitive | Unchanged |
| Oxidative phosphorylation in reconstituted system | Abolished | Abolished |
| ATP-driven reversed electron transfer in reconstituted system | Abolished | Abolished |

[a] Soluble $F_1$ was treated with 5 $\mu$g trypsin/mg for 3 min at 25°C. Soybean trypsin inhibitor prevented all trypsin effects.
[b] Soluble $F_1$ was passed through Sephadex G25 in 50 mM tris $SO_4$–1 mM EDTA–60 mM $K_2SO_4$, pH 8.0.
[c] Native soluble $F_1$ contained zero "tightly bound" ATP and 1.8 mole "tightly bound" ADP per mole $F_1$.

driven by ATP or oxidative phosphorylation.[48] These experiments are summarized in Table I. The nucleotides were removed either by treatment with low levels of trypsin or by gel filtration at high ionic strength, and the ability of the soluble $F_1$ to "reload" the nucleotides was very limited. The nucleotide-depleted $F_1$ retained all the other activities which we tested (Table I) and in fact we have not at the time of writing been able to pinpoint any structural change in the $F_1$, even after the trypsin treatment (techniques used were SDS gel electrophoresis in 10% or 12% gels and gel electrophoresis in 8 M urea containing buffer at pH 3.9, 4.3, and 8.9).

Our experiments therefore suggested that in some crucial way the "tightly bound" nucleotide in $F_1$ was involved in energy coupling. We then performed the same kinds of experiments on ETPH and got exactly the same kinds of results (see Table II and ref. 49). Both trypsin treatment and high ionic strength gel filtration of ETPH depleted the particles of "tightly bound" ATP and ADP and caused loss of energy coupling without loss of several other properties of the ATPase complex. Thus these experiments confirmed and extended the experiments on soluble $F_1$ and greatly underscored the importance of these "tightly bound" nucleotides. From our work, we conclude that a cluster or series of tight nucleotide-binding sites must be present on $F_1$ in order for energy-linked ATP hydrolysis and ATP synthesis to occur. Non-energy-linked ATP hydrolysis does not require these "tightly bound" nucleotides.

In further recent experiments we have found that $F_1$ depleted of nucleotide by high ionic strength gel filtration can in fact be induced to bind both ATP and ADP tightly after the $F_1$ is reconstituted with OSCP and depleted membranes under energized conditions at low ionic strength, opening up the possibility that the structural requirements of these binding sites can be studied by use of analogues of ATP and ADP. It may even be possible to insert "affinity labels" or the "photoaffinity label" described above to localize these sites.[50]

Our own data do not yet answer the question as to whether these "tightly bound" nucleotides act catalytically. Both Boyer and Slater have recently proposed speculative schemes to account for the catalytic action of these nucleotides and have provided evidence which suggests that "tightly bound" ADP is phosphorylated, is the initial Pi acceptor, is "tightly bound" as ATP after a largely energy-independent condensation, and is released in an energy-dependent fashion to the medium.[42, 44, 51-53] Strotmann and McCarty have also provided evidence which suggests a possible catalytic role for the nucleotides.[45 47] However, the question is by no means finally settled; it has still to be shown satisfactorily that the rates of phosphorylation of "tightly bound"

*Table II. Effects of Removal of "Tightly Bound" Nucleotides from Membrane-Bound $F_1(ETP_H)$*[a]

| Property | Trypsin treatment[b] | High ionic strength treatment[c] |
|---|---|---|
| Content of tightly bound ATP | ATP → 0–23% | ATP → zero |
| Content of tightly bound ADP | ADP → zero | ADP → 32–48% |
| Content of tightly bound (ATP + ADP) | → 0–12% | → 14–22% |
| ATPase activity | Enhanced 50% | Enhanced 40–160% |
| ATP-driven reversed electron transfer | Abolished | → 15–23% |
| NADH oxidase | Unchanged | Unchanged |
| Succinate oxidase | Unchanged | Unchanged |

[a] $ETP_H$ is phosphorylating submitochondrial particles. Our preparations contained 1.55–2.4 nmoles "tightly bound" ATP/mg protein and 1.25–1.9 nmoles "tightly bound" ADP/mg protein.
[b] $ETP_H$ was treated with 125 μg trypsin/mg protein for 5 min. at 25°C. Soybean trypsin inhibitor prevented all trypsin effects.
[c] $ETP_H$ was passed through Sephadex G25 in 0.25 M sucrose–10 mM tris acetate–1 mM dithiothreitol ± 4 mM $MgCl_2$, pH 7.5, plus $K_2SO_4$ up to 500 mM.

ADP are adequate to account for the overall phosphorylation process, and it has to be decided whether all the "tightly bound" nucleotides are catalytic or whether some are "structural."

## Conclusion

I hope that I have been able to show how resolution and reconstitution of the ATPase complex have been so important. First, they have enabled us to build up, piece by piece, a reasonably clear picture of the structure of this extremely complicated enzyme. Second, they have given us clear insights into the biosynthesis and assembly of a membrane enzyme, under both nuclear and mitochondrial genetic control. Third, we have already been able to make some statements about the mechanism of ATP synthesis and we are able to make finite choices on how to proceed experimentally. I would like to expand just a few more moments on this third point. The concept of this ATPase complex as a unit is experimentally verified. Mitchell's idea that this unit has the ability to generate an electrical gradient through proton pumping coupled to ATP hydrolysis, and to accomplish ATP synthesis by reversing the process, is experimentally verified. Of course, the main question now is, how is this done? We know from the evidence presented above that "tightly bound" nucleotides on the $F_1$, whose binding affinities may change markedly as a result of ion-gradient-induced conformational changes, are intimately involved in the overall process, although exactly how is not yet clear. We have some clues as to which amino acid groups in the $F_1$ are involved in the action. I think most would agree that by pursuing these lines of inquiry, and extending them to other individual components of the ATPase complex such as OSCP, $F_6$, and the membrane sector subunits we are likely to progress toward an answer.

This enzyme complex embodies in it some features which manmade machines do not; notably, it transduces electrical into chemical energy at fairly low temperatures in a fairly efficient manner. It is nonexplosive and nonpolluting, and we know that it works. There is the possibility that if the principles were understood we could apply them to our own energy problems.

ACKNOWLEDGMENTS

Work described here and done in our laboratory in Rochester was supported by Grants GB-38350 and PCM76-04991 from NSF and AM-16366 from NIAMDD, USPHS. I thank Richard Leimgruber for his valuable suggestions and contributions on the "tightly bound" nucleotide work and Mrs. Christine Whitman for her continued energetic technical assistance.

## References

1. Senior, A. E. (1973) *Biochim. Biophys. Acta 301*, 249–277.
2. Pederson, P. L. (1975) *J. Bioenergetics 6*, 243–275.
3. Nelson, N. (1976) *Biochim. Biophys. Acta 456*, 314–388.
4. Abrams, A., and Smith, J. B. (1974) *Enzymes 10*, 395–429.
5. Simoni, R. D., and Postma, P. W. (1975) *Ann. Rev. Biochem. 44*, 523–554.
6. Senior, A. E. (1978) in *Membrane Proteins in Energy Transduction* (Capaldi, R. A., ed.), Marcel Dekker, New York, in preparation.
7. Kanner, B. I., Serrano, R., Kandrach, M. A., and Racker, E. (1976) *Biochem. Biophys. Res. Commun. 69*, 1050–1056.
8. Shankaran, R., Sani, B. P., and Sanadi, D. R. (1975) *Arch. Biochem. Biophys. 168*, 394–403.
9. Higashiyama, T., Steinmeier, R. C., Serrianne, B. C., Knoll, S. L., and Wang, J. H. (1975) *Biochemistry 14*, 4117–4121.
10. Sou, K. S., and Hatefi, Y. (1976) *Biochim. Biophys. Acta 423*, 398–412.
11. Joshi, S., Shaikh, F., and Sanadi, D. R. (1975) *Biochem. Biophys. Res. Commun. 65*, 1371–1377.
12. Chen, C., Yang, M., Durst, H. D., Saunders, D. R., and Wang, J. H. (1975) *Biochemistry 14*, 4122–4126.
13. Capaldi, R. A. (1973) *Biochem. Biophys. Res. Commun. 53*, 1331–1337.
14. Serrano, R., Kanner, B. I., and Racker, E. (1976) *J. Biol. Chem. 251*, 2453–2461.
15. Simoni, R. D., and Shandell, A. (1975) *J. Biol. Chem. 250*, 9421–9427.
16. Ryrie, I. J. (1975) *Arch. Biochem. Biophys. 168*, 712–719.
17. Tzagoloff, A., Rubin, M. R., and Sierra, M. F. (1973) *Biochim. Biophys. Acta 301*, 71–104.

18. Satre, M., DeJerphanian, M. B., Huet, J., and Vignais, P. V. (1975) *Biochim. Biophys. Acta* 241–255.
19. Landry, Y., and Goffeau, A. (1975) *Biochim. Biophys. Acta 376*, 470–484.
20. Foury, F., and Tzagoloff, A. (1976) *Eur. J. Biochem. 68*, 113–121.
21. Raff, R. A., and Mahler, H. R. (1972) *Science 177*, 575–582.
22.. Vadineanu, A., Berden, J. A., and Slater, E. C. (1976) *Biochim. Biophys. Acta 449*, 468–479.
23. Bertina, R. M., Schrier, P. I., and Slater, E. C. (1973) *Biochim. Biophys. Acta 305*, 503–518.
24. Ferguson, S. J., Lloyd, W. L., and Radda, G. K. (1976) *Biochem. J. 159*, 347–353.
25. Dawson, A. P., and Selwyn, M. J. (1975) *Biochem. J. 152*, 333–339.
26. Montecucco, C., and Azzi, A. (1975) *J. Biol. Chem. 250*, 5020–5025.
27. Bragg, P. D., Davies, P. L., and Hou, C. (1973) *Arch. Biochem. Biophys. 159*, 664–670.
28. Nelson, N., Kanner, B. I., and Gutnick, D. L. (1974) *Proc. Natl. Acad. Sci USA 71*, 2720–2724.
29. Futai, M., Sternweiss, P. C., and Heppel, L. A. (1974) *Proc. Natl. Acad. Sci. USA 71*, 2725–2729.
30. Smith, J. B., Sternweiss, P. C., and Heppel, L. A. (1975) *J. Supramol. Struct. 3*, 248–255.
31. Stockdale, M., and Selwyn, M. J. (1971) *Eur. J. Biochem. 21*, 416–423.
32. Senior, A. E., and Tometsko, A. M. (1975) in *Electron Transfer Chains and Oxidative Phosphorylation* (Quagliariello, E., et al., eds.), pp. 155–160, North-Holland, Amsterdam.
33. Cantley, L. C., and Hammes, G. G. (1973) *Biochemistry 12*, 4900–4904.
34. Russell, J., Jeng, S. J., and Guillory, R. J. (1976) *Biochem. Biophys. Res. Commun. 70*, 1225–1234.
35. Ferguson, S. J., Lloyd, W. J., Lyons, M. H., and Radda, G. K. (1975) *Eur. J. Biochem. 54*, 117–126.
36. Ferguson, S. J., Lloyd, W. J., and Radda, G. K. (1975) *Eur. J. Biochem. 54*, 127–133.
37. Ferguson, S. J., Lloyd, W. J., Radda, G. K., and Slater, E. C. (1976) *Biochim. Biophys. Acta 430*, 189–193.
38. Marcus, F., Shuster, S. M., and Lardy, H. A. (1976) *J. Biol. Chem. 251*, 1775–1780.
39. Cheng, T. M., and Penefsky, H. S. (1974) *J. Biol. Chem. 248*, 1090–1098.
40. McCarty, R. E., and Fagan, J. (1973) *Biochemistry 12*, 1503–1507.
41. Harris, D. A., Rosing, J., Van de Stadt, R. J., and Slater, E. C. (1973) *Biochim. Biophys. Acta 314*, 149–153.
42. Harris, D. A., and Slater, E. C. (1975) *Biochim. Biophys. Acta 387*, 335–348.
43. Slater, E. C., Rosing, J., Harris, D. A., Van de Stadt, R. J., and Kemp, A. (1974) in *Membrane Proteins in Transport and Phosphorylation* (Azzone, G. F., et al., eds.), pp. 137–147, North-Holland, Amsterdam.
44. Harris, D. A., and Slater, E. C. (1975) in *Electron Transfer Chains and Oxidative Phosphorylation* (Quagliariello, E., et al., eds.), pp. 379–384, North-Holland, Amsterdam.
45. Strotmann, H., Bickel, S., and Hurchzermeyer, B. (1976) *FEBS Lett. 61*, 194–198.
46. Bickel-Sandkötter, S., and Strotmann, H. (1976) *FEBS Lett. 65*, 102-106.
47. Magnusson, R. P., and McCarty, R. E. (1976) *J. Biol. Chem. 251*, 7417-7422.
48. Leimgruber, R. M., and Senior, A. E. (1976) *J. Biol. Chem. 251*, 7103–7109.
49. Leimgruber, R. M., and Senior, A. E. (1976) *J. Biol. Chem. 251*, 7110–7113.
50. Leimgruber, R. M., and Senior, A. E. (1977) *Biophys. J. 17*, 66a.
51. Cross, R. L., and Boyer, P. D. (1975) *Biochemistry 14*, 392–398.
52. Boyer, P. D., Smith, D. J., Rosing, J., and Kayalar, C. (1975) in *Electron Transfer Chains and Oxidative Phosphorylation* (Quagliariello, E., et al., eds.), pp. 361–372, North-Holland, Amsterdam.
53. Smith, D. J., Stokes, B. O., and Boyer, P. D. (1976) *J. Biol. Chem. 251*, 4165–4177.

## Chapter 6

# Organization and Role of Lipids in Membranes

## Giorgio Lenaz

### Introduction

Since the pioneering studies of Gortel and Grendel[1] and Danielli and Davson,[2] it was postulated that lipids in biomembranes are organized in the form of bilayers; the same picture is still considered valid, and uncertainties in membrane structure have been related in past years to the structural relations existing between proteins and the lipid bilayer.[3-6] The importance of hydrophobic interactions between lipids and proteins was stressed by Green and Fleischer[7] and subsequently confirmed in reconstitution studies of the inner mitochondrial membrane by Lenaz *et al.*[8-11]

It is now recognized[12,13] that two types of proteins are linked to the lipid bilayer. Extrinsic or peripheral proteins are bound to the membrane surfaces mainly by polar interactions, directed perpendicularly to the plane of the bilayer, while intrinsic or integral proteins penetrate into the bilayer to various extents and may even span the whole membrane thickness: these latter proteins are bound to lipids mainly by hydrophobic interactions parallel to the plane of the lipid bilayer.[14,15]

Lipids in membranes are in a liquidlike state.[16,17] Pure amphipathic lipids are known to undergo transitions to an ordered crystalline arrangement at temperatures depending on the chemical nature of the phospholipid molecules. Such transitions for complex lipid mixtures are accompanied by phase separations,[18] so that fluid and solid regions

*Giorgio Lenaz* • Istituto di Biochimica, University of Ancona, Ancona, Italy.

may coexist in the same bilayer. However, true endothermic phase changes do not occur at physiological temperatures for most membranes, such as the inner mitochondrial membrane.[19, 20] On the other hand, EPR* techniques have demonstrated transitions in the motion parameters of "spin labels" at temperatures near the physiological range[21,22]; such transitions result from formation of quasicrystalline clusters of lipid molecules within a fluid bilayer,[23,24] having largely increased lifetimes in comparison with the surrounding fluid medium. Therefore, under physiological conditions, lipid bilayers in natural membranes, although not in a crystalline state to significant extents, do not have the properties of a true liquid phase but are in a state of heterogeneity in relation to the anisotropic motion of the lipid molecules.

The fluidity of a membrane may be modified by several agents, which, interestingly, can vary under physiological conditions (e.g., pH, ionic strength, concentration of divalent cations) inducing isothermal fluidity changes or phase separations.[17]

Not all the lipids in a membrane can be described as a bilayer undergoing transitions independently of the proteins.[6] A fraction of the lipids (about 20%) do not show a thermal transition by differential scanning calorimetry or other techniques, and this is the fraction believed to be directly in contact with the intrinsic protein molecules. These lipids are strongly immobilized and represent one layer of molecules directly surrounding the protein[25]; these are called "boundary lipids"[25] or "lipid annulus"[26] and are responsible for the lower fluidity observed in natural membranes, containing proteins, in comparison with the extracted lipids (Table I). The lipids in the annulus appear of importance in the function of the associated protein molecule.

### Roles of Lipids in Membranes

Lipids in biomembranes may have several roles at the same time.[3, 6, 27, 28] They are involved in permeability and transport, in the lateral diffusion of membrane-associated receptors, in the determination of hormonal effects (for polypeptide hormones), and in membrane

---

* Abbreviations: ANS, 1-anilinonaphthalene-8-sulfonate; CD, circular dichroism; CoQ, coenzyme Q; $E_A$ activation energy; EPR, electron paramagnetic resonance; ETP, electron transfer particle (submitochondrial membranes obtained by sonic disruption of mitochondria); $F_0$, maximal fluorescence of a probe at infinite membrane concentration; NMR, nuclear magnetic resonance; NPN, N-phenyl naphthylamine; 5-NS, 5-doxylstearic acid; 16-NS, 16-doxylstearic acid.

Table I. Mobility of Stearic Acid Spin Labels in Different Membranes

| Membrane | Mobility of spin label (ratio $h_0/h_{-1}$)[a] | | |
|---|---|---|---|
| | 5-NS | 12-NS | 16-NS |
| Lipid vesicles | 6–7 | 4–5 | 1.8–2.5 |
| ETP | 15–20 | 9–10 | 3–4 |
| ETP + pronase | 9–12 | 6–8 | 3–4 |

[a] The ratio $h_0/h_{-1}$ (heights of the medium and high field peaks in the EPR spectra) is roughly indicative of membrane fluidity and is used to calculate the rotational correlation times $\tau_c$. The ratio ranges from 1 for a nitroxide freely tumbling in solution to infinite for an immobilized nitroxide. For the limitations in expression of fluidity by the $\tau_c$, see Kivelson.[74]

fusion and fusion-related processes, and are also required for the activity of a large number of membrane-bound enzymes.

There may be several reasons why lipids are involved in enzymic activity, and often more than one reason may be operative for the same enzyme.[6, 27, 28]

a. Lipids may be required for establishing a binding area or a hydrophilic surface for enzymes which are extrinsic proteins. In some cases, partial penetration of the enzyme assures a firm anchorage.[29] Enzymes bound to a membrane surface may have optimized catalytic possibilities because of restrictions of random movements and also because of the changed physicochemical environment in comparison with the same enzyme in solution (e.g., the $\zeta$ potential, the hydration of the phospholipid polar heads).[30]

The membrane-bound enzymes do not necessarily have higher activity than the solubilized enzyme; sometimes binding to lipids may inhibit activity, as shown for glutamate dehydrogenase.[31] This could represent an important regulatory mechanism for enzymes which can alternatively be bound or released from a membrane.

b. When enzymic activities are associated with transport phenomena, as in $Ca^{2+}$-ATPase from sarcoplasmic reticulum, ($Na^+$ + $K^+$)ATPase from plasma membranes, or electron transport complexes in mitochondria (which are linked to asymmetrically disposed proton pumps), an optimal lipid composition may be necessary to separate the two compartments and to prevent backflow of the solutes which are moved across the membrane. Warren *et al.*[32] have demonstrated that certain lipids are optimal in restoring $Ca^{2+}$-ATPase, but other lipids are preferred for assuring ATPase-linked $Ca^{2+}$ transport into the sarcoplasmic reticulum. Since the lipid regions near the protein are those where passive permeability increases because of perturbation of the bilayer by the irregular protein surface,[32] it is the composition of the annulus that controls compartmentation. The composition of the annulus is determined by the protein structure, and there is a marked specificity of association. [32]

c. Lipids represent a bidimensional medium [33] to dissolve lipid-soluble substrates and cofactors. Enzymes involved in lipid metabolism have a phospholipid requirement. [34] Also, mitochondrial electron transport requires lipids to dissolve coenzyme Q, a lipid-soluble redox component (see later).

Membrane proteins are dissolved in the bidimensional lipid medium, where they undergo processes of lateral diffusion and rotation [6, 35]; protein diffusion may be critical for concerted multienzymic processes involving several proteins individually dissolved in the lipid milieu.

d. Green and Tzagoloff [36] first recognized a membrane-forming role for the phospholipids. The assembly of mitochondrial membranes from their dissociated subunits can take place only if lipids are present in the reconstitution medium, by allowing bidimensional polymerization along a plane, contrasted to random aggregation when lipids are not included. The lipid requirement of membrane-bound enzymes was explained by a better availability of substrates and coenzymes to the active sites in bidimensional lamellae (membranes) rather than in tridimensional aggregates.

e. A conformational role for lipids has been suggested by Lenaz. [3,6,17,27,28] In much the same way as a polar aqueous medium assures the folding of water-soluble proteins by hiding hydrophobic groups in the polypeptide interior, membrane proteins, which are very rich in hydrophobic residues, will assume the conformation having the minimal free energy in a hydrophobic medium by optimizing hydrophobic interactions with the lipid alkyl chains; such conformation will usually give to the enzymic protein the optimal activity. In addition, it may be postulated that the lipid milieu allows the active site to assume the correct conformational cycle accompanying catalytic activity with formation and breakage of the transition state. [28]

### Lipids and Protein Conformation

Although it is known that interaction with lipids of synthetic polypeptides or serum lipoprotein apoproteins affects profoundly their conformation, [17, 27] little is known in the case of membrane proteins.

London *et al.* [37] found that myelin proteolipid becomes completely $\alpha$ helical in hydrophobic media or in lipid bilayers. Large extents of $\alpha$ helix in membrane proteins have been observed by circular dichroism

(CD) in bulk membranes[38] and also in some purified proteins such as the proteolipid from sarcoplasmic reticulum.[39] The extent of helical structure is different in different membranes, but is usually large, in contrast to $\beta$ structure.[17]

We have investigated the effect of lipids on the conformation of the total protein in mitochondrial membranes by performing CD of lipid-depleted mitochondria and mitochondria reconstituted with different types of phospholipids.

The CD spectra of membranes are characteristic in being largely distorted by artifacts arising from the large size of the membrane particles (light scattering, absorption dampening, differential light scattering)[40]; we have employed, when necessary, the corrections devised by Urry and Long[41] and Masotti *et al.*[42] and have obtained corrected spectra indicative of the real conformation of the membrane proteins.

The corrected CD spectra of intact mitochondria and sub-mitochondrial particles ETP show large extents of $\alpha$ helical structure,[43] but lipid extraction produces CD spectra indicative of increased disordered conformation (decrease of the 208-nm band).[44] Readdition of different phospholipids restores CD spectra, approaching the original conformation observed in mitochondria.[44] The optimal restoration of the spectra is obtained with cardiolipin, in agreement with its specific mitochondrial location (Fig. 1). It is to be noted that phospholipid addition to lipid-depleted mitochondria restores some of the enzymic activities lost by delipidation, such as succinate oxidation[9, 45]; cardiolipin was found the best phospholipid in restoring mitochondrial respiratory activities,[46] and experiments in Racker's laboratory[47] have established the optimal ratios of PE:PC:cardiolipin for electron transfer complexes and ATPase.

In sarcoplasmic reticulum the optimal activity of $Ca^{2+}$-ATPase is restored by unsaturated lecithins, but the activity is inhibited by substituting the native phospholipids with anionic phospholipids.[32] Using the lipid substitution method, we have found that a conformation similar to that of sarcoplasmic reticulum ATPase with its native lipids was obtained with dioleyl lecithin, whereas different CD spectra were given by anionic phospholipids (A. Spisni, G. Lenaz, and L. Masotti, unpublished). In all cases available, including the studies in lipoproteins and model polypeptides, the optimal conformation of the protein in the presence of lipids consisted largely of $\alpha$ helix.

Thermodynamic reasons favor $\alpha$ helix in membrane proteins in their natural environment.[48] The $\alpha$ helix is the conformation maximizing intrapeptide hydrogen bonds; on the surface of a water-soluble protein in water, the solvent molecules compete with peptide bonds for hydrogen bonding, whereas the hydrophobic core may have helical

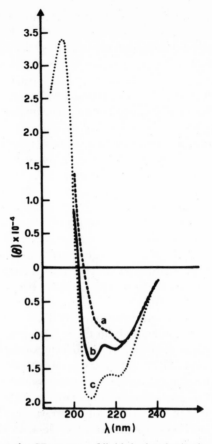

Fig. 1. Effect of lipids on the CD spectra of lipid-depleted mitochondria. From Masotti *et al.*[44] Curve a: Lipid-depleted mitochondria. Curve b: Lipid-depleted mitochondria + cardiolipin. Curve c: Original mitochondria. CD spectra were obtained with a Cary-60 spectropolarimeter modified to obtain simultaneous reading of ellipticity and absorbance in order to correct the CD spectra.[42]

regions favored thermodynamically. In hydrophobic media, the intrapeptide hydrogen bonding will be maximal; this of course is possible if the protein has a sufficient conformational flexibility and a sufficient number of nonpolar residues to direct to the exterior. Membrane proteins, which have a large number of nonpolar residues,[49] are favored in the conformation where they can direct such residues to the exterior in a hydrophobic medium.

Sometimes intrinsic proteins may be extracted from membranes in a water-soluble form[50]; in such case, a profound conformational alteration must occur upon extraction.[6] In the case of lipid removal from mitochondrial membranes, proteins still remain in an insoluble form resembling the native membrane[51]; in this latter case, although a different conformation has been shown between native and extracted membrane,[44] no such dramatic change in conformation is necessary as with solubilized proteins.[6] Accordingly, the interaction of proteins extracted from erythrocytes in soluble form requires an initial ionic interaction followed by hydrophobic stabilization (penetration),[50] whereas in the case of lipid-depleted mitochondria the extracted residue directly binds phospholipids by exclusively hydrophobic interactions.[10, 11]

## Kinetics of Membrane-Bound Enzymes

Kinetic changes in membrane-bound enzymes have been observed after solubilization, or when lipids are removed or perturbed with organic solvents[6, 17] (Table II). In many instances, removing or altering the lipids (e.g., by organic solvents) resulted in stabilization of an enzyme–substrate complex; in mitochondrial ATPase[52] and in other enzymes,[6] an uncompetitive type of inhibition was observed, with decrease of both $V_{max}$ and $K_m$. This is the case for $(Na^+ + K^+)$ATPase, where alcohols stabilize a phosphorylated intermediate formed in presence of $Na^+$ and prevent its dissociation in the presence of $K^+$.[53] In $Ca^{2+}$-ATPase, phospholipids are not required to form the phosphorylated intermediate but are necessary for its hydrolysis.[54]

In mitochondrial membranes, ATPase activity is decreased by solvent addition or by phospholipase $A_2$ treatment, and there is a decrease of both $V_{max}$ and $K_m$ for ATP, suggesting that a similar

Table II. *Lipids and the Kinetics of Some Membrane-Bound Enzymes*[a]

| Enzyme | Treatment | $K_m$ | $V_{max}$ |
|---|---|---|---|
| β-Hydroxybutyrate dehydrogenase | Delipidation | Decreased | Decreased |
| Glucose-6-phosphatase | Delipidation | Decreased | Decreased |
| Glucose-6-phosphatase | Solubilization (detergent) | Decreased | Decreased |
| Succinate-cytochrome $c$ reductase | Delipidation | Decreased | Decreased |
| Cytochrome $c$ oxidase | Delipidation | Decreased | Decreased |
| ATPase (mitochondrial) | Delipidation | Decreased | Decreased |
| ATPase (mitochondrial) | Butanol, solvents | Decreased | Decreased |
| $(Na^+ + K^+)$ATPase | Delipidation | Unchanged | Decreased |
| $(Na^+ + K^+)$ATPase | Butanol | Decreased | Decreased |
| Pyruvate oxidase | Delipidation | Increased | Decreased |

[a] From Lenaz[6, 28] and Fourcans and Jain.[75]

mechanism (stabilization of an intermediate) may be operative[52, 55] (Fig.2).

The kinetic changes may be related to an alteration of the catalytic cycle, possibly for a changed conformation of the active site, preventing the normal structural events from taking place during catalysis.

### Arrhenius Plots of Membrane-Bound Enzymes

Membrane-associated activities, linked to either enzymes or carriers, present anomalous temperature dependence. Whereas most soluble enzymes have constant activation energies in large ranges of temperatures, the lipid-dependent activities exhibit discontinuities in the Arrhenius plots, with breaks in activation energies at well-defined temperatures.[3,6,17,27,56] For example, mitochondrial ATPase shows a sharp break at about 16–18°C, but the temperature dependence becomes smoother after detergent treatment[57] (Fig. 3). The breaks have been related to changes occurring in the lipid phase. In the case of membranes reconstituted with pure phospholipids, the breaks often correspond to the real (calorimetric) transition from a fluid to a crystalline phase. Sometimes two breaks were found,[22] related to the onset and the termination of the solidification process during a phase separation. The correspondence is often lacking in the case of natural membranes, where the lipids usually have transitions at low temperatures ($-4$°C to $-12$°C in the inner mitochondrial membrane[20]); however, a correspondence exists with fluid phase separations and formation of quasicrystalline clusters[23] observed in fluid lipids by spin labeling, fluorescence, and NMR techniques. Sharp increases in anisotropy of the lipid motion by decreasing the temperature may correspond to the breaks in Arrhenius plots of enzymic activities.[28]

Organic solvents modify the Arrhenius plots of membrane-bound enzymes[17] (Table III). In mitochondrial ATPase we have found breaks in Arrhenius plots to be lowered by diethyl ether whereas *n*-butanol abolishes the discontinuity by increasing the activation energy above the break temperature.[27] Pentrane, a general anesthetic, induces both a decrease of break temperature and an increase of activation energy. An increased activation energy above the break is observed also after phospholipase $A_2$ treatment of mitochondrial particles.[52]

The increased activation energies are interpreted as the loss of a correct lipid environment of the enzyme, as also shown for detergent solubilization.

The mechanism whereby the physical state of phospholipids affects catalytic power is obscure. There may be two principal reasons for that phenomenon. In the case of concerted activities requiring a series of

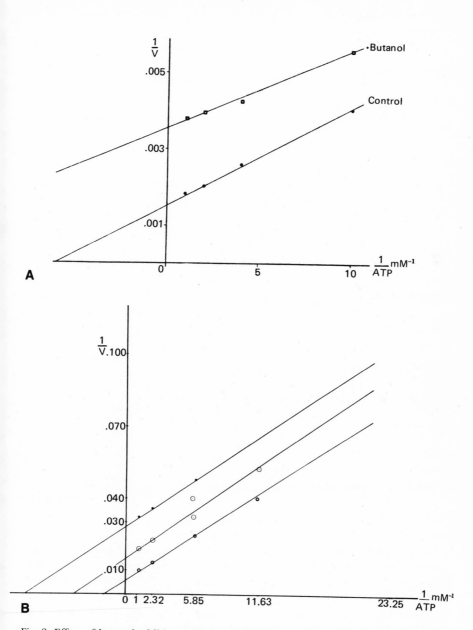

Fig. 2. Effect of butanol addition and phospholipase $A_2$ treatment on Lineweaver–Burk plots of mitochondrial ATPase in ETP. A: $\bigcirc$, no addition; $\square$, butanol, 0.1 M. B: $\bigcirc$, no treatment; $\odot$, phospholipase $A_2$, 5 min hydrolysis; $\bullet$, phospholipase $A_2$, 10 min hydrolysis.

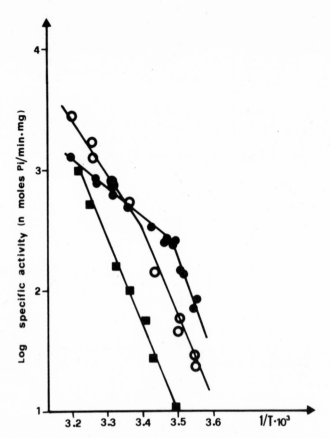

Fig. 3. Arrhenius plots of mitochondrial ATPase. ●, ± No treatment; ○, Triton X-100, 2.5 mg/ml; ■, butanol, 0.32 M.

enzymic proteins which have to undergo collisions in the lipid matrix (e.g., NADH-cytochrome $b_5$ reductase flavoprotein and cytochrome $b_5$[29]), the break may be linked to a sharp decrease in the rate of protein diffusion in a solidifying medium, preventing fast interaction of the proteins.[58] This does not hold in activities not involving protein motion in the membrane. In sarcoplasmic reticulum ATPase it was shown that, below the break, an increased $\Delta H$ of activation is compensated by an increased $\Delta S$ so that the activation state is entropically favored but reached with more difficulty.[54] The change in thermodynamic parameters when lipids become more viscous may be explained by an increased protein rigidity in the pretransition state, so that a nonfunctional conformation is favored; alternatively, a conformational change involving the enzyme–substrate complex and necessary to the catalytic

*Table III. Effect of Various Treatments on the Temperature Dependence of Mitochondrial ATPase* [a]

| Treatment | Break temperature (°C) | $E_A$ above break (kcal/mole) | $E_A$ below break (kcal/mole) |
|---|---|---|---|
| 1. None | 16.0 | 11.6 | 36.3 |
| Triton X-100 (1 mg/ml) | 21.0 | 19.3 | 36.3 |
| Triton X-100 (2.5 mg/ml) | 22.0 | 22.6 | 30.9 |
| Butanol (0.35 M) | — | 31.3 | |
| Diethyl ether (two-phase system) | 12.0 | 2.7 | 40.0 |
| 2. None | 25.5 | 4.5 | 24.2 |
| Pentrane (0.5%) | 18.5 | 9.1 | 31.5 |
| 3. None | 19.0 | 9.7 | 25.2 |
| Phospholipase $A_2$ incubation (10 min) | 19.0 | 14.7 | 26.0 |

[a] From Lenaz *et al.*,[17, 57, 61] Landi *et al.*,[52] and Parenti-Castelli *et al.*[55]

cycle will take place with increased difficulty. Such possibility, which may also account for the effects of lipids on $V_{max}$ and $K_m$, is depicted schematically in Fig. 4. The demonstration that a sudden conformational change occurs at the temperature where a break is observed in the Arrhenius plot was given for a soluble enzyme, D-amino acid oxidase.[59]

The average conformation of mitochondrial membranes has been preliminarily studied in our laboratory by CD at different temperatures; the changes observed are indicative of different conformational states at low and high temperatures (A. Spisni, L. Masotti, and G. Lenaz, unpublished observations).

## Drugs and Membrane Fluidity: A Conformational Mechanism for Anesthesia

Several exogenous compounds—*viz*, different kinds of drugs—are capable of affecting membrane lipid fluidity.[17, 60] In the course of our

$$E + S \underset{K_2}{\overset{K_1}{\rightleftarrows}} ES \overset{K_3}{\rightleftarrows} ES^* \rightarrow E + P$$

Lipid Removal Induces:
a. Higher $E_A$ (decreased formation of ES*)
b. Lower $V_{max}$ (decreased formation of ES*)
c. Lower $K_m$ (increased accessibility of S to E or increased stability of ES?)

Fig. 4. Scheme of the possible role and sites of involvement of phospholipids in enzyme activity.

studies of agents employed to perturb the lipid bilayer and lipid–protein interactions,[61, 62] we have found that monohydric alcohols and other compounds known as general anesthetics[17] have the property of weakening lipid–protein interactions in membranes. The fluidization of the lipid bilayer induced by general and local anesthetics in model lipid membranes[62] has been confirmed by us in natural membranes, such as erythrocyte ghosts and the inner mitochondrial membrane, by means of EPR and fluorescence techniques.[17]

The fluorescence of anilinonaphthalene derivatives (ANS and NPN), when they are bound to membranes, decreases as a function of the fluidity of the lipid bilayer[64]; ANS probes the membrane surface, while NPN is located into the membrane interior (Fig. 5).

By double reciprocal plots of ANS and NPN fluorescence against membrane concentration,[17] it is possible to extrapolate maximal fluorescence $(F_0)$ of the probes at infinite membrane concentration when all the probe added is bound to the membrane.

Alcohols and anesthetics decrease $F_0$ of both ANS and NPN bound to mitochondrial membranes or phospholipid vesicles (Figs. 6 and 7) without shifting the emission maxima of the fluorescence spectra (Fig. 8), indicating that the effects are due to increased fluidity and not to increased polarity of the probe environment (a polar medium, besides quenching the fluorescence of the probes, induces a red shift of the emission maximum[17]).

The spin label experiments have confirmed the fluidization of the membranes by anesthetics (Fig. 9) but have revealed in addition that the disordering effect is much more pronounced in mitochondrial mem-

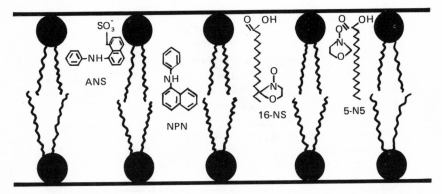

Fig. 5. Schematic localization of spin labels and fluorescent probes in a lipid bilayer.

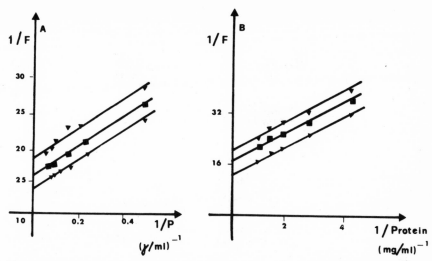

Fig. 6. Double reciprocal plots of NPN fluorescence in phospholipid vesicles (A) and mitochondrial membranes (B) in the presence of *n*-butanol. For theoretical details on fluorescent probes, see Lenaz *et al.*[17] Fluorescence was recorded with a Perkin-Elmer MPF-2 spectrophotofluorometer; excitation at 380 nm, emission at 420 nm. △, No butanol; ■, butanol, 0.27 M; ▲, butanol, 0.41 M.

branes than in lipid vesicles (Fig. 10). The data have been interpreted by assuming that anesthetics weaken lipid–protein interactions and release the immobilization of the lipid molecules induced in the bilayer by the hydrophobic penetrating proteins.

## A Working Hypothesis for Anesthesia

The idea that anesthetics act at the membrane level is rather old (*cf*. refs. 17, 63); Hill[65] has theoretically shown that anesthesia is produced when the free energy of the membrane is decreased by a critical amount, independently of the method used. The results discussed previously that organic solvents and anesthetics not only increase lipid fluidity but also disrupt lipid–protein interactions, and that lipids are required for the correct conformation of membrane-bound enzymes, allow the postulation of a conformational hypothesis for general anesthesia (Fig. 11).

According to this hypothesis, the changes in fluidity and lipid–protein interactions induced by anesthetics will initiate a series of

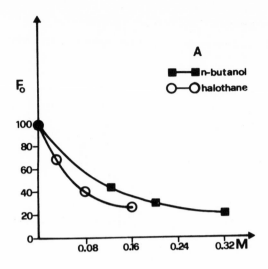

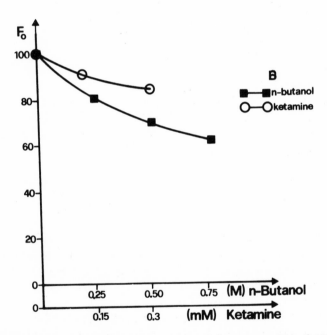

Fig. 7. Effect of different anesthetics on the maximal fluorescence ($F_0$) of ANS(A) and NPN(B) in mitochondria. $F_0$ was calculated from the intercept on the ordinates in double reciprocal plots as in Fig. 6. The values of $F_0$ are in percent of control values without anesthetics.

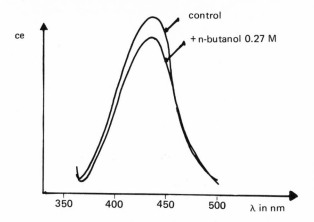

Fig. 8. Effect of *n*-butanol on the fluorescence spectra of NPN in mitochondria. Excitation wavelength at 360 nm.

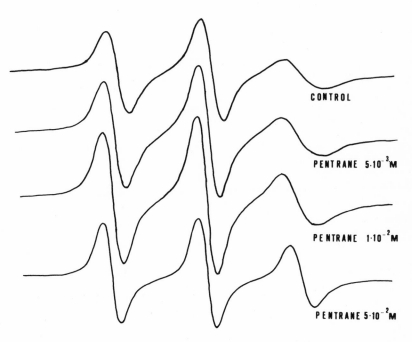

Fig. 9. EPR spectra of 16-NS in mitochondrial membranes in the presence of increasing amounts of pentrane.

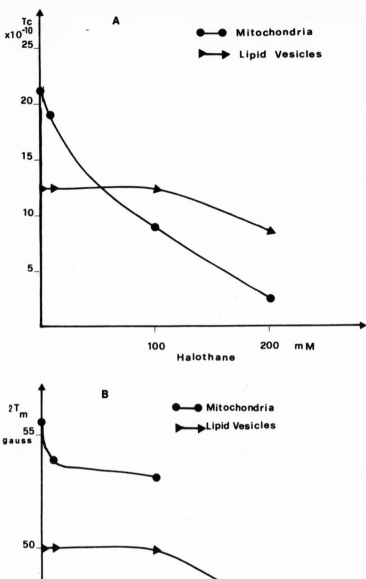

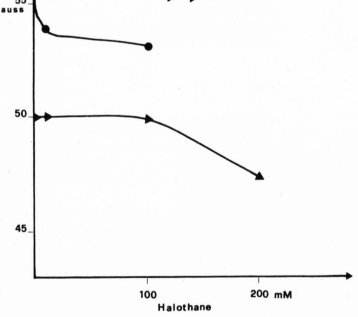

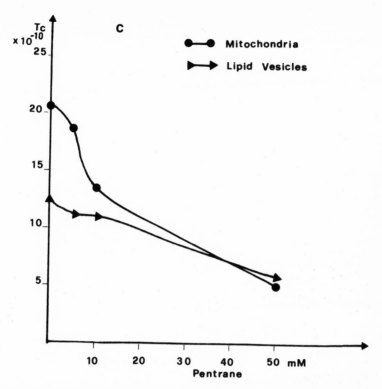

Fig. 10. A: Effect of halothane on the $\tau_c$ of 16-NS in lipid vesicles and mitochondrial membranes. B: Effect of halothane on the $2T_\parallel$ of 5-NS in lipid vesicles and mitochondrial membranes. C: Effect of pentrane on the $\tau_c$ of 16-NS in lipid vesicles and mitochondrial membranes.

conformational changes in membrane proteins which will change their functions: catalytic proteins, either enzymes or carriers, will modify their kinetic properties. In the case of carriers, changes in transmembrane ion transport will be possible; such changes in ionic channels ($Na^+$ channels) in neuronal membranes will abolish the transmission of the action potential and give rise to anesthesia.

Preliminary data that solvents and anesthetics at low concentrations induce conformational changes, measured by CD, in model membranes (mitochondrial and sarcoplasmic reticulum ATPase) have been obtained in our laboratory.[17] The discussed changes in breaks in Arrhenius plots and other kinetic parameters induced by organic solvents and anesthetics are best considered as resulting from conformational changes of membrane proteins.

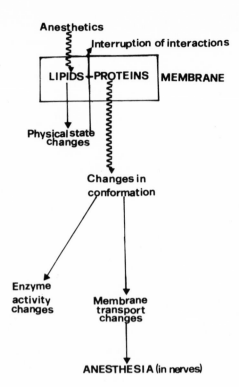

Fig. 11. A working hypothesis for the mechanism of action of general anesthetics at the membrane level.


## Involvement of Lipid Fluidity in the Motion of Coenzyme Q in the Mitochondrial Membrane

In mitochondria, lipids are also required to dissolve the lipid-soluble electron carrier, coenzyme Q (CoQ, ubiquinone). In view of the important role of CoQ in the mitochondrial respiratory chain and of the assumption that it is as a mobile electron carrier[65] between reduced flavoproteins and oxidized cytochrome(s) *b*, we have studied the effect of different quinones in mitochondrial electron transport. We have observed that short-chain quinones, markedly $CoQ_2$ and $CoQ_3$, are very poor in restoration of NAD-linked electron transfer in ubiquinone-depleted mitochondria, in contrast to the long-chain physiological quinones[67] (Fig. 12); furthermore, $CoQ_2$ and $CoQ_3$ are inhibitors of NADH oxidation but not of succinate oxidation in submitochondrial particles, and the inhibition is competitive with respect to the natural $CoQ_{10}$ (Fig. 13). It was also found that exogenously added $CoQ_3$ can be reduced by both NADH and succinate in submitochondrial particles

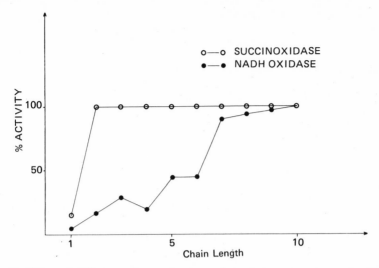

Fig. 12. Effect of different quinones in restoration of electron transport in pentane-extracted mitochondria. From Lenaz *et al.*[67]

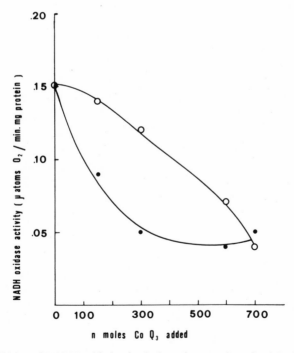

Fig. 13. Inhibition of NADH oxidation by CoQ$_3$ and restoration of activity by CoQ$_7$ in the presence of deoxycholate (0.5 mg/mg protein). ●, CoQ$_3$; ○, CoQ$_3$ in the presence of CoQ$_7$ (134 $\mu$M).

(Table IV), suggesting that its reoxidation may be impaired in the case that its reduction is sustained by NADH through complex I (G. Lenaz and P. Pasquali, unpublished observations).

We have tentatively interpreted these findings as the result of some kind of compartmentation, so that short-chain quinones reduced by NADH cannot reach their site of reoxidation.[69] In view of the higher polarity of short-chain quinones in comparison with the very hydrophobic $CoQ_{10}$, such compartmentation could be the result of sidedness of the redox cycle of the quinone pool in the membrane. If reduced ubiquinone must cross the membrane to be reoxidized, it is plausible that the polar $CoQ_3$ may not traverse the bilayer with a sufficient rate in comparison to the hydrophobic long-chain quinones. This idea is substantiated by the finding that $CoQ_3$ is a much more powerful inhibitor of NADH oxidation at lower temperatures, indicating that when the membrane is more fluid $CoQ_3$ may serve as an electron carrier in the chain (G. Lenaz and L. Cabrini, unpublished) (Fig. 14).

Additional support for this hypothesis originates from the difference observed in hydroquinone oxidation in intact mitochondria and submitochondrial particles, which are known to have opposite polarity or sidedness (Table V). It was found that reduced $CoQ_3$ is oxidized at a much higher rate in submitochondrial particles ETP than in mitochondria, whereas reduced $CoQ_{10}$ is oxidized at similar rates in intact mitochondria and in ETP (G. Lenaz, T. Ozawa, and P. Pasquali, unpublished). Also, fast kinetics in a stopped-flow apparatus (in collaboration with S. Papa from the University of Bari) show a much higher oxidation rate and a lower $t/2$ for $CoQ_3$ in ETP than in mitochondria, suggesting that the site of oxidation of the hydroquinone is located in the inner side of the mitochondrial membrane.

Figure 15 tentatively suggests a scheme of electron transport

*Table IV. Reduction of Exogenous $CoQ_3$ by NADH and Succinate in Antimycin A Inhibited Submitochondrial Particles*

|  | $CoQ_3$ reductase activity (nmoles/min-mg protein) | |
| --- | --- | --- |
| $CoQ_3$ (mM) | NADH | Succinate |
| — | 0 | 0 |
| 0.05 | 184 | 56 |
| 0.1 | 199 | — |
| 0.17 | — | 134 |
| 0.2 | 227 | 219 |
| 0.3 | 241 | — |

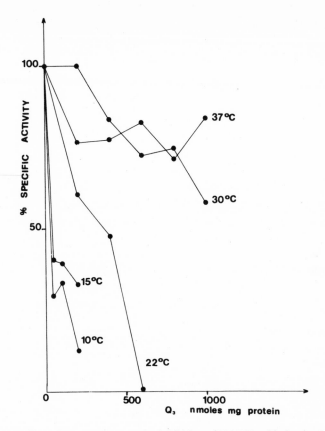

Fig. 14. Effect of temperature on the inhibition of NADH oxidation by CoQ$_3$.

mediated by CoQ that explains the observed phenomena. It must be noted that reduced CoQ moves in this scheme in a way opposite to that proposed by Mitchell[70, 71] for the chemioosmotic hypothesis; for this reason, it would be necessary to postulate a proton carrier other than CoQ (*cf.* ref. 72) for the observed outward movement of H$^+$ during electron transfer. According to this interpretation, reduced CoQ$_3$ is unable to cross the bilayer to reach its reoxidation site on the inner side of the membrane. No such movement is apparently necessary in the case of CoQ reduced by succinate.

Experiments on the effects of quinones on the physical state of membrane lipids are in agreement with this interpretation (Fig. 16 and Table VI). Quinones (CoQ$_3$ and CoQ$_{10}$) were added to phospholipid vesicles containing fatty acids spin-labeled either in the 5-position or in the 16-position. The results show that oxidized quinones have a fluidiz-

*Table V. Ubiquinol-Cytochrome c Reductase in Intact Mitochondria and Submitochondrial Particles*[a]

| Substrate (reduced CoQ) | Mitochondria | | ETP (nmoles/min-mg protein) | |
|---|---|---|---|---|
| | Total | AA sensitive | Total | AA sensitive |
| $CoQ_{10}$ | 24 | 14 | 100 | 80 |
| $CoQ_3$ | 17 | 3 | 205 | 187 |

[a] Ubiquinol-cytochrome *c* reductase was assayed by following the increase in absorbance of cytochrome *c* at 550 nm upon reduction. Assay temperature was 20°C. Antimycin A (AA) when added was 2 μg/ml. Under the same conditions, intact mitochondria have a NADH-cytochrome *c* reductase of 14 nmoles cytochrome *c* reduced/min-mg protein (11 in the presence of antimycin A). Antimycin-sensitive NADH-cytochrome c reductase (3 nmoles/min-mg) corresponds to antimycin-sensitive $CoQ_3$-cytochrome *c* reductase. From G. Lenaz, T. Otawa, and P. Pasquali (unpublished).

*Table VI. Effect of Oxidized and Reduced Quinones on the Mobility of 5-NS and 16-NS in Egg Lecithin Vesicles*[a]

| Quinone (0.4 mM) | 5-NS $T_{\parallel}$ (gauss) | 16-NS $\tau_c\ 10^{-9}$ sec |
|---|---|---|
| — | 25.6 | 0.8 |
| $CoQ_3$ oxidized | 25.2 | 0.7 |
| $CoQ_3$ reduced | 25.9 | 1.0 |
| $CoQ_{10}$ oxidized | 24.6 | 0.7 |
| $CoQ_{10}$ reduced | 25.0 | 0.8 |

[a] The quinones were reduced with $NaBH_3CN$ rather than with $NaBH_4$ in order to prevent reduction of the spin labels.

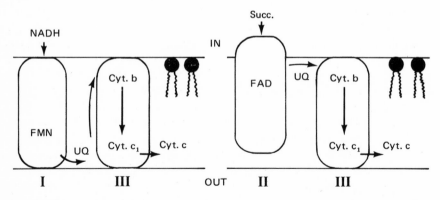

Fig. 15. A tentative scheme of the movements of reduced CoQ in the mitochondrial membrane.

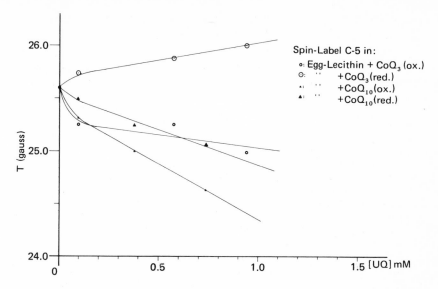

Fig. 16. Effect of reduced and oxidized quinones on the mobility of 5-NS in sonicated vesicles of egg lecithin.

ing effect on the membrane lipids; a similar disordering effect is shared by $CoQ_{10}$ also in its reduced form. However, reduced $CoQ_3$ has a strong ordering effect on the lipid bilayer. This effect is interpreted as stacking of the isoprenoid chain of the short quinone with the lipid alkyl chains; it is to be noted that short-chain quinones have lengths approximating the width of a half-bilayer. Possible formation of hydrogen bonds with the polar heads of the phospholipids, as suggested by Cain et al.,[73] would stabilize the ordered situation and prevent diffusion of the short hydroquinone across the bilayer.

Studies of the polarization of the fluorescence of perylene in lipid bilayers are qualitatively in accord with the spin label studies (unpublished observations).

ACKNOWLEDGMENTS

Part of the unpublished data reported here were obtained in collaboration with the Istituto di Chimica Biologica, University of Bologna, Italy. I wish to thank the Hoffmann La Roche Co., Basel, for the kind gift of quinones. Thanks are due to Dr. Griffiths, University of Warwick, U. K., and Dr. Rotilio, University of Camerino, Italy, for allowing the use of their EPR facilities. Some of the recent CD data were obtained with a Juan spectropolarimeter, in the Department of Intermediates Chemistry of the University of Bologna.

The experiments described in this chapter have been supported in part by grants from the CNR, Rome, Italy.

## References

1. Gortel, E., and Grendel, F., (1925) *J. Exp. Med. 41*, 439.
2. Danielli, J. F., and Davson, H. (1935) *J. Cell. Comp. Physiol. 5*, 495.
3. Lenaz, G. (1974) *SubCell. Biochem. 3*, 167.
4. Vandenheuvel, F., (1971) *Adv. Lipid Res. 9*, 161.
5. Hendler, R. A. (1971) *Physiol. Rev. 51*, 66.
6. Lenaz, G. (1977) in *Membrane Proteins and Their Interaction with Lipids* (Capaldi, R. A., ed.), p. 47, Marcel Dekker, New York.
7. Green, D. E., and Fleischer, S. (1972) *Horizons in Biochemistry* (Kasha, M., and Pullamn, B., eds.), p. 381, Academic Press, New York.
8. Lenaz, G., Sechi, A. M., Parenti-Castelli, G., and Masotti, L. (1970) *Arch. Biochem. Biophys. 141*, 79.
9. Lenaz, G., Sechi, A. M., Masotti, L., and Parenti-Castelli, G. (1970) *Arch. Biochem. Biophys. 141*, 89.
10. Lenaz, G., Parenti-Castelli, G., Sechi, A. M., and Masotti, L. (1972) *Arch. Biochem. Biophys. 148*, 391.
11. Lenaz, G., Parenti-Castelli, G., Monsigni, N., and Silvestrini, M. G. (1971) *J. Bioenergetics 2*, 119.
12. Vanderkooi, G., and Green, D. E. (1970) *Proc. Natl. Acad. Sci. USA 66*, 615.
13. Glaser, M. H., Simpkins, H., Singer, S. J., Sheetz, M., and Chan, S. I. (1970) *Proc. Natl. Acad. Sci. USA 65*, 721.
14. Singer, S. J., and Nicolson, G. L. (1970) *Science 175*, 720.
15. Green, D. E. (1972), *Ann. N. Y. Acad. Sci. 195*, 150.
16. Chapman, D. (1969) *Lipids 4*, 251.
17. Lenaz, G., Curatola, G., and Masotti, L., (1975) *J. Bioenergetics 7*, 223.
18. Shimshick, E. J., and McConnell, H. M., (1973) *Biochemistry 12*, 2351.
19. Reinert, J. C., and Steim, J. M. (1974) *Science 168*, 1580.
20. Höchli, M., and Hackenbrock, C. R. (1976) *Proc. Natl. Acad. Sci. USA 73*, 1636.
22. Raison, J. K., and McMurchie, E. J. (1974) *Biochim. Biophys. Acta 363*, 135.
21. Raison, J. K., Lyons, J. M., Mehlhorn, R. J., and Keith, A. D. (1971) *J. Biol. Chem. 246*, 4036.
23. Lee, A. A., Birdsall, N. J. M., Metcalfe, J. C., Tson, P. A., and Warren, G. B. (1974) *Biochemistry 13*, 3699.
24. Davis, D. G., Inesi, G., and Gulik-Krzywicki, T. (1976) *Biochemistry 15*, 1271.
25. Jost, P. C., Griffith, O. H., Capaldi, R. A., and Vanderkooi, G. (1973) *Biochim. Biophys. Acta 311*, 141.
26. Warren, G. B., Birdsall, N. J. M., Lee, A. G., and Metcalfe, J. C. (1974) *Membrane Proteins in Transport and Phosphorylation* (Azzone, G. F., Klingenberg, M. E., Quagliariello, E., and Siliprandi N., eds.), p. 1, North-Holland, Amsterdam.
27. Lenaz, G. (1973) *Acta Vitamin. Enzymol. (Milano) 27*, 62.
28. Lenaz, G. (1978) *SubCell. Biochem.*, in press.
29. Spatz, L., and Strittmatter, P. (1973) *J. Biol. Chem. 248*, 793.
30. Triggle, D. J. (1970) *Recent Progr. Surface Sci. 3*, 273.

31. Dodd, G. H., (1973) *Eur. J. Biochem. 33*, 418.
32. Warren, G. B., Bennett, J. P., Hesketh, T. R., Houslay, M. D., Smith, G. A., and Metcalfe, J. C., (1975) *Proc. 10th FEBS Meet.,* p. 3.
33. Vanderkooi, G., (1974) *Biochim. Biophys. Acta 344*, 307.
34. Dawson, R. M C. (1973) in *Form and Function of Phospholipids*, 2nd ed. (Ansell, G. B., Dawson, R. M. C., and Hawthorne, J. N., eds.), Chap. 5, Elsevier, Amsterdam.
35. Cherry, R. J. (1975) *FEBS Lett. 35*, 1.
36. Green, D. E., and Tzagoloff, A. (1966) *J. Lipid Res. 7*, 587.
37. London, Y., Demel, R. A., Geurts van Kessel, W. S. M., Zahler, P. and Van Deenen, L. L. M. (1974) *Biochim. Biophys. Acta 332*, 69.
38. Urry, D. W., Masotti, L., and Krivacic, J. R. (1971) *Biochim. Biophys. Acta 241*, 600.
39. Laggner P. (1975) *Nature 255*, 427.
40. Urry, D. W. (1972) *Biochim. Biophys. Acta 265*, 15.
41. Urry, D. W., and Long, M. M. (1975) in *Methods in Membrane Biology* (Korn, E. D., ed.), Plenum, New York.
42. Masotti, L., Urry, D. W., Krivacic, J. R., and Long, M. M. (1972) *Biochim. Biophys. Acta 266*, 7.
43. Urry, D. W., Masotti, L., and Krivacic, J. R. (1970) *Biochem. Biophys. Res. Commun. 41*, 521.
44. Masotti, L., Lenaz, G., Spisni, A., and Urry, D. W. (1974) *Biochem. Biophys. Res. Commun. 56*, 892.
45. Fleischer, S., and Fleischer, B. (1967) *Methods Enzymol. 10*, 406.
46. Fleischer, S., Brierley, G. P., Klouwen, H., and Slautterback, G. (1962) *J. Biol. Chem. 237*, 3264.
47. Ragan, C. I., and Racker, E. (1973) *J. Biol. Chem. 248*, 5263.
48. Singer, S. J. (1971) in *Structure and Function of Biological Membranes* (Rothfield, L. I., ed.), p. 145, Academic Press, New York.
49. Capaldi, R. A., and Vanderkooi, G. (1972) *Proc. Natl. Acad. Sci. USA 69*, 930.
50. Zwaal, R. F. A., and Van Deenen, L. L. M. (1970) *Chem. Phys. Lipids 4*, 311.
51. Fleischer, S., Fleischer, B. and Stoeckenius, W., (1967) *J. Cell Biol. 32*, 193.
52. Landi, L., Olivo, G., Parenti-Castelli, G., Sechi, A. M., and Lenaz, G. (1976) *Bull Mol. Biol. Med. 1*, 29.
53. Hegivary, C. (1973) *Biochim. Biophys. Acta 311*, 272.
54. Hidalgo, C., Ikemoto, N., and Gergely, J. (1976) *J. Biol. Chem. 251*, 4224.
55. Parenti-Castelli, G., Sechi, A. M., Landi, L., Cabrini, L., Mascarello, S., and Lenaz, G., *Arch. Biochem. Biophys.,* submitted.
56. Raison, J. K. (1973) in *Membrane Structure and Mechanisms of Biological Energy Transduction* (Avery, J., ed.), p. 559, Plenum, London.
57. Lenaz, G., Parenti-Castelli, G., Sechi, A. M., Landi, L., and Bertoli, E. (1972) *Biochem. Biophys. Res. Commun. 49*, 536.
58. Strittmatter, P., and Rogers, M. J. (1975) *Proc. Natl. Acad. Sci. USA 72*, 2658.
59. Massey, V., Curti, B., and Gauther, H. (1966) *J. Biol. Chem. 241*, 2347.
60. Chapman, D., Urbina, J., and Keough, K. M. (1974) *J. Biol. Chem. 249*, 2512.
61. Lenaz, G., Parenti-Castelli, G., and Sechi, A. M. (1975) *Arch. Biochem. Biophys. 169*, 1499.
62. Lenaz, G., Bertoli, E., Curatola, G., Mazzanti, L., and Bigi, A. (1976) *Arch. Biochem. Biophys. 172*, 278.
63. Seeman, P. (1972) *Pharmacol. Rev. 24*, 503.
64. Trauble, H., and Overath, P. (1973) *Biochim. Biophys. Acta 307*, 491.
65. Hill, M. W. (1974) *Biochim. Biophys. Acta 356*, 117.
66. Green, D. E. (1966) in *Comprehensive Biochemistry, Vol. 4* (Florkin, M., and Stotz, E. H., eds.), p. 309, Elsevier, Amsterdam.

67. Lenaz, G., Daves, G. D., and Folkers, K. (1968) *Arch. Biochem. Biophys. 123*, 539.
68. Lenaz, G., Pasquali, P., Bertoli, E., Parenti-Castelli, G., and Folkers, K. (1975) *Arch. Biochem. Biophys. 169*, 217.
69. Lenaz, G., Pasquali, P., and Bertoli, E. (1975) in *Electron Transfer Chains and Oxidative Phosphorylation* (Quagliariello, E., Papa, S., Palmieri, F., Slater, E. C., and Siliprandi, N., eds.), p. 251, North-Holland, Amsterdam.
70. Mitchell, P. (1974) *FEBS Lett. 43*, 189.
71. Mitchell, P. (1975) in *Electron Transfer Chains and Oxidative Phosphorylation* (Quagliariello, E., Papa, S., Palmieri, F., Slater, E. C., and Siliprandi, N., eds.), p. 305, North-Holland, Amsterdam.
72. Papa, S. (1976) *Biochim. Biophys. Acta 456*, 39.
73. Cain, J., Santillan, G., and Blasie, J. K. (1972) in *Membrane Research* (Fox, C. F., ed.), p. 3, Academic Press, New York.
74. Kivelson, D. (1960) *J. Chem. Phys. 33*, 1099.
75. Fourcans, B., and Jain, K. M. (1974) *Adv. Lipid. Res. 12*, 147.

Chapter 7

# Studies on Nicotinamide Nucleotide Dehydrogenation and Transhydrogenation by Mitochondria

## Youssef Hatefi and Yves M. Galante

### Introduction

Nicotinamide nucleotides constitute the major vehicle for the transfer of reducing equivalents to the mitochondrial electron transport system. The mitochondrial respiratory chain is capable of oxidizing NADH and NADPH as well as catalyzing transhydrogenation from NAD(P)H to NAD(P). In this chapter, we should like to describe the nicotinamide nucleotide dehydrogenase and transhydrogenase properties of mitochondria at three levels of structural complexity, i.e., submitochondrial particles, complex I, and the soluble NADH dehydrogenase isolated from complex I.

### Results

#### Submitochondrial Particles

Submitochondrial particles prepared by sonication and differential centrifugation of beef heart mitochondria in the presence of ATP and $Mg^{2+}$ are capable of coupled oxidation of NADH, succinate, and NADPH. Until recently, it was believed that the mitochondrial respira-

*Youssef Hatefi and Yves M. Galante* • Department of Biochemistry, Scripps Clinic and Research Foundation, La Jolla, California 92037.

Table I. Characteristics of NADPH Oxidation by Submitochondrial Particles

| | |
|---|---|
| Rate of NADPH oxidation (30°C, pH $\leq$ 6.0) | $\geq$ 250 nmoles/min/mg protein |
| Respiratory chain inhibitors | Rotenone, piericidin A, antimycin A, cyanide |
| P/O in phosphorylation | 2.4–2.9 |
| Distribution of NADPH dehydrogenase in complexes | Complex I |
| Stereospecificity for NADPH hydrogen | 4B |
| pH optimum | < pH 6.0 |

tory chain was incapable of direct oxidation of NADPH and that the oxidation of this nucleotide occurred only by way of transhydrogenation to NAD followed by oxidation of the NADH so formed through the normal NADH oxidase pathway of the respiratory chain. Our studies showed that the NADPH → NAD transhydrogenase activity of sub-mitochondrial particles could be completely inhibited by treatment of the particles with trypsin or butanedione (in the presence of borate buffer) with little (<10%) or no effect on the oxidation rates of NADH and NADPH.[1] Such particles were also capable of undiminished transhydrogenase activity from NADH to NAD (or 3-acetylpyridine adenine dinucleotide, AcPyAD). The oxidation of NADPH by sub-mitochondrial particles is linked to ATP synthesis with P/O of 2.4–2.9, and at appropriate pH values (< 6.0) the oxidation rate is $\geq$ 250 nmoles $min^{-1}$ mg $protein^{-1}$ at 30°C.[2] Similar to the oxidation of NADH and transhydrogenation from NADPH to NAD, the oxidation of NADPH by submitochondrial particles involves the abstraction of hydrogen 4-B.[3] The NADPH oxidase properties of submitochondrial particles are summarized in Table I. Data regarding NADH and NADPH oxidation

Table II. Effects of Treatment of Submitochondrial Particles with Trypsin or Butanedione on the Nicotinamide Nucleotide Dehydrogenase and Transhydrogenase Activities[a]

| | Percent activity of untreated ETP[b] | |
|---|---|---|
| Reaction | Trypsin-treated ETP | Butanedione-treated ETP |
| NADH → $O_2$ | 100 | 100 |
| NADPH → $O_2$ | 92 | 90 |
| NADH → AcPyAD[c] | 90 | 90 |
| NADPH → AcPyAD | 0.0 | 5 |

[a] From Figs. 1, 2, and 6 of Djavadi-Ohaniance and Hatefi.[1]
[b] ETP, Submitochondrial particles prepared by sonication.
[c] 3-Acetylpyridine adenine dinucleotide.

NADPH ⟶ AcPyADP
(400 – 450 nm)

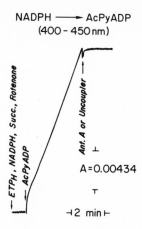

Fig. 1. Energy-linked transhydrogenation from NADPH to 3-acetylpyridine adenine dinucleotide phosphate (AcPyADP). For details, see Hatefi and Galante.[8]

and NADPH → NAD and NADH → NAD transhydrogenation by particles treated with trypsin or butanedione are given in Table II.

As has been shown by Ernster *et al.*[4,5] and Kaplan *et al.*,[6] the reduction of NADP by NADH is energy linked in mitochondria. In the absence of an energy supply this reaction is slow, but when energy is provided, either by oxidation of a substrate (e.g., succinate) or by hydrolysis of ATP, then transhydrogenation from NADH to NADP is facilitated severalfold, and the apparent equilibrium constant of the reaction can go far beyond the equilibrium constant of the non-energy-linked transhydrogenase reaction (NADH + NADP $\rightleftharpoons$ NAD + NADPH; $K \simeq 0.79$; $\Delta G^0 = +0.14$ kcal/mole) to $K' \simeq 500$.[7] In practice, it has been shown that one high-energy bond is utilized per molecule of NADP reduced when ATP is the energy source.[5]

For reasons that will be explained below, we predicted that submitochondrial particles should also catalyze transhydrogenation from NADPH to NADP, and that this reaction should also be energy linked. The results of an experiment demonstrating energy-linked transhydrogenation from NADPH to the NADP analogue 3-acetylpyridine adenine dinucleotide phosphate (AcPyADP) are shown in Fig. 1. Rutamycin-treated phosphorylating submitochondrial particles (ETP$_H$) were placed in a cuvette containing buffer, NADPH, succinate, and rotenone. Then AcPyADP was added and its reduction was monitored in a dual-wavelength spectrophotometer at 400 *minus* 450 nm. It is seen that AcPyADP was reduced by NADPH. That this reaction is energy linked is demonstrated by the fact that antimycin A, which inhibited succinate oxidation and energy production, and uncouplers inhibited the reduction of AcPyADP.[8] Since submitochondrial particles contain $\leq$ 0.2 nmoles NAD per milligram of protein, the possibility was considered that AcPyADP reduction might have occurred as follows:

$$\text{NADPH} \longrightarrow \text{bound NAD} \xrightarrow{\text{energy}} \text{AcPyADP}$$

To test for this possibility, the source of reducing power was changed from NADPH to 3-hydroxybutyrate. It was found that no AcPyADP reduction took place when NADPH was replaced with 3-hydroxybutyrate unless NAD was also added to the reaction mixture. In this system, a rate of AcPyADP reduction comparable to that shown in Fig. 1 required the addition of 3.2 $\mu$M NAD, i.e., 50 times as much as might have been bound to the added $\text{ETP}_H$.

*Complex I*

Preparations of complex I (Table III) catalyze the oxidation of NADH and NADPH by ferricyanide or ubiquinone as electron acceptor.[9, 10] They also catalyze transhydrogenation from NADH or NADPH to NAD.[10] Furthermore, Ragan and Widger[11] have shown that phospholipid vesicles inlaid with complex I and a mitochondrial protein fraction designated "hydrophobic proteins," then supplemented with $F_1$ (ATPase), are capable of catalyzing ATP-supported energy-linked transhydrogenation from NADH to NADP. Thus it appears that the enzymes concerned with NADH and NADPH dehydrogenation and non-energy-linked and energy-linked transhydrogenation are present in complex I preparations.

*NADH Dehydrogenase*

The resolution of complex I (Table III) with chaotropic agents at $\geq 15°C$ yields a soluble fraction containing essentially all the flavin

Table III. Molecular and Enzymic Properties of Complex I and the Soluble, Low Molecular Weight NADH Dehydrogenase

| Parameter | Complex I | Dehydrogenase |
|---|---|---|
| g protein/mole flavin | $7 \times 10^5$ | $7-8 \times 10^4$ |
| Flavin:Fe:S | 1:16–18:16–18 | 1:4:4 |
| Turnover number (per mole flavin) (NADH → ferricyanide) | $5 \times 10^5$ | $2.9 \times 10^4$ |
| $K_m^{NADH}$ ($\mu$M) | ~45 | ~80 |

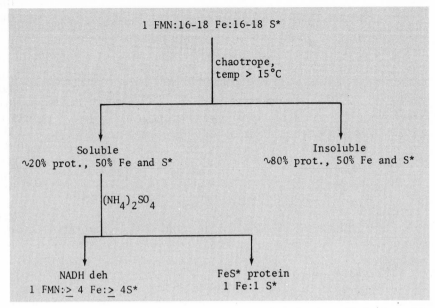

Fig. 2. Scheme showing the resolution of complex I into three fractions with the chaotropic salt $NaClO_4$.

(FMN), 15–20% of the protein, and approximately 50% of the nonheme iron and acid-labile sulfide (S*) content of complex I (Fig. 2). This soluble fraction is composed mainly of two proteins, an iron-sulfer protein with the EPR spectral characteristics of iron-sulfur center 2 of complex I[12] and an iron-sulfur flavoprotein with NADH dehydrogenase and NADH → NAD (or AcPyAD) transhydrogenase activities.

The soluble NADH dehydrogenase, as now prepared in our laboratory (Fig. 2), appears to be pure. It moves as a single symmetrical peak when chromatographed on Sephadex G-100, and shows a single protein band when subjected to polyacrylamide gel electrophoresis in the absence of sodium dodecylsulfate (SDS). In the presence of SDS, the enzyme shows three polypeptide bands upon gel electrophoresis, with $M_r$ values of approximately 51,000, 24,000, and $\leq 10,000$. The molecular weight calculated from Sephadex G-100 chromatography is, however, 69,000 ± 5%, which agrees with the flavin content of the preparation (14.0–14.7 nmoles/mg protein).[13]

As compared to complex I, the soluble NADH dehydrogenase has a low NADH dehydrogenase activity per mole of flavin and a higher $K_m$ for NADH. In addition, it was found that under the same assay conditions as applied to complex I, the resolved fractions of the complex were essentially devoid of NADPH dehydrogenase and NADPH →

NAD transhydrogenase activities.[8, 10] These differences appeared to be associated in part, with a somewhat more relaxed conformation of the NADH dehydrogenase active site, and were largely corrected when the assays were performed in the presence of guanidine HCl. Thus, as seen in Fig. 3, addition of guanidine HCl up to about 150 mM decreased the apparent $K_m^{NADH}$ and increased the $V_{max}$ for NADH oxidation by ferricyanide in the direction of values obtained for complex I.[8] Various alkylguanidines, including arginine and arginyl methyl ester, had similar but, on a molar basis, lower effects than guanidine, while phospho-arginine was without effect. The reason for the guanidine effect is probably related to the fact that in a large number of nicotinamide nucleotide and adenine nucleotide linked enzymes, arginyl residues have been found to act at the enzyme active site as positively charged groups (i.e., the guanido moiety) for recognition or neutralization of the negatively charged substrates (probably the phosphoryl groups in the above cases).

Figure 4 shows the effect of 75 mM guandine HCl on the NADH and NADPH dehydrogenase activities of the soluble NADH dehy-

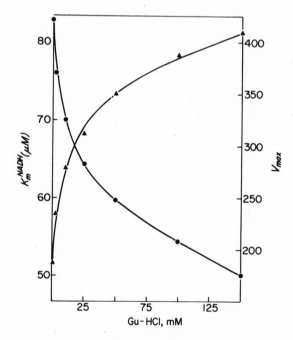

Fig. 3. Effect of guanidine HCl on the $K_m^{NADH}$ (●) and $V_{max}$ (▲) of the reaction NADH → $K_3Fe(CN)_6$ catalyzed by NADH dehydrogenase. $V_{max}$ is expressed as $\mu$moles NADH oxidized per minute per milligram of protein at 38°C. For conditions, see Hatefi and Galante.[8]

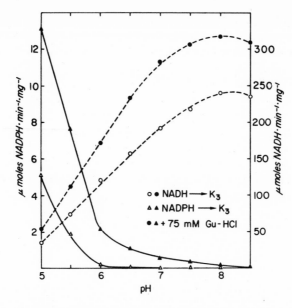

Fig. 4. Effects of 75 mM guanidine HCl and pH on the NADH and NADPH dehydrogenase activities of NADH dehydrogenase in the presence of 2-methylnaphthoquinone (K$_3$) as electron acceptor. For conditions, see Hatefi and Galante.[8]

drogenase as a function of pH, and Fig. 5 shows a similar plot for the NADPH → NAD transhydrogenase activity of the enzyme. It is seen in both figures that in the absence of guanidine the soluble dehydrogenase has negligible NADPH dehydrogenase and NADPH → NAD transhydrogenase activities at pH ≥ 6.0. However, in the presence of 75 mM guanidine HCl, the enzyme exhibits appreciable activities, which increase with lowering of pH. Thus, at pH 5.0 and in the presence of 75 mM guanidine HCl, the NADPH dehydrogenase activity of NADH dehydrogenase was 13.1 $\mu$moles min$^{-1}$ mg protein$^{-1}$, and its NADPH → NAD transhydrogenase, activity at pH 5.5 was >1.5 $\mu$moles min$^{-1}$ mg protein$^{-1}$. It should be noted in Figs. 4 and 5 that in the absence of added guanidine lowering of pH below pH 6.0 also elicited considerable NADPH dehydrogenase and NADPH → NAD transhydrogenase activities in the NADH dehydrogenase. This effect, as will be discussed below, has been interpreted as being associated with protonation of substrate phosphoryl groups at pH ≤ 6.0, and suggests that guanidine and alkylguanidines (but not phosphoarginine) serve a similar function by electrostatic interaction with, and neutralization of, the substrate phosphoryl groups. As might be expected from the data of Fig. 4, it was found that at pH 6.0 the enzyme was very sluggishly bleached (because of reduction of the flavin and iron-sulfur centers) by NADPH (25 min from trace 1 to trace 6; see Fig. 6 and inset). However, in the presence of 50 mM guanidine HCl, the same degree of bleaching occurred within 50 sec (i.e., the time required to record the spectrum from 600 to 450 nm) (Fig. 6). The above results show that under appropriate conditions the soluble NADH dehydrogenase isolated from complex I is capable of rapid oxidation of NADH by ferricyanide, oxidation of NADPH by ferricyanide and quinoid structures, and transhydrogenation from NADH and NADPH to NAD. Even in the presence of guanidine HCl and at low pH, the enzyme exhibited only marginal NADH → NADP and undetectable NADPH → NADP transhydrogenase activities. These results are summarized in Table IV.

## Discussion

In the preceding section we have traced NAD(P)H dehydrogenase and NAD(P)H → NAD transhydrogenase activities from submitochondrial particles to complex I, and from complex I to a soluble iron–sulfur flavoprotein preparation which appears to be a pure enzyme composed of three subunits. The results presented here and elsewhere[8, 13, 14] suggest strongly that this purified dehydrogenase is the

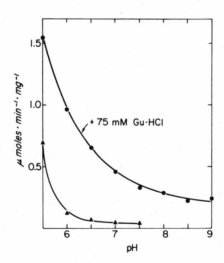

Fig. 5. Effects of 75 mM guanidine HCl and pH on the NADPH → AcPyAD (3-acetylpyridine adenine dinucleotide) transhydrogenase activity of NADH dehydrogenase. For conditions, see Hatefi and Galante.[8]

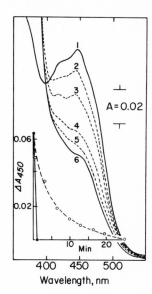

Fig. 6. Reduction of NADH dehydrogenase chromophores by NADPH in the absence (traces 1 to 6) and presence (traces 1 and 6) of 50 mM guanidine HCl at pH 6.0. Inset: Decrease of 450-nm absorbance ($\Delta A_{450}$) as a function of time in the absence (dashed line) and presence (solid line) of 50 mM guanidine HCl. For details, see Hatefi and Galante.[8]

moiety in complex I and submitochondrial particles responsible for the NAD(P)H dehydrogenase activities of these particles. The NADH → NAD transhydrogenase activities of submitochondrial particles, complex I, and the soluble enzyme appear also to be related. However, there seem to be at least two enzymes in submitochondrial particles for catalyzing NADPH → NAD transhydrogenation. Although not yet purified, the membrane-bound NADPH → NAD transhydrogenase

*Table IV. Dehydrogenase and Transhydrogenase Activities of NADH Dehydrogenase Isolated from Complex I*

| Reaction | pH | Activity ($\mu$moles min$^{-1}$ mg$^{-1}$ protein at 38°C) | |
|---|---|---|---|
| | | $-$Gu$-$HCl | +Gu $-$ HCl (75 mM) |
| NADH → ferricyanide | 8 | 242 | 485 |
| NADH → AcPyAD | 8 | 12 | 30 |
| | 5.5 | 6 | 8 |
| NADH → AcPyADP | 5.5 | <0.04 | <0.04 |
| NADPH → ferricyanide | 8 | nil | 0.3 |
| | 5.5 | 4.2 | 20.8 |
| NADPH → AcPyAD | 8 | <0.05 | 0.3 |
| | 5.5 | 0.7 | 1.54 |
| NADPH → AcPyADP | 5.5 | nil | nil |

appears to be different from the soluble NADH dehydrogenase in several respects:

a.   Many nicotinamide nucleotide linked dehydrogenases are known which also exhibit transhydrogenase activity. The soluble NADH dehydrogenase is capable of catalyzing the non-energy-linked transhydrogenations [NAD(P)H → NAD] and exhibits negligible activity for transhydrogenations that are energy linked in the particles [NAD(P)H → NADP]. However, whether its membrane-bound form can catalyze energy-linked transhydrogenation is not known.

b.   Under appropriate conditions, NADH dehydrogenase activity in both the particles and the soluble enzyme is strongly inhibited at high NADH concentrations. NAD and adenine nucleotides, but not adenosine, also inhibit to a lesser extent. The characteristics of these inhibitory effects, as studied elsewhere,[15, 16] suggest that both the membrane-bound and the soluble dehydrogenase can bind more than one nucleotide molecule at or near the enzyme active site. Since as compared to NADH the $K_m$ of the enzyme for NADPH is very high (~500 $\mu$M in submitochondrial particles), we predicted that in the case of NADPH oxidation addition of other nucleotides might have an activating effect at low concentrations followed by an inhibitory effect at higher levels. This was tested in the presence of 2′-phosphoadenosine-5′-diphosphoribose and adenosine-2′, 5′-diphosphate and found to be true.[17] The possibility that both the membrane-bound and the soluble dehydrogenase are capable of binding more than one molecule of nucleotide at or near the enzyme active site is in agreement with an active site design for the dual functions of dehydrogenation and transhydrogenation. However, the transhydrogenase reactions NADPH + NAD ⇌ NADP + NADH in submitochondrial particles involve direct hydride ion transfer from one nucleotide to another. Whether the transhydrogenase reactions catalyzed by complex I and the soluble dehydrogenase also involve direct hydride ion transfer or indirect transfer of reducing equivalents by way of an enzyme moiety has yet to be determined.

c.   The pH optimum of the membrane-bound NADPH → NAD transhydrogenase reaction is at pH 6.0–6.5, whereas the same reaction catalyzed by the soluble enzyme exhibits no pH optimum from pH 9 to pH 5.0 (the lowest possible pH at which the reaction could be measured without complications by rapid enzyme denaturation; see ref. 8). This is not necessarily a serious difference, because pH values below 6.0 might cause other changes in the membrane with secondary inhibitory effects on the membrane-bound transhydrogenase activity, whereas the soluble enzyme could be free of such secondary effects.

d.   Rydström *et al.*[18] believe that the mitochondrial transhydrogenase enzyme is not a flavoprotein. This point has yet to be unequivocally established, however. We have prepared theirs[18,19] and Kaplan's

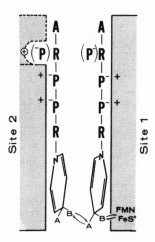

Fig. 7. Proposed arrangement of nicotinamide adenine dinucleotides at the active sites of NADH dehydrogenase for dehydrogenation (site 1) and transhydrogenation (sites 1 and 2). The trypsin/butanedione-susceptible arginyl residue of the transhydrogenase enzyme is shown by a plus sign inside a circle. The dashed lines indicate that the above arginyl residue may not be present in the soluble, low molecular weight dehydrogenase. For simplicity, the carbamyl groups have been deleted from the nicotinamide rings, and both nucleotides have been shown in reduced form to indicate the stereospecificities of hydrogen abstraction (curved arrows) in dehydrogenation and transhydrogenation. A, R, and P⁻ stand for adenine, ribose, and phosphate, respectively. The $P^-$ in parentheses is 2′-phosphate when the nucleotide is NADP or NADPH. The shaded areas represent portions of the enzyme around the active sites. For details, see text.

preparations of NADPH → NAD transhydrogenase.[20] These preparations are all detergent-solubilized membrane fragments containing a large number of polypeptides, and those prepared in our laboratory were not free of flavin and contained polypeptides in the molecular weight range of those found in complexes I and V.

However, regardless of whether or not the soluble NADH dehydrogenase is in any way related to the membrane-bound energy-linked transhydrogenase, it appears to serve as an interesting model for study of the mechanism of nicotinamide nucleotide transhydrogenation. Regarding the dehydrogenase and the transhydrogenase properties of the soluble preparation, our current working hypothesis is that the enzyme contains two active sites, site 1 for NADH or NADPH dehydrogenation and site 2 for attachment of a second nucleotide for transhydrogenation (Fig. 7).* Since site 1 appears to have a low affinity for NADPH, it is possible that NADH or NAD would preferentially bind to site 1 during the transhydrogenation reactions involving also NADPH or NADP.

---

\* Attachment of NADH to both sites at high substrate concentration would result in inhibition of NADH dehydrogenase activity as described above.

Consequently, hydrogen abstraction for dehydrogenation and trans-hydrogenation would take place as depicted by the curved arrows in Fig. 7. These modes of hydrogen abstraction are in agreement with the results found for membrane-bound nicotinamide nucleotide dehydro-genation and transhydrogenation.[3, 21] Should the soluble enzyme con-tain an essential arginyl residue analogous to the trypsin/butanedione-susceptible arginyl residue of the membrane-bound transhydrogenase, then this arginyl residue would have to be located at site 2, possibly for binding of the 2'-phosphate of NADP (H). This is because modification of the membrane-bound arginyl residue inhibits the transhydrogena-tions involving NADP (H) but not NAD(P)H dehydrogenation or NADH → NAD transhydrogenation.[1]

Regarding energy-linked transhydrogenation, Table IV shows that the soluble enzyme can catalyze dehydrogenation and transhydrogena-tion reactions with all the reduced and oxidized nucleotides, except when NADP is used in a transhydrogenation reaction. In the latter case, the reaction catalyzed by the soluble enzyme is either extremely slow (NADH → AcPyADP) or undetectable (NADPH → AcPyADP). How-ever, as discussed above, these reactions are energy linked in sub-mitochondrial particles. Therefore, it is possible that the problem in mitochondrial transhydrogenation concerns hydride ion transfer to NADP and that somehow membrane energization facilitates this reaction.

A comparison of the structure of the four oxidized and reduced nucleotides shows that NADP differs from the others by having both a negatively charged 2'-phosphate and a positively charged nitrogen in the nicotinamide ring. The folded structure of $\beta$-NADP in solution, as depicted in Fig. 8, shows that the negatively charged oxygen (or the hydroxyl group) of the 2'-phosphate can be very close to the C-4 of the nicotinamide ring, which through ring resonance carries a partial positive charge. Therefore, it is possible that electrostatic attraction between the above groups results in greater stabilization of the folded structure of NADP and diminished electrophilic property of the nicotinamide C-4.* These differences might be a major contributing factor in making the reduction of NADP by the transhydrogenase en-zyme sluggish as compared to the reduction of NAD.

---

* This should result in a somewhat lower reduction potential for NADP as compared to NAD. The equilibrium constant of the non-energy-linked transhydrogenase reaction suggests that this difference might be only about 3 mV at 30°C. Data of Olson and Anfinsen[24] and Kaplan *et. al.*[25] obtained for two different systems and at two different temperatures (27°C and 37°C) indicate a difference of 5 mV between the reduction potentials of NAD and NADP.

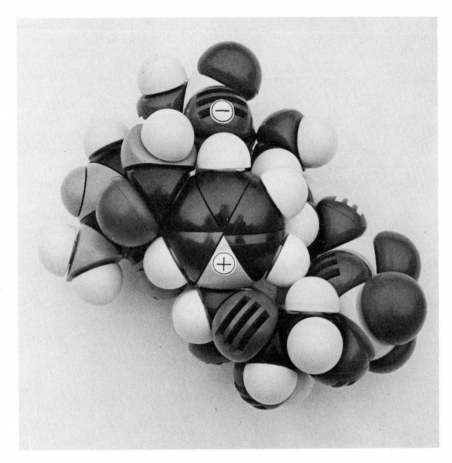

Fig. 8. Molecular model of $\beta$-NADP in folded (stacked) conformation. The positively charged nicotinamide ring nitrogen and the negatively charged oxygen of 2'-phosphate are marked with (+) and (−) signs, respectively. Note the close proximity of the negatively charged oxygen (or the hydroxyl group) of 2'-phosphate to the C-4 of the nicotinamide ring, which through the ring resonance carries a formal positive charge. The model shown is based on the structure of the stacked conformation of $\beta$-NADH proposed by Miles and Urry[26] and Kaplan and Sarma.[27]

In support of the above hypothesis, the following results might be mentioned: (a) The studies of Ryström *et al.*[22] show that in going from non-energy-linked to energy-linked transhydrogenation the $K_m$ values of all the interacting nucleotides remain essentially unchanged, except the $K_m$ of NADP, which diminishes from 41 $\mu$M to 6.5 $\mu$M (Table V). These results agree with our conclusion regarding NADP as the focus of the problem in nonenergized transhydrogenation. (b) As shown in Fig.

9, at acid pH (5.0–5.5), where the protonated form of the nucleotide
phosphoryl groups is favored, transhydrogenation catalyzed by submitochondrial particles from NADH to NADP is quite efficient (70% of
the rate of energy-linked transhydrogenation at the optimum pH) and
very little improved (~10%) under energized conditions. These results
are also in agreement with our hypothesis, and could be interpreted to
mean that protonation of the 2′-phosphate group of NADP by conducting the assay at a low pH will to a large extent obviate the structural
problem discussed above and allow rapid transhydrogenation from
NADH to NADP in the absence of an energy supply.

An important feature of the energy-linked transhydrogenation
reaction NADH + NADP $\rightleftharpoons$ NAD + NADPH is the "overshoot" of the
equilibrium constant from $K = 0.79$ to an apparent value of $K' \simeq 500$. In
the forward direction 1 mole of ATP (or energy equivalent) is consumed
per mole of NADP reduced, but the reverse reaction catalyzed by
coupled particles does not appear to yield much energy. Indeed, as far
as ATP production coupled to hydride ion transfer from NADPH to
NAD is concerned, the results of Van de Stadt et al.[23] show only
marginal amounts of ATP synthesis (22 nmoles min$^{-1}$ mg protein$^{-1}$).
Whether even this low level of ATP synthesis was indeed coupled to
transhydrogenation or to a slow oxidation of NADH and NADPH due
to leaks in the inhibited electron transport system is not absolutely

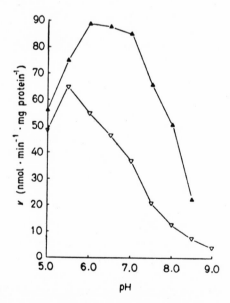

Fig. 9. Effect of pH on the rates of non-energy-linked ($\triangledown$) and energy-linked ($\blacktriangle$)
transhydrogenation from NADH to NADP. From Rydström.[28]

Table V. $K_m$ Values of Nicotinamide Nucleotides in the Non-Energy-Linked and Energy-Linked Transhydrogenase Reactions NADH + NADP $\rightleftharpoons$ NAD + NADPH[a]

| Reaction | $K_m$ ($\mu$M) | | | |
|---|---|---|---|---|
| | NADH | NAD | NADPH | NADP |
| Non-energy/linked | 7–9 | 28–31 | 20 | 40–41 |
| Energy-linked | 12.6 | 43.5 | 20 | 6.5 |

[a] From Rydström *et al.*[22, 29]

certain. Therefore, the available data (including the maximum apparent equilibrium constant of $K'$ $\simeq$ 500 calculated for the energy-linked transhydrogenase reaction) indicate irretrievable consumption of considerable amounts of energy for NADP reduction. These considerations, the apparent problem of NADP reduction because of its unique structure (specifically the extra 2'-phosphate as compared to NAD) which is largely overcome at pH < 6.0, and the essentiality of the trypsin/butanedione-susceptible arginyl residue for transhydrogenations involving NADP (H) suggest the possibility that energy-linked transhydrogenation might involve an endergonic interaction between the 2'-phosphate of NADP and the essential arginyl residue of the enzyme to form a transitory intermediate, such as enzyme-guanido . . . 2'-phospho-NAD. Such an intermediate would be expected to facilitate the reduction of NADP because of its sequestered 2'-phosphate.

As regards the thermodynamics of the energy-linked system, the available values are a maximum apparent equilibrium constant of $K'$ $\simeq$ 500, and the stoichiometric utilization of ATP for NADP reduction. The value of $K'$ $\simeq$ 500 was obtained for a reaction energized by succinate oxidation. However, if we assume a comparable equilibrium for energy-linked transhydrogenation energized by ATP hydrolysis ($\Delta G^0$ $\simeq$ −8.7 kcal/mole), then in the reaction NADH + NADP + ATP $\rightleftharpoons$ NAD + NADPH + ADP + Pi, 3.7 kcal/mole is utilized to achieve a $K'$ $\simeq$ 500. The remainder, i.e., 8.7 − 3.7 = 5 kcal/mole, could be the amount of energy utilized by the transhydrogenase enzyme to increase its affinity for NADP. This may involve a change in enzyme conformation or surface charge, especially as pertains to its recognition of, and interaction with, the 2'-phosphate of NADP.

ACKNOWLEDGMENT

This work was supported by USPH Grant AM 08126 to Youssef Hatefi.

NOTE ADDED IN PROOF

For a recent report on purification of the mitochondrial transhydrogenase enzyme, see Höjeberg, B., and Rydström, J. (1977) *Biochem.
Biophys. Res. Commun. 78*, 1183–1190.

## References

1. Djavadi-Ohaniance, L., and Hatefi, Y. (1975) *J. Biol. Chem. 250*, 9397–9403.
2. Hatefi, Y., Djavadi-Ohaniance, L., and Galante, Y. M. (1975) in *Electron Transfer Chains and Oxidative Phosphorylation* (Quagliariello, E., Papa, S., Palmieri, F., Slater, E. C., and Siliprandi, N., eds.), pp. 257–263, North-Holland, Amsterdam.
3. Hatefi, Y. (1974) in *Dynamics of Energy-Transducing Membranes* (Ernster, L., Estabrook, R. W., and Slater, E. C., eds.), pp. 125–141, Elsevier, Amsterdam.
4. Danielson, L., and Ernster, L. (1963) *Biochim. Biophys. Acta 10*, 91–96.
5. Ernster, L., Lee, C.-P., and Torndal, U. B. (1969) in *The Energy Level and Metabolic Control in Mitochondria* (Papa, S., Tager, J. M. Quagliariello, E., and Slater, E. C., eds.), pp. 439–451, Adriatica Editrice, Bari.
6. Kawasaki, T., Satoh, K., and Kaplan, N. O. (1964) *Biochem. Biophys. Res. Commun. 17*, 648–654.
7. Rydström, J., Teixeira da Cruz, A., and Ernster, L. (1970) *Eur J. Biochem. 17*, 56–62.
8. Hatefi, Y. and Galante, Y. M. (1977) *Proc. Natl. Acad. Sci. USA 74*, 846–850.
9. Hatefi, Y., and Stiggall, D. L., (1976) in *The Enzymes*, Vol. 13, 3rd ed. (Boyer, P. D., ed.), pp. 175–297, Academic Press, New York.
10. Hatefi, Y. and Hanstein, W. G. (1973) *Biochemistry 12*, 3515–3522.
11. Ragan, C. I., and Widger, W. R. (1975) *Biochem. Biophys. Res. Commun. 62*, 744–749.
12. Orme-Johnson, N. R., Hansen, R. E., and Beinert, H. (1974) *J. Biol. Chem. 249*, 1922–1927.
13. Galante, Y. M., and Hatefi, Y., manuscript in preparation.
14. Hatefi, Y., and Bearden, A. J. (1976) *Biochem. Biophys. Res. Commun. 69*, 1032–1038.
15. Hatefi, Y., and Stempel, K. E. (1969) *J. Biol. Chem. 244*, 2350–2357.
16. Hatefi, Y., Stempel, K. E., and Hanstein, W. G. (1969) *J. Biol. Chem. 244*, 2358–2365.
17. Galante, Y. M., and Hatefi, Y., unpublished results.
18. Rydström, J., Hoek, J. B., and Hundal, T. (1974) *Biochem. Biophys. Res. Commun. 60*, 448–455.
19. Rydström, J. (1976) *Biochim. Biophys. Acta. 455*, 24–35.
20. Kaplan, N. O. (1967) *Methods Enzymol. 10*, 317–322.
21. Lee, C.-P., Simard-Duquesne, N., Ernster, L., and Hoberman, H. D. (1965) *Biochim. Biophys. Acta 105*, 397–409.
22. Rydström, J., Teixeira da Cruz, A., and Ernster, L. (1971) *Eur. J. Biochem. 23*, 212–219.
23. Van de Stadt, R. J., Nieuwenhuis, F. J. R. M., and Van Dam, K. (1971) *Biochim. Biophys. Acta 234*, 173–176.
24. Olson, J. A., and Anfinsen, C. B. (1953) *J. Biol. Chem. 202*, 841–856.
25. Kaplan, N. O. Coldwick, S. P., and Neufeld, E. F. (1953) *J. Biol. Chem. 205*, 1–29.
26. Miles, D. W., and Urry, D. W. (1968) *J. Biol. Chem. 243*, 4181–4188.
27. Kaplan, N. O., and Sarma, R. H. (1970) in *Pyridine Nucleotide Dependent Dehydrogenases* (Sund, H., ed.), pp. 39–56, Springer-Verlag, Berlin.
28. Rydström, J. (1974) *Eur. J. Biochem. 45*, 67–76.
29. Teixeira da Cruz, A., Rydström, J., and Ernster, L. (1971) *Eur. J. Biochem. 23*, 203–211.

*Chapter 8*

# Presence and Function of Metals in the Mitochondrial Electron Transfer System: Facts and Speculations

## Helmut Beinert

### Introduction

When I saw the first tentative program for this meeting provided by our organizers, and when I noticed that my lecture was proposed to be on "The Role of Metals in the Electron Transport System," my first reaction was to say: "No!" True, we know a fair amount about metals associated with the system, but it is their role about which our knowledge is most deficient, or close to nonexistent in some instances.

However, this occasion today is not one to desert from. After all, we did not assemble here only to show each other the latest hard experimental results. At this point we are equally interested as to how knowledge and progress come about and how individual scientists of unusual qualities may be able to foresee, to influence, and to shape the course of events.

What I propose, therefore, is to go back some decades in history, search for the roots of some of the thoughts and developments we witness today, and then try to project into the future and in between give you some sprinkling of reality, namely, some of the hard facts on metals in the electron transport system, as I know them today. Since most of us

*Helmut Beinert* • Institute for Enzyme Research, University of Wisconsin, Madison, Wisconsin 53706.

probably have become sufficiently specialized, a brief survey of facts may not be redundant.

### Some Early Ideas and Concepts Concerning Metals in Electron Transport

With their concern to conserve space, primary scientific journals are no longer the best sources for tracking ideas, speculations, and development of concepts. Proceedings of meetings are a more rewarding source of such information, and so I will present a few sentences from meetings long ago and probably largely forgotten.

There was a Symposium at the Henry Ford Hospital at Detroit in 1955 on "Enzymes: Units of Biological Structure and Function" at which David Green presented a paper on "Structural and Enzymatic Pattern of the Electron Transfer System."[1] There we read (italics added):

> There is a gap in our knowledge of the mechanism by which the various cytochromes interact with one another. The nub of the uncertainty might be expressed in terms of the following questions. First, are we to think of the different components in the chain as separate entities which react with one another by molecular collision, or as *parts of a single unit which are linked to one another structurally* as well as functionally? Second, *are there other oxidation-reduction components in the electron transfer process besides those which are observed by spectroscopic means? These* are *questions* which *can be answered only by* application of *the methods of isolation and direct examination.* (p. 465)

The questions asked there are now partly answered by the isolation of the intimate complexes of the electron transfer system, by the establishment of membranes as scaffolding for the mounting of the components of the chain, and by the discovery of a multitude of new components of the chain, indeed not or barely detectable by optical spectroscopy.

To elaborate further on this last theme, we may recall what we read on p. 470: "For each molecule of flavin there are in round numbers 7 to 8 molecules of total heme, 61 atoms of nonheme iron, and 8 atoms of copper. *Both the nonheme iron and the copper are very firmly bound in ETP*." Although the stoichiometry of components is somewhat off from what is accepted today, the firm binding of nonheme iron and copper was clearly recognized. And then a picture of the whole system:

> On the basis of the properties of ETP and its derivative particulate fragmentation products, we have the impression that the mitochondrial electron transfer system is not a mixture of discrete entities which react with one another by molecular collisions. As far as we can ascertain, *the system behaves like a single structure* in which all the oxidation-reduction elements are bound one to the others in firm linkage. The classical notions of kinetics would appear, therefore, to be inapplicable to conduction of electrons within such a system though applicable to the interaction of the system with external reactants such as substrate and oxygen.

Our own preference is expressed in terms of the representation shown in Fig. 9, which is purely hypothetical. It is intended to express the notion that there is no simple molecular unit of electron transfer but rather *a network arrangement*. Whether metal atoms do indeed connect neighboring hemes and do cross-link parallel chains is at best an inspired guess based on the very high concentration of metallic constituents in ETP. (p. 479)

We find here expressed the idea of a network—rather than a chain—to which I very much subscribe and the impression that we are dealing with "a single structure"—what we know today as the membrane. In 1961, at the International Congress of Biochemistry at Moscow, focusing was somewhat closer[2]:

The above results have led us to propose a role for $Fe_{NH}$ and copper as obligatory electron carriers in terminal electron transport. Although at this point in our inquiries we know little about the binding of these metals to the components of the electron transport system, *there seems to be little doubt that we are dealing with a class of biologically active metal chelates different from the well known iron porphyrin complexes characteristic of the cytochromes*. The possible *interaction between heme a and copper* remains an attractive hypothesis. In this respect, $Fe_{NH}$ analyses and EPR studies of highly purified segments of the electron transport chain should enable us to define more precisely the location and function of these complexes with respect to other electron transport components. (p. 189)

Here we see mapped out what many of us have been trying to do ever since, and, to my delight, even copper–heme interaction in cytochrome *c* oxidase, which Bob van Gelder and I postulated 8 years ago[3] on the basis of our EPR results, is foreshadowed.

Since I will be presenting to you later some thoughts on the function of iron-sulfur components in electron transfer, it may also be worthwhile to recall some of the ideas on the function of metals in metalflavoproteins particularly put forth by Henry Mahler at about the same time[4]:

On the contrary, their [cytochromes] interaction with a two-electron donor such as a reduced flavoprotein presents formidable mechanistic complications. Simultaneous transfer by collison with two acceptor molecules in solution is kinetically forbidden, binding of two acceptor molecules in a stepwise manner seems unlikely for lack of a satisfactory binding site, while stepwise one electron transfer [the Michaelis postulate] requires the existence of long-lived flavoprotein semiquinones at physiological pH values and at relatively high concentrations.

*The presence of the metal closely linked spatially and electronically to the flavin site eliminates these difficulties*. (pp. 591–592)

After the unpairing of the electrons *all subsequent changes are considered to occur rapidly and spontaneously. Valence changes of the metal component are transitory*, and the net result is direct electron transfer within one molecule (the reduced metalloflavoprotein-acceptor complex) rather than intermolecular electron transport. (p. 594)

If I have given the impression that there are more unanswered questions in the field of metalloflavoprotein catalysis than there are settled ones, then I have left exactly the opinion I wanted to convey. The work that has been done so far certainly can do no more than suggest possible new approaches and points of view, which, suitably modified, might be the starting point of a real understanding of some electron transport functions. (p. 596)

Some of the experimental work on which these early ideas were based may not have stood the test of time, but in a way, now in 1977, we seem to come back full circle pretty close to Mahler's original views. One of the objections to the general applicability of Mahler's postulates was the finding, which I and Fred Crane reported at the same meeting in 1955,[5] that the acyl CoA dehydrogenases were linked by a metal-free electron-transferring flavoprotein (which one would probably now call a flavodoxin) to the electron transport system. Although the enzyme, as we had it, was able to reduce cytochrome $c$, it has since become clear that the pure flavoprotein does not react significantly with cytochrome $c$[6] and thus does, in fact, not bring about an efficient translation of 2 $e^-$ into 1 $e^-$ transfer. It was indeed fortuitous that 2 years ago, in our characterization of mitochondrial Fe-S proteins by EPR, we stumbled onto a thus far unknown iron-sulfur flavoprotein[7] which has not shown any activity other than being rapidly (<100 msec) reduced by acyl CoA dehydrogenase plus acyl CoA plus ETF and rapidly reoxidized by $Q_1$. Thus it seems that this is the missing link between the flavoproteins of $\beta$ oxidation of fatty acids and the respiratory chain; and this protein is indeed a Fe-S flavoprotein, as are the better-known dehydrogenases of the electron transfer system, namely those for succinate and NADH.

### Nonheme Metal Components of the Electron Transfer System

Since I have recently summarized the present status concerning Fe-S proteins on several occasions,[8-10] I would mainly like to refer you to those presentations and very briefly state only the best present information here.

Succinate dehydrogenase, when associated with the respiratory chain, has three distinct Fe-S centers, approximately at a 1:1 ratio to bound flavin. Two of these centers are of the ferredoxin type with [2Fe-2S] centers. Only one of these is reduced by succinate. A third center is of the type found in the "high-potential" Fe-S protein of chromatium,[11] with a [4Fe-4S] center. This is also reduced by succinate. On solubilization this last center is either lost (4Fe/flavin enzyme) or inactivated (8Fe/flavin enzymes), unless special precautions are taken.

NADH dehydrogenase almost certainly has at least four distinct Fe-S centers of widely varying oxidation-reduction potentials, again approximately stoichiometric to flavin. Two additional centers have been postulated, but they have not been shown to be present in meaningful quantities.

In the cytochrome $b\text{-}c_1$ particle there seems thus far to be only one Fe-S component, namely the "Rieske" Fe-S center, although what is presently known about the stoichiometry of EPR-detectable Fe-S with respect to the cytochromes does not completely rule out additional undetected components.

The Fe-S center mentioned above, which is thought to be associated with the fatty acid $\beta$-oxidation pathway,[7] has always been found in mammalian mitochondrial-type electron transfer systems. The corresponding protein contributes one [4Fe-4S] center, stoichiometric to its acid-extractable FAD. In addition, there are at least three Fe-S centers, clearly characterized by their distinct EPR signals. The two more abundant ones are the outer membrane [2Fe-2S] protein, discovered by Bäckström *et al.*,[13] which seems to be involved in an electron transfer pathway including cytochrome $b_5$, and the "high-potential"-type Fe-S protein[13] (probably with a [4Fe-FeS] center) which is solubilized on sonication of mitochondria and is therefore not present in ETP. Unless special precautions are taken, considerable amounts of outer membrane fragments are associated with ETP. The Fe-S protein may serve as a marker for contamination of ETP with outer membrane. There is also an unidentified ferredoxin-type Fe-S protein associated with this "high-potential"-type Fe-S protein which can be solubilized from mitochondria. The function and association of these two Fe-S centers, *viz.*, the soluble high-potential Fe-S protein and the last mentioned ferredoxin-type protein, are unknown at this time.

This is what we can detect with certainty. There may be additional Fe-S components not detectable by EPR. If Fe-S centers are very closely adjacent, their EPR signals may not be detectable. An example for this is found in the Mo-Fe protein of bacterial nitrogenase, where only a fraction of the iron and labile sulfur found by analysis is accounted for by EPR. A balance of labile sulfide in ETP vs. known Fe-S centers, which I have tried to construct,[9, 10] indicates that there may well be room for additional Fe-S components, particularly if they should occur at concentrations as low as those of the substrate-linked dehydrogenases. Among nonheme heavy metal components we must, of course, also count the copper atoms of cytochrome $c$ oxidase, but it seems redundant to enlarge on this well-known topic here.

## *Ideas on the Function of Nonheme Heavy Metal Components in the Electron Transfer System*

It is obvious that the electron transfer system contains a greater variety of Fe-S components than of cytochromes and, with all the attention that the cytochromes have received through the decades, it is time that we pay similar attention to the properties and function of the nonheme components. We may take it as a clue that the majority—with the Rieske Fe-S center as the only exception—of Fe-S centers are clustered at the substrate end of the system, and all those in association with flavin. The example of the three flavoproteins in a row, as apparently required in fatty acid oxidation, seems to argue that for entrance into the Q-$b$-$c_1$ region of the electron transfer system, intervention of a Fe-S-containing component is obligatory. Here we may recall the early ideas of Henry Mahler (4), to which I alluded above: a transition from $2e^-$ transfer from substrate to $1e^-$ transfer in the cytochrome system proper has to be accomplished. Since flavins and flavoproteins have been shown to be able to occur in semiquinoid forms, one may, at first sight, argue that this transition might be accomplished by the flavin prosthetic group itself. However, at least two flavins acting in concert would be required to properly shuffle and dispose of electrons in such a reaction. Fortunately, the required events have been thought through more rigorously than most biochemists would be capable of doing, by the initiator of modern flavin chemistry, P. Hemmerich. According to his views, either two flavin and one Fe-S group acting in concert or a flavin and two Fe-S centers are required to accomplish an effective transition from $2e^-$ to $1e^-$ transfer. A detailed scheme, the $2/1e^-$ transformase complex cycle, is shown in Fig. 1 (*cf*. ref. 11). The component indicated by the large circle is the dehydrogenase flavoprotein (which cannot be replaced by a Fe-S protein), which has $2e^-$ input and $1e^-$ output. The small circles represent the "$1e^-$ collectors," either two separate Fe-S centers or one flavodoxin-type flavoprotein and one Fe-S center. Only the low-potential $1e^-$ carrier can be replaced by a flavodoxin. This component transfers a single electron into the electron transfer chain, while the Fe-S center of higher potential has the function of an electron sink for reoxidation of the dehydrogenase flavoprotein, so that $2e^-$ uptake from the substrate may once more occur. The required electron shifts within the complex are conditioned by conformation-linked proton shifts between the 1- and 5-positions on the flavin.

Recalling now the schemes of Mahler,[4] we find one of the differences in the ideas of the role of the iron, or, as we know now, the Fe-S centers, in that Mahler, according to his schemes, thought of the iron mainly as the next member in an $e^-$ transfer series, whereas we now tend more to see it as sitting on the side, doing some shuffling with the flavin but presumably in many cases letting the flavin do the communicating with the outside world. This may not hold for all Fe-S centers of multi-Fe-S center proteins such as succinate or NADH dehydrogenase, but it is true for some of them.

A second clue to the function of Fe-S centers is provided by their widely varying oxidation-reduction potentials, which may span a range of as much as 400 mV. In NADH dehydrogenase, for example, there are

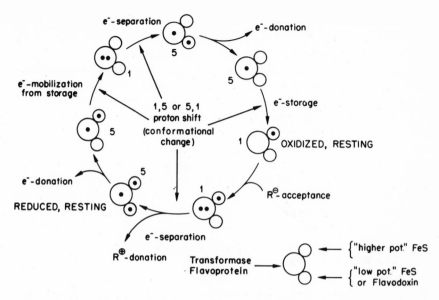

Fig. 1. Scheme proposed by Dr. Peter Hemmerich, University of Konstanz, Germany, for the transition from $2e^-$ to $1e^-$ transfer, brought about by the enzyme complex called "2/1 transformase." From Hemmerich (personal communication).

Fe-S centers at the potential of the $e^-$ donor and there is one with a potential close to that of Q, the acceptor. This means that the oxidative step is not actually occurring between donor and enzyme or between enzyme and acceptor but within the enzyme. A similar situation holds for cytochrome $c$ oxidase, where the midpoint redox potential of $c$ is as high as that of components of the oxidase. Such an arrangement may well be of critical importance for the tapping off of energy from the ongoing oxidations. The occurrence of Fe-S centers with different oxidation-reduction potentials is also an important ingredient in Hemmerich's hypothesis, as discussed above.

Just as the Fe-S groups at the substrate end of the electron transfer system can be thought of as largely being internal electron shufflers and temporary reservoirs, so the copper atoms of cytochrome $c$ oxidase probably fulfill a similar function in a higher redox potential range. Here it becomes necessary to gear a $1e^-$ intake to a $4e^-$ or rapid $2 \times 2e^-$ requirement for the reduction of $O_2$. Again, it is likely that the classical prosthetic groups, namely cytochromes $a$ and $a_3$, communicate directly with the incoming donor and acceptor, while the copper atoms stand by in the internal shuffle, which makes repeated $1e^-$ acceptance from cytochrome $c$ possible and provides the needed quantity of electrons for $O_2$ reduction.

Finally, it is healthy to return on safe grounds and look at some experimental examples. I want to mention two here, namely milk xanthine oxidase and bacterial trimethylamine dehydrogenase. In an impressive series of papers, Olson *et al.*[15, 16] have elaborated and quantitatively documented a plausible mechanism for the oxidation of xanthine by oxygen. The details will have to be assimilated from their papers, but, concerning the two [2Fe-2S] centers of xanthine oxidase, Olson *et al.* come to the conclusion that they fulffill the function of internally keeping the electron balance such that molybdenum is restored to the $Mo^{VI}$ state, for efficient interaction with the $2e^-$ donor substrate, and FAD is converted to $FADH_2$ for efficient reduction of $O_2$ to $H_2O_2$.

The second example is from unpublished work in which Drs. D. J. Steenkamp, T. P. Singer, and I are currently involved (*cf.* ref. 17). Trimethylamine dehydrogenase has a single flavin prosthetic group and one [4Fe-4S] center and oxidizes its substrate to dimethylamine and formaldehyde. The reaction catalyzed by this enzyme may not be of the greatest interest to mitochondriologists, but the way in which it is brought about may have a message to convey concerning mechanisms of Fe-S flavoprotein function in general. As it often happens, when one looks at a large variety of examples, eventually one will be found which for some reason shows some sought-for salient features more clearly than all others. Trimethylamine dehydrogenase may well be that parade example. The turnover of this enzyme is of the order of 2 per second at 20°C, when measured with PMS as acceptor. The natural acceptor is not known. Our studies to date by stopped-flow-optical and by freeze-quench EPR and reflectance spectroscopy show the following features: There is a very rapid bleaching of the flavin-type chromophore within the dead time of our rapid-mixing equipment (~3 msec) (Fig. 2). Since neither optical signs nor EPR signals of semiquinone appear, we assume that this indicates rapid $2e^-$ uptake and reduction of the flavin-type component to a dihydro state (this state may not be simple dihydroflavin but may in fact include some substrate such as a covalent adduct, but the overall state of the flavin is thought to be "dihydro"). Since the reoxidation of reduced (iron-) flavoproteins by PMS is usually very rapid and not rate determining, we expected to find a slow, rate-determining step in the reductive half-reaction. EPR spectra of TMA dehydrogenase are shown in Fig. 3. With an excess of dithionite, we find the rhombic signal of a ferredoxin-type Fe-S center (Fig. 3, bottom curve) and no other features. When we reduce with substrate, however, a complex pattern of signals arises with time (Fig. 3, top and center spectra), and we see only traces of the reduced ferredoxin signal superimposed on a novel, complex signal. If this signal by itself were not interpretable, a

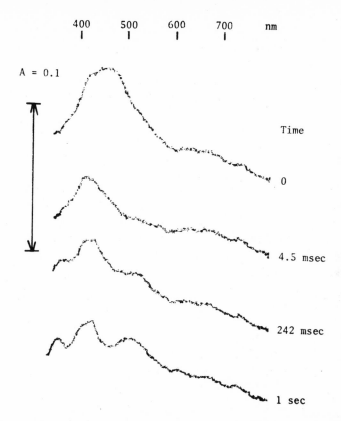

Fig. 2. Low-temperature (100°K) reflectance spectra of trimethylamine dehydrogenase reacting with trimethylamine. The enzyme (60 μM) was dissolved in 0.1 M sodium pyrophosphate buffer, pH 7.7, and mixed at 15°C with an equal volume of 2 mM trimethylamine hydrochloride in the same buffer in the rapid freeze-quench apparatus. The samples shown were frozen after the reaction times indicated.

signal always appearing in association with it at $g = 4$ (Fig. 4) would clearly tell us that we are dealing here with a fairly strong interaction of two closely associated paramagnetic species, i.e., we have a triplet state signal. From all that we know from the enzyme, these species must be the flavin semiquinone and the reduced Fe-S center. If we follow the formation of the signal which indicates interaction, most clearly expressed at $g = 4$, we find a half-time of ~300 msec for this reaction at 15°C (Fig. 5). On mixing with PMS, this signal disappears within the resolution of our apparatus. The reflectance spectra of Fig. 2 show the appearance of absorption at 520 and 360 nm simultaneous with the development of the interaction signal. It is possible that these bands originate from the semiquinone. However, it appears from stopped-flow

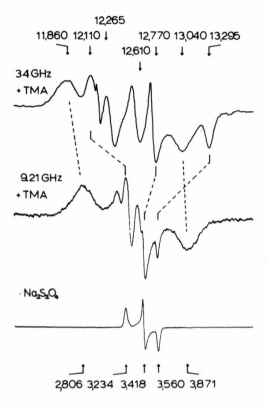

Fig. 3. EPR spectra of trimethylamine dehydrogenase reduced with an excess of trimethylamine (top and center curves) or dithionite (bottom curve). Enzyme, 32 μM, in 0.1 M pyrophosphate buffer, pH 7.7, was mixed anaerobically with trimethylamine hydrochloride (final concentration 5 mM) for the center spectrum and at approximately 10 times the enzyme concentration for the top spectrum and was frozen approximately 1 min after mixing. The bottom spectrum was obtained after addition of a tenfold excess of dithionite to 32 μM enzyme. The conditions of EPR spectroscopy were as follows: top spectrum, 34 GHz, ~3 mW, 5 G modulation amplitude, and 13°K; center and bottom spectra 9.2 GHz, 2.7 mW, 8 G modulation amplitude, and 13°K. The amplification of the center spectrum was 20 times that of the bottom spectrum. The corresponding resonances at 34 and 9.2 GHz are connected by broken lines, to show the large spread of the (sharp) lines of the Fe-S center as compared to the broad lines indicating interaction. Two new, sharp lines in the center in the top and center spectra, which are not seen in the bottom spectrum (reduced with dithionite), are also due to interaction.

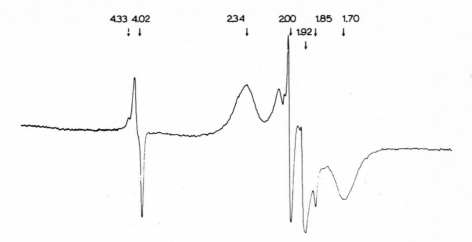

Fig. 4. EPR spectrum at 9.2 GHz, showing field scan from 200 to 4000 G, of trimethylamine dehydrogenase reduced with trimethylamine. The conditions of the reaction and of spectroscopy were analogous to those of the center spectrum of Fig. 3. Note the strong signal at $g$ = 4.0 ("half-field signal"), which is an infallible sign of spin–spin interaction. A small signal of contaminating high-spin iron is seen at $g$ = 4.3.

spectrophotometry that there may be other steps intervening with rates intermediate between the fast initial reduction and the slow development of the Fe-S semiquinone interaction.

Without going into further detail, I would like to focus on the two reactions we think we have reasonably well identified, namely a very fast $2e^-$ transfer to the flavin-type chromophore and then a slow electron redistribution to generate the semiquinone with electron donation to a very closely associated Fe-S center. This latter reaction appears to be the rate-determining step for the overall turnover with PMS, which includes a rapid $1e^-$ transfer to this acceptor, probably via flavin. Thus with this enzyme we seem to see in slow motion what probably happens in a less transparent fashion in the better-known but more complex Fe-S flavoproteins of the electron transfer system.

If the flavin of TMA dehydrogenase is indeed the exit port for electron transfer to PMS, we would have to postulate that the flavin returns to the fully oxidzed state in this reaction and then rapidly transfers the second electron from the Fe-S group, since we cannot see any reduced Fe-S center in rapid reaction studies with PMS as acceptor. In Hemmerich's scheme, as shown in Fig. 1, the flavin shuttles between reduced and semiquinone forms during the "splitting" of the electron pair. In this case, however, two Fe-S centers are required in conjunction

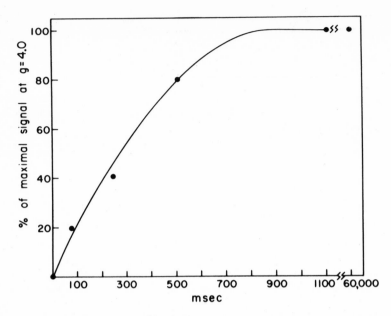

Fig. 5. Plot of progress of formation of the EPR signal at $g = 4.0$ (see Fig. 4) with time of reaction. The experiment was that of Fig. 2.

with the flavin, as postulated by Hemmerich, not only one center, as we find in trimethylamine dehydrogenase.

Maybe at this point I should repeat Henry Mahler's statement of 22 years ago (see above) that we might be "at the starting point of a real understanding of some electron transport functions."

ACKNOWLEDGMENTS

I am indebted to Dr. Peter Hemmerich, University of Konstanz, Germany, for permitting me to show his scheme of Fig. 1 and to Drs. D. J. Steenkamp and T. P. Singer for permission to discuss our preliminary unpublished results on trimethylamine dehydrogenase.

### References

1. Green, D. E. (1956) in *Enzymes: Units of Biological Structure and Function* (Gaebler, O. H. ed.), p. 465, Academic Press, New York.
2. Green, D. E., Griffiths, D. E., Doeg, K. A., and Wharton, D. C. (1963) in *Proceedings of the Fifth International Congress of Biochemistry*, Vol. IV, p. 182, Pergamon Press, New York.

3. van Gelder, B. F., and Beinert, H. (1969) *Biochim. Biophys. Acta 189*, 1–24.
4. Mahler, H. R., and Glenn, J. L. (1956) in *Inorganic Nitrogen Metabolism* (McElroy, W. D., and Glass, B., eds), p. 575, Johns Hopkins Press, Baltimore.
5. Beinert, H., and Crane, F. L. (1956) in *Inorganic Nitrogen Metabolism* (McElroy, W. D., and Glass, B., eds.), p. 601, Johns Hopkins Press, Baltimore.
6. Hall, C. L., Heijkenskjold, L., Bartfai, T., Ernster, L., and Kamin, H. (1976) *Arch. Biochim. Biophys. 177*, 402–414.
7. Ruzicka, F. J., and Beinert, H. (1975) *Biochem. Biophys Res. Commun. 66*, 622–631.
8. Beinert, H., and Ruzicka, R. J. (1975) in *Electron Transfer Chains and Oxidative Phosphorylation* (Quagliariello, E., *et al.*, eds.), pp. 37–42, North-Holland, Amsterdam.
9. Beinert, H. (1977) in *The Iron-Sulfur Proteins*, Vol. III (Lovenberg, W., ed.), Academic Press, New York.
10. Beinert, H. (1976) in *Iron and Copper Proteins*, Vol. 74 (Yasunobu, K. T., Mower, H. F. and Hayaishi, O., eds.), p. 137, Plenum, New York.
11. Hemmerich, P. (1977) *Adv. Chem. Ser., 162* (Kenneth N. Raymond, ed.), 312–329.
12. Dus, K., De Klerk, H., Sletten, K., and Bartsch, R. G. (1967) *Biochim. Biophys. Acta 140*, 291.
13. Bäckström, D., Hoffström, I., Gustafsson, I., and Ehrenberg, A. (1973) *Biochem. Biophys. Res. Commun. 53*, 596.
14. Ruzicka, F. J., and Beinert, H. (1974) *Biochem. Biophys. Res. Commun. 58*, 556–563.
15. Olson, J. S., Ballou, D. P., Palmer, G., and Massey, V. (1974) *J. Biol. Chem. 249*, 4350–4362.
16. Olson, J. S., Ballou, D. P., Palmer, G., and Massey, V. (1974) *J. Biol. Chem.* 4363–4382.
17. Steenkamp, D. J., and Singer, T. P. (1976) *Biochim. Biophys. Res. Commun. 71*, 1289–1295.

## Chapter 9

# Microsomal Oxidases

## Daniel M. Ziegler

### Introduction

Most eukaryotic cells contain a membranous network—the endoplasmic reticulum—that forms vesicular fragments when the cells are homogenized. These vesicles, separated from homogenates by differential centrifugation, make up the major membrane components[1] in the fraction usually referred to as "microsomes." Microsomal fractions derived from mammalian tissues contain oxidoreductases that vary in concentration and specificity not only in different tissues but also in the same tissue of a single species as a function of development, sex, or exposure to different environmental conditions. Changes in the concentration of specific enzymes can be especially marked in liver microsomes since the endoplasmic reticulum of hepatocytes apparently readily adapts to meet changing physiological or environmental conditions.

Oxidative enzymes located in liver microsomes play a major role in cholesterol biosynthesis,[2] desaturation of fatty acids,[3] and oxidative metabolism of a variety of nonnutritive compounds including drugs.[4] The microsomal membrane fragments contain several hemoproteins, flavoproteins, and other oxidoreduction components. The organization, sequence of electron transfer, and nature of the terminal oxidases involved in the various catalytic reactions have been studied extensively over the past decade. Progress in defining the structure and paths of electron transfer among the cytochromes and their reductases in liver microsomes is described in detail in several recent reviews[5, 6] and will be

***Daniel M. Ziegler*** • Clayton Foundation Biochemical Institute and Department of Chemistry, University of Texas at Austin, Austin, Texas 78712.

discussed only briefly. This chapter will describe in more detail the properties and probable physiological functions of a flavin-containing mixed-function oxidase also present in substantial amounts in hepatic microsomes.

### Cytochrome P-450 Monooxygenases

While cytochrome P-450 is present in the microsomal fraction of several organs, monooxygenases of the cytochrome P-450 type in hepatic tissue have been the most extensively characterized. These microsomal oxidases catalyze reduced pyridine nucleotide- and oxygen-dependent oxygenation of remarkably diverse substrates. In addition to endogenous substrates such as steroids,[7] the cytochrome P-450 system catalyzes hydroxylation of nonnutritive alkyl and aryl hydrocarbons and dealkylation of ethers, thioethers, and N-substituted amines or amides.[5] Hepatic microsomes contain multiple forms of cytochrome P-450 distinguished by spectral differences of the reduced —CO complexed hemoproteins and by their different substrate specificities. The concentration of these hemoproteins changes dramatically in animals exposed to a variety of drugs and other foreign compounds.[8, 9] Selective induction can be achieved by pretreating animals with different nonnutritive compounds. For example, the reduced hemoprotein CO spectrum of the phenobarbital-induced species absorbs at 450 nm whereas the maximum of the methylcholanthrene-induced hemoprotein is at 448 nm. Spectral and substrate specificities characteristic of microsomes from animals exposed to differential inducing agents are retained in highly purified preparations of these hemoproteins.[10] Two forms of the hemoprotein have been purified to homogeneity,[11, 12] and the studies of Coon et al.[12] suggest that four more species may be present in liver microsomes.

Terminal oxidases of the cytochrome P-450 type in hepatic microsomes are reduced by a NADPH-specific flavoprotein.[13] This flavoprotein, originally isolated and characterized as a NADPH-cytochrome c reductase,[14] has more recently been separated from detergent extracts of microsomes.[15] Antisera to the purified reductase specifically inhibit both reduction of cytochrome P-450 in microsomal membranes and catalytic cytochrome P-450 dependent hydroxylations.[13] While the NADPH-cytochrome reductase can supply reducing equivalents required for hydroxylations catalyzed by purified cytochrome P-450, the studies of Cohen and Estabrook[16] suggest a more complex electron flow in the intact membrane. In carefully controlled experiments, these authors demonstrated that both NADPH and NADH serve as reductants in cytochrome P-450 dependent hydroxylations. After reduction of the hemoprotein substrate complex by transfer of one electron from the NADPH-specific flavoprotein, NADH is the preferred reductant for the second electron required for reduction of the oxygenated complex. The route for the electron transfer from NADH to cytochrome P-450 has not been completely resolved, but it could occur via cytochrome $b_5$ and its reductase.

The probable sequence of electron transfer among various redox components of the microsomal membrane and intermediate states of cytochrome P-450 are summarized in Fig. 1. As indicated, binding of substrate precedes reduction of the hemoprotein by the NADPH-cytochrome reductase.[17] The reduced hemoprotein substrate complex combines with oxygen.[18] The ternary complex accepts a second electron

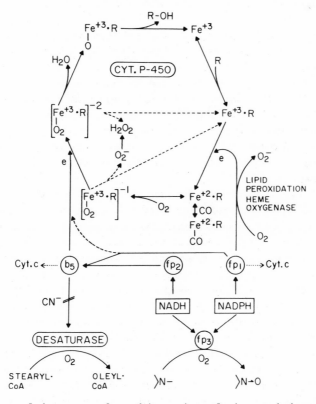

Fig. 1. Routes of electron transfer and interactions of microsomal electron transfer components. Abbreviations: $b_5$, cytochrome $b_5$; $fp_1$, cytochrome reductase; $fp_2$, NADH-cytochrome $b_5$ reductase; $fp_3$, mixed-function amine oxidase. Courtesy of R. W. Estabrook.

from either NADH or NADPH, but NADH is apparently the preferred reductant. [16] Oxygenation of substrate occurs within the fully reduced ternary complex, followed by dissociation of products and regeneration of the fully oxidized hemoprotein. Although there are unresolved problems on the exact mechanism of cytochrome P-450 mediated hydroxylations, these uncertainties should be quickly resolved by further studies on the recently available homogeneous hemoproteins [11, 12] correlated with spectral and catalytic properties of the membrane-bound cytochromes.

### Cytochrome $b_5$

Cytochrome $b_5$, isolated by Strittmatter and Velick [19] in 1956, was the first hemoprotein purified to homogeneity from microsomal membranes. Descriptions of the isolation and properties of this hemoprotein are given in detail in reviews. [20, 21] The roles of cytochrome $b_5$ in microsomal-catalyzed reactions are also described in a recent review. [22] The cytochrome is reduced by a NADH-specific flavoprotein first isolated from protease-treated bovine liver homogenates. [23] Later the reductase was solubilized and purified from bovine liver microsomes by the use of detergents. [24, 25] The catalytic and spectral properties of the flavoprotein isolated by the different methods were similar, but amino acid analyses indicated that the detergent-extracted flavoprotein contained an additional 98 amino acids. Sixty percent of these were hydrophobic and are presumed to play a role in attachment of the reductase to the membrane.

Physiological functions of the microsomal cytochrome $b_5$ and its reductase have only recently been defined. In erythrocytes, they constitute an electron transfer system capable of reducing methemoglobin, and, as indicated in Fig. 1, the microsomal cytochrome $b_5$ may function as an electron donor for cytochrome P-450 dependent hydroxylations. Cytochrome $b_5$, its NADH-specific reductase, and a third poorly defined microsomal protein are also required to reconstitute the NADH-$N$-hydroxyamine reductase of pig liver microsomes. [26] However, the major function of this microsomal electron transfer system is its role in the desaturation of fatty acids. The studies of Sato's laboratory [27] demonstrated that the microsomal stearyl CoA 9-desaturase required cytochrome $b_5$, its reductase, and a cyanide-sensitive terminal oxidase. The terminal oxidase, purified to homogeneity, is a nonheme iron protein. [28] The desaturase system apparently requires lipid, [29] but the function of lipid in this system has not been completely defined. In intact microsomes, NADH and NADPH are equally effective reductants. [30] The probable sequence of electron transfer in the membrane-bound de-saturase is also illustrated in Fig. 1. Electrons from either NADH or NADPH are transferred by way of their respective dehydrogenase to cytochrome $b_5$ and the reduced cytochrome supplies reducing equivalents to the terminal oxidase which catalyzes desaturation of stearyl CoA.

## Flavin-Containing Microsomal Oxidase

Early reports from Kiese's laboratory[31] demonstrated marked differences in the NADPH-dependent oxidation of aniline and *N*-alkylanilines catalyzed by liver microsomes. *N*-Oxidation of *N*-alkylanilines in contrast to aniline was insensitive to carbon monixide, and higher concentrations of oxygen were also required to saturate the microsomal di- and trisubstituted amine *N*-oxidase.[31–33] By the use of inhibitors specific for cytochrome P-450 dependent hydroxylations, several laboratories[32, 34, 35] demonstrated that, unlike the case with aniline, *N*-oxidation of most secondary and tertiary amines was not catalyzed by the microsomal cytochrome P-450 system. However, the nature of the microsomal monooxygenase catalyzing *N*-oxidation of these amines was not resolved until the oxygenase was solubilized and extensively purified.[36] It was subsequently purified to homogeneity[37] and shown to be a FAD-containing flavoprotein distinctly different from other microsomal flavoproteins.[38] This flavin-containing mono-oxygenase is usually referred to by its trivial name, "microsomal amine oxidase," or simply as the "amine oxidase."

Progress in isolating and characterizing the microsomal amine oxidase was greatly facilitated by the development of a sensitive and specific colorimetric method for measuring the *N*-oxide of *N, N*-dimethylaniline.[39] In virtually all species and tissues studied, the *N*-oxide of this tertiary amine is produced solely by the amine oxidase, and the species and tissue distribution follow the trimethylamine *N*-oxidase activity described by Baker *et al.*[40] In all species studied, liver contains the highest amounts of amine oxidase, and it is unusually concentrated in hog[37] and human liver tissue.[41] Rodent hepatocytes apparently contain somewhat lower amounts of this enzyme, and it is especially low in the rat. Table I summarizes activities of this monooxygenase in hepatic tissue from adults of several species.

In hog liver the amine oxidase is a major component of the endoplasmic reticulum. Based on activities of homogenates, isolated microsomes, and turnover of the purified enzyme,[37] this enzyme accounts for 3–5% of microsomal protein of hog liver, and higher amounts, up to 10% of total microsomal protein, are not unusual. If the

*Table I. Hepatic Mixed-Function Amine Oxidase Activity in Several Species*[a]

| Species | Activity (mg liver protein) |
| --- | --- |
| Pig | 3.4 |
| Human | 2.0 |
| Guinea pig | 1.7 |
| Hamster | 1.7 |
| Rabbit | 0.6 |
| Rat | 0.5 |

[a] Activities based on rate of dimethylaniline $N$-oxide formation at 37°C. Values listed are averages obtained with 15 or more different preparations from adult animals. Considerable interindividual variations with both pig and human liver values are encountered. For example, activities in the 15 different human liver samples varied from 0.7 to 5.3 (*cf.* ref. 41).

turnover of the human liver enzyme is similar to that from the hog, it is also a major protein component of human liver microsomes.

In contrast to a number of other microsomal enzymes, induction of amine oxidase by pretreating animals with drugs cannot be demonstrated.[38] However, hepatic amine oxidase activity changes markedly during the life cycle of rodents. The amount of enzyme present may be under hormonal control, but direct evidence that this is the case has not been demonstrated.

Changes in hepatic microsomal amine oxidase activity during the life cycle of the rat have been extensively studied by Uehleke *et al.*[42] Activity barely detectable in the fetus increases about tenfold during the first 2 days after delivery. Then the activity declines somewhat, but after 8–10 days increases dramatically and reaches a distinctly higher peak just prior to weaning at 27 days. The activity then declines in young animals, and the decline is more marked in females than in males. In male rats the activity increases slowly but progressively as the animals age and can reach rather high levels in 1-year-old males.[42] The activity of this enzyme also exhibits a characteristic "weaning spike" in hamsters,[44] and changes in microsomal amine oxidase activity during the life span of this species are illustrated in Fig. 2. The activity rises sharply after delivery and reaches a distinct peak in 4-week-old hamsters. Immediately following weaning the activity drops by about one-half and then remains relatively constant during the life span of both male and female adults. The development of hepatic amine oxidase in rodents is similar to that of other enzymes described by Greengard[45] as the "late suckling cluster" of enzymes.

Whether the amine oxidase also undergoes similar changes in humans is not known. The studies of Rane[46] demonstrate that it is detectable in human fetal liver as early as 13 weeks' gestation and increases rapidly during development of the fetus, and, as indicated earlier, this enzyme is present in substantial amounts in adults. However, critical data on concentration of amine oxidase in the human neonate are not available. The function of this microsomal oxidase during development remains obscure, but the consequences of the increased activity in the neonate on metabolism of amine drugs and some carcinogenic amines disposed by alternate routes in the adult have been pointed out by Uehleke *et al.*[42] The microsomal amine oxidase develops earlier and reaches its maximum activity before other microsomal mixed-function drug oxidases. The role of *N*-oxidation in the metabolism and disposition of drugs and other nonnutritive compounds is discussed at length in several excellent reviews.[47–49]

## Properties of the Isolated Amine Oxidase

The oxidase purified to homogeneity from hog liver contains close to 15 nmoles FAD per milligram of protein. SDS-treated preparations migrate as a homogeneous protein upon electrophoresis in polyacrylamide gels, and molecular weight of the SDS-treated preparations estimated by migration rate on polyacrylamide is close to the minimum molecular weight (65,000) based on flavin. However, the sedimentation rate of the native enzyme suggests that the enzymically active species exists as an octomer in equilibrium with small quantities of tetrameric forms. The purified oxidase appears free from metals, hemoproteins, and lipid.[37] Spectra of both the oxidized and reduced forms are similar

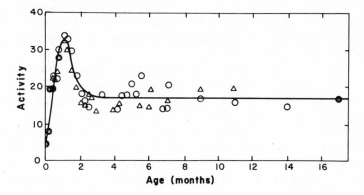

Fig. 2. Changes in hepatic amine oxidase during life cycle of hamster. △, Females; ○, males. Each point represents averages of three or more animals. Activity is expressed as nmoles of substrate oxidized per minute per milligram of microsomal protein. Compiled from data of Wiginton.[44]

to those of other flavoproteins. Anaerobically the flavin prosthetic group is completely reduced by either NADPH or NADH, but spectral changes characteristic of flavin semiquinones are not observed. Aerobically, the reduction of flavin by reduced nucleotides cannot be detected, which suggests that in the absence of an oxidizable substrate the reduction of the flavin is considerably slower than its reoxidation by oxygen. However, the endogenous NADPH oxidase activity is quite low and usually less than 10% of the substrate-dependent rate. Kinetic studies still in progress suggest that addition of substrates is ordered-sequential, and the kinetic order appears to be NADPH, the oxidizable substrate, and oxygen.

At saturating levels both NADPH and NADH are equally effective reductants, but NADPH saturates the enzyme at less than one-tenth the concentrations required with NADH. Phosphate also markedly affects reduction of the enzyme by NADH, whereas activity of the oxidase with NADPH as reductant is quite independent of the type of buffer used. The apparent $K_m^{NADPH}$ at 38°C in solutions at equilibrium with atmospheric oxygen is $5 \times 10^{-6}$ M. This value does not change over a wide range in pH, and appears relatively constant regardless of the nature of the substrate hydroxylated.

The purified oxidase catalyzes the oxidation of diverse types of amines[36, 50] and hydrazines.[51] Types of amines oxidized and products formed are illustrated in Table II. Specificity of the enzyme for tertiary amines has been the most thoroughly investigated, and in general tertiary amines with N-methyl, N-isopropyl, or N-benzyl substituents containing an alkyl side chain branched three or four carbons from the nitrogen are preferred substrates. The tricyclic phenothiazine drugs meet these requirements and are among the best substrates, with apparent $K_m$'s in the micromolar range.[50] N-Oxides are the only products formed, and, in compounds containing more than one tertiary amine group, only one position is oxidized.[52]

Secondary amines are also N-oxidized, but at a somewhat slower rate than tertiary amines. N-Hydroxyamines produced by N-hydroxylation of secondary amines do not differ significantly in physical properties from the parent compound, and the enzyme catalyzes their further oxidation to nitrones.[53] The nitrones are apparently formed by dehydration of very unstable intermediate N-hydroxyamine oxides. With the exception of α-phenyl-substituted nitrones, all other nitrones are quite unstable and readily hydrolyze to the corresponding aldehydes and primary N-hydroxyamines. That this sequence of secondary amine oxidation can occur in intact microsomes is supported by recent studies.[53] Work from Beckett's laboratory[55, 56] also suggests that metabolites of chlorphenteramine and phenteramine in human subjects are apparently produced by this route.

*Table II. Oxidation of Nitrogen Compounds Catalyzed by the Amine Oxidase*[a]

| Reaction[b] | Example |
|---|---|
| *tert*-Amines<br><br>$X-N\begin{smallmatrix}R_1\\R_2\end{smallmatrix} \longrightarrow X-\overset{+}{\underset{R_2}{N}}(OH)-R_1$ | Chlorpromazine, ethylmorphine, atropine, nicotine, meperidine, etc. |
| *sec*-Amines<br><br>$X-NHR_2 \longrightarrow X-\overset{OH}{N}R_2$<br>*sec*-N-Hydroxyamines | Desipramine, desmethylmeperidine, methamphetamine |
| $X-\overset{OH}{N}R_1 \begin{cases} \longrightarrow X=\overset{O}{N}-R_1 \\ \longrightarrow X-\overset{O}{N}=R_1 \end{cases}$ | N-Hydroxy-N-benzylamphetamine, N-methyl-N-hydroxybenzylamine |
| Imines and arylamines<br><br>$\rangle{=}NH \longrightarrow \rangle{-}NHOH$ | Rosaniline, 2-naphthylamine |
| Hydrazines<br><br>$X-N\begin{smallmatrix}NH_2\\R_1\end{smallmatrix} \longrightarrow X-\overset{+}{\underset{R_1}{N}}(OH^c)-NH_2$ | 1-Methyl-1-phenylhydrazine, N-aminopiperidine |

[a] X, Any lipophilic alkyl or aryl side chain free from negatively charged groups and groups more polar than a hydroxyl within a two-carbon radius of the nitrogen; $R_1$, methyl, isopropyl, or benzyl groups preferred substrates; $R_2$, any alkyl side chain or can be part of ring as in guanethidine or desmethylmeperidine.

[b] All reactions require 1 mole NADPH and 1 mole $O_2$/mole amine.

[c] Suggested product from stoichiometry measurements. Too reactive to be isolated.

Primary alkylamines, aromatic amines, and most primary arylamines are not N-oxidized by the purified amine oxidase. However, 2-naphthylamine and a few other arylamines (i.e., 2-aminoazulene and rosaniline) that readily form imine tautomers are rapidly N-oxidized by the microsomal amine oxidase. The studies of Poulsen *et al.*[57] suggest that this is the only route for N-hydroxylation of 2-naphthylamine in hog liver homogenates.

In addition to amines, recent studies[58] have shown that the microsomal amine oxidase also catalyzes oxidation of a variety of nucleophilic sulfur compounds. Sulfur substrates are competitive with amine substrates, which suggests that they interact at the same catalytic site. Although the substrate specificity of this enzyme for sulfur-containing compounds has not been thoroughly examined, types of compounds tested so far and the products formed are listed in Table III.

*Table III. Oxidation of Sulfur-Containing Compounds Catalyzed by the Amine Oxidase*[a]

| Reaction[b] | Examples |
|---|---|
| **Thioureylenes** $$X-NH \\\ \phantom{}\rangle -SH \longrightarrow \phantom{}\rangle -SO_2H \\\ R_2-N \phantom{XXXXXX} N$$ | Methimazole, methylthiourea, propylthiouracil |
| **Thioamides** $$X-\overset{S}{\underset{\parallel}{C}}-NH_2 \longrightarrow X-\overset{S=O}{\underset{\parallel}{C}}-NH_2$$ | Thioacetamide, thiobenzamide |
| **Thiols** $$2\,R-SH \longrightarrow R-S-S-R$$ | Mercaptoethanol, dithiothreitol |
| **Aminothiols** $$2\,NH_2CH_2CH_2SH \longrightarrow (NH_2CH_2CH_2S-)_2$$ | Cysteamine, homocysteamine |

[a] Nature of X and R substituents similar to those for amines listed in Table II.
[b] All reactions require 1 mole NADPH and 1 mole $O_2$ for every oxygen atom incorporated into substrate. Formation of the disulfides appears to involve the intermediate sulfenic acid, and only 1 mole of nucleotide and oxygen is consumed for every mole of disulfide formed.

In all cases, the products identified are those expected as a result of an initial oxidative attack on the sulfur atom.

Substrates containing the thiocarbamide moiety are oxidized initially to their respective, extremely reactive, formamidine sulfenic acid derivatives. In the absence of thiols, the formamidine sulfenic acid disproportionates to yield the somewhat more stable formamidine sulfinic acid and the parent thiocarbamide. However, the formamidine sulfenic acid preferentially reacts with glutathione or protein thiols, and by thiol–disulfide exchange a second mole of glutathione regenerates the parent thiocarbamide with formation of glutathione disulfide. This series of reactions can deplete cells of glutathione and appears to be the molecular basis for toxicity of these compounds.

Thioamides form stable monooxygenated derivatives, and, with the few simple thioamides tested, the corresponding sulfoxide is the only oxidation product detected.[59] While dithiothreitol and a few alkyl thiols[57] are substrates for this microsomal oxidase, low molecular weight thiols with one or more negatively charged functional groups (i.e., glutathione and cysteine) do not interact with the oxidase.

Cysteamine, the decarboxylated metabolite of cysteine, may be the physiological substrate for this microsomal enzyme, and it is specifically oxidized only to the disulfide cystamine.[59] The significance of this reaction as a source of disulfide for synthesis of protein disulfide bonds has been described,[60] and evidence that this reaction can support renaturation of reduced ribonuclease in the presence of glutathione and glutathione reductase has been presented.[59]

## References

1. Palade, G. E., and Sickevitz, P. (1956) *J. Biophys. Biochem. Cytol. 2*, 171.
2. Goad, L. J. (1970) in *Natural Substances Formed from Mevalonic Acid* (Goodwin, T. W., ed.), p. 35, Academic Press, New York.
3. Oshimo, N., Imai, Y., and Sato, R. (1971) *J. Biochem. 69*, 155.
4. Estabrook, R. W., Gillette, J. R., and Leibman, K. D. (eds.) (1972) *Microsomes and Drug Oxidation*, Williams and Wilkins, Baltimore.
5. Orrenius, S., and Ernster, L. (1974) *Molecular Mechanisms of Oxygen Activation* (Hayaishi, O., ed.), pp. 215–244, Academic Press, New York.
6. Gunsalus, I. C., Pederson, T. C., and Sligar, S. G. (1975) *Ann. Rev. Biochem. 44*, 377–406.
7. Hamberg, M., Samuelsson, B., Bjorkhern, J., and Danielsson, H. (1974) in *Molecular Mechanism of Oxygen Activation* (Hayaishi, O., ed.), pp. 69–76, Academic Press, New York.
8. Conney, A. H., Miller, E. C., and Miller, J. A. (1956) *Cancer Res. 16*, 450.
9. Conney, A. H. (1967) *Pharmacol. Rev. 19*, 317.
10. Lu, A. Y., Levin, W., West, S., Jacobson, M., Ryan, D., Kuntzman, R., and Conney, A. H. (1973) *J. Biol. Chem. 248*, 456.
11. Imai, Y., and Sato, R. (1974) *Biochem. Biophys. Res. Commun. 60*, 8.
12. Coon, M. J., Ballou, D. P., Guengerich, F. P., Nordblom, C. D., and White, R. E. (1976) in *Iron and Copper Proteins* (Yasunobu, K. T., Mower, H. F., and Hayaishi, O., eds.), pp. 270–280, Plenum, New York.
13. Masters, B. S. S., Kamin, H., Gibson, Q. H., and Williams, C. H. (1965) *J. Biol. Chem. 240*, 921.
14. Williams, C. H., and Kamin, H. (1962) *J. Biol. Chem. 237*, 587.
15. Coon, M. J., Strobel, H. W., and Boyer, R. F. (1973) *Drug Metab. Disposition 1*, 92.
16. Cohen, B. S., and Estabrook, R. W. (1971) *Arch. Biochem. Biophys. 143*, 54.
17. Tsai, R., Yu, C. A., Gunsalus, I. C., Peisach, J., Orme-Johnson, W. E., and Beinert, H. (1970) *Proc. Natl. Acad. Sci. USA 66*, 1157.
18. Estabrook, R. W., Hildebrandt, A. G., Baron, J., Netter, K. J., and Leibman, K. (1971) *Biochem. Biophys. Res. Commun. 42*, 132.
19. Strittmatter, P., and Velick, S. F. (1956) *J. Biol. Chem. 221*, 253.
20. Strittmatter, P. (1963) *Enzymes 8*, 114.
21. Ozols, J., and Strittmatter, P. (1968) in *Structure and Function of Cytochromes* (Okunuki, K., Kamen, M. D., and Sekuza, I., eds.), pp. 578-580, University Park Press, Baltimore.
22. Schenkman, J. B., Jansson, I., and Robie-Suh, K. M. (1976) *Life Sci. 19*, 611.
23. Strittmatter, P., and Velick, S. F. (1957) *J. Biol. Chem. 228*, 785.
24. Panfili, E., Sottocasa, G. L., and De Bernard, B. (1972) *Biochim. Biophys. Acta 253*, 51.
25. Spatz, L., and Strittmatter, P. (1972) *Fed. Proc. 31*, 411.
26. Kadlubar, F. F., and Ziegler, D. M. (1974) *Arch. Biochem. Biophys. 162*, 83.
27. Oshimo, N., Imai, Y., and Sato, R. (1966) *Biochim. Biophys. Acta 128*, 13.
28. Strittmatter, P., Spatz, L., Corcoram, D., Rogers, M., Setlow, B., and Redline, R. (1974) *Proc. Natl. Acad. Sci. USA 71*, 4565.
29. Jones, P. D., Holloway, P. W., Peluffo, R. O., and Wakil, S. J. (1969) *J. Biol. Chem. 244*, 744.

30. Hollowway, P. W., Peluffo, R., and Wakil, S. J. (1963) *Biochem. Biophys. Res. Comm. 12*, 300.
31. Kampffmeyer, H., and Kiese, M. (1964) *Naunyn-Schiedebergs Arch. Exp. Pathol. Pharmakol. 246*, 397.
32. Kampffmeyer, H., and Kiese, M. (1964) *Biochem. Z. 339*, 454.
33. Beije, B., and Hulten, T. (1971) *Chem. Biol. Interact. 3*, 321.
34. Hlavica, P. H., and Kiese, M. (1969) *Biochem. Pharmacol. 18*, 1501.
35. Uehleke, H. (1966) *Life Sci. 5*, 1489.
36. Ziegler, D. M., Mitchell, C. A., and Jollow, D. (1969) in *Microsomes and Drug Oxidations* (Gillette, J. R., Conney, A. H., Cosmides, G. J., Estabrook, R. W., Fouts, J. R., and Mannering, G. J. eds.), pp. 173–188, Academic Press, New York.
37. Ziegler, D. M., and Mitchell, C. A. (1972) *Arch. Biochem. Biophys. 150*, 116.
38. Masters, B. S. S., and Ziegler, D. M. (1971) *Arch. Biochem. Biophys. 145*, 358.
39. Ziegler, D. M., and Pettit, F. H. (1964) *Biochem. Biophys. Res. Commun. 15*, 188.
40. Baker, J., Struempler, A., and Chaykin, S. (1963) *Biochim. Biophys. Acta 71*, 58.
41. Gold, M. S., and Ziegler, D. M. (1973) *Xenobiotica 3*, 179.
42. Uehleke, H., Reiner, O., and Hellmer, K. H. (1971) *Res. Commun. Chem. Pathol. Pharmacol. 2*, 793.
43. Das, M., and Ziegler, D. M. (1970) *Arch. Biochem. Biophys. 140*, 300.
44. Wiginton, D. (1976) Master's thesis, University of Texas.
45. Greengard, O. (1971) in *Essays in Biochemistry* (Campbell, P. N., and Dickens, F., eds.), p. 159, Academic Press, New York.
46. Rane, A. (1973) *Clin. Pharmacol. Ther. 15*, 32.
47. Bickel, M. H. (1969) *Pharmacol. Rev. 21*, 325.
48. Kiese, M. (1974) *Methemoglobinemia: A Comprehensive Treatise*, CRC Press, Cleveland.
49. Gorrod, J. W. (1973) *Chem. Biol. Interact. 7*, 289.
50. Ziegler, D. M., Jollow, D., and Cook, D. (1971) in *Flavins and Flavoproteins: Third International Symposium*, (Kamin, H., ed.), pp. 507–522, University Park Press, Baltimore.
51. Prough, R. A. (1973) *Arch. Biochem. Biophys. 158*, 442.
52. Sofer, S. S., Ziegler, D. M., and Popovich, R. (1975) *Biotech. Bioengin. 17*, 107.
53. Poulsen, L. L., Kadlubar, F. F., and Ziegler, D. M. (1974) *Arch. Biochem. Biophys. 164*, 774.
54. Prough, S. S., and Ziegler, D. M. (1977) *Arch. Biochem. Biophys. 180*, 363.
55. Beckett, A. H., and Belanger, P. M. (1974) *J. Pharm. Pharmacol. 26*, 206.
56. Beckett, A. H., and Belanger, P. M. (1974) *J. Pharm. Pharmacol. 26*, 560.
57. Poulsen, L. L., Masters, B. S. S., and Ziegler, D. M. (1976) *Xenobiotica 6*, 481.
58. Poulsen, L. L., Hyslop, R., and Ziegler, D. M. (1974) *Biochem. Pharmacol. 23*, 3431.
59. Poulsen, L. L., and Ziegler, D. M. (1977) *Arch. Biochem. Biophys. 183*, 563.
60. Ziegler, D. M., and Poulsen, L. L. (1977) *Trends Biochem. Sci. 2*, 79.

## Chapter 10

# Molecular Mechanism of Energy Coupling in Mitochondria and the Unity of Bioenergetics

## David E. Green and George A. Blondin

### Introduction

The mitochondrion is a cellular organelle specialized for coupling the oxidation of citric cycle intermediates, fatty acids, certain amino acids, etc., to the synthesis of ATP. The nature and molecular basis of this coupling have intrigued and baffled three generations of investigators. There have to be four steps in arriving at a solution to the coupling problem. First, the source of the energy has to be specified and defined. In this case, the energy is chemical and is released by oxidation. The second is the recognition of the principle by which the free energy released during mitochondrial oxidation can be conserved and utilized (the principle of energy coupling). The third is the development of a model that specifies how the mitochondrion translates the principle of energy coupling. The model we are proposing for this translation is the paired moving charge model of energy coupling.[1] Finally, if the principle and the model are valid and relevant, these should lead step by step to the solution of all the phenomena of mitochondrial energy coupling. This complete rationalization is the only infallible test of such validity. The structural, functional, and control aspects of the mitochondrion must be consistent with and illuminated by the principle and the

*David E. Green and George A. Blondin* • Institute for Enzyme Research, University of Wisconsin, Madison, Wisconsin 53706.

model. If such is not the case, the principle and the model cannot be correct. When the rationalization is across the board, the possibility that the principle and the model are invalid becomes vanishingly small.

## Principle of Energy Coupling

Energy coupling is an interaction between two chemical processes—one releasing free energy and the other requiring free energy for consummation. Since it is chemical energy that is involved in such cooperativity during energy coupling, logically the first order of business should be the definition of chemical energy. When the energy of an electron drops in the transition of bond $A—B$ to bond $A'—B'$, where the letters represent the atoms participating in these two bonds, free energy is released and this particular form of free energy can be equated with chemical energy. The decrease in electronic energy as an electron acquires a new set of bonding atoms is what chemical energy is all about. Given this definition, it is obvious that there is only one way in which this energy can be tapped. The original bond must be ruptured and a new bond must be formed. This means charge separation of the atoms constituting the original bond and charge reuniting of the atoms that are partners in the new bond. It is also obvious that the rupture of bond $A—B$ and formation of bond $A'—B'$ has to go parallel with the rupture of bond $C—D$ and the formation of bond $C'—D'$. Thus the unit of chemical change is always a pair of bonds in the stage of rupture and a pair of bonds in the stage of formation.

In order to conserve chemical free energy, the fall in electronic energy of the electron undergoing a change in bonding partner must be coupled to an increase in energy of an electron undergoing an energy-requiring change in bonding partner. This is the fundamental principle of energy coupling in chemical systems. Coupling involves an energy-releasing bond change in one molecule cooperatively linked to an energy-requiring bond change in a second molecule, and in this cooperativity there has to be interplay between an electron moving down the free energy scale and an electron moving up the free energy scale (Fig. 1).

## Model of Energy Coupling

There is in fact no way of achieving direct electron–electron coupling when this coupling would have to depend on an interaction between molecules that are separated in space (this separation in space is

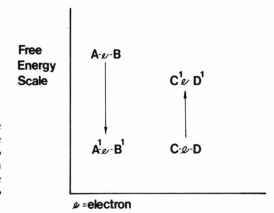

Fig. 1. Coupling of the fall in free energy of an electron (in the transition from bond *A—B* to bond *A'—B'* to the increase in free energy of an electron (in the transition from bond *C—D* to *C'—D'*).

the biological condition that has to be satisfied). However, we can construct a physical model for achieving electron–electron coupling by carrying out the coupling in two stages (see Fig. 2). Let us assume that an energy-releasing chemical sequence takes place within a center that spans a membrane separating an internal from an external aqueous phase; this exergonic center is cheek by jowl with another membrane-spanning center (the endergonic center) where an energy-requiring chemical sequence takes place. We now have the two interacting centers in close apposition within a membrane, and we shall further assume that the chemical reactions taking place in these two centers have a vectorial character, proceeding in parallel and in synchrony from one side of the membrane to the other.

By virtue of paired charge separation in the two centers, and the ejection of the positively charged species in the exergonic center and of the negatively charged species in the endergonic center, the flow of the electron in the exergonic center can be coupled to the flow of an ionophore*–$Me^+$ complex in the endergonic center ($Me^+$ stands for a

---

* Ionophores are relatively small molecules (molecular weight 300–2000) with the following characteristics: an oily exterior that permits passage through nonpolar membranes and a polar interior with groups capable of coordination with cations. In the movement of a cation from an aqueous phase to the ionophore in the membrane phase, the cation sheds its water of hydration and forms instead coordination links with the polar groups in the interior of the ionophore. There are two main types of ionophores—doughnut shaped, like valinomycin, which encapsulate the cation (the ionophore opens and closes during this encapsulation) and open, a pair of which encapsulate the cation. The ionophore can be neutral (valinomycin) or negatively charged (A23187). Several excellent review articles as well as a monograph on ionophores are available.[2-6]

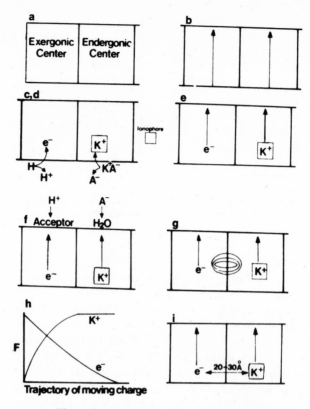

Fig. 2. Ground rules for energy coupling.

positively charged metal ion). The coupling depends on the electrostatic interaction between the negatively charged electron and the positively charged ionophore–$Me^+$ complex.[1] The interaction of the electron with the final acceptor in the chain of the exergonic center involves a simultaneous uptake of a proton from the aqueous phase on the side of the membrane opposite to the side where charge separation took place. This reuniting of an electron and a proton in the acceptor is paired to the transfer of $Me^+$ from the ionophore to an anion ($A^-$) also derived from the aqueous phase on the terminus side of the membrane. In effect, we have thereby coupled the downhill flow of an electron in the electron transfer sequence to the uphill flow of an electron in the cation-transporting sequence (Fig. 3). The electron in the latter case is associated with the anion ($A^-$) that combines with $Me^+$ released from the ionophore. One has to bear in mind that $Me^+$ is moved from one side of the membrane to the other, and, in consequence of this positional change, the chemical potential of $Me^+$ is increased so that the free energy of the electron in $Me^+A^-$ in the aqueous phase on the terminus side of the membrane is higher than that of the electron in $Me^+A^-$ in the aqueous phase on the other side of the membrane.

The above model solves the problem of electron–electron coupling by substituting such coupling by electron–$Me^+$ coupling actuated by the attractive force of electrostatic interaction.[17] There is no such comparable force for electron–electron coupling. Work is done on ionophore-linked $Me^+$, and when the electron (in the form of $A^-$) reunites with $Me^+$ the energy level of the new species ($Me^+A^-$) reflects this work performance.

Essential for electron–$Me^+$ coupling is paired charge separation, paired charge movement, and, finally, paired charge elimination. Unpaired charge separation is an energy-requiring process, but paired charge separation catalyzed by an enzyme can be a spontaneous process.[1, 7] The maneuver by which an electron is separated from a proton or by which the ionophore–$Me^+$ complex is separated from its anion ($A^-$) depends on the separation of one charge species into the membrane phase and of the other charged species into the aqueous phase in contact with the membrane phase. In charge elimination, two charged species in the exergonic and endergonic centers, respectively, are reunited, and since the reuniting of charges is paired and synchronous, in these two centers charge elimination can be a spontaneous process. One must bear in mind that the formation of the species $A^-$ and $H^+$,

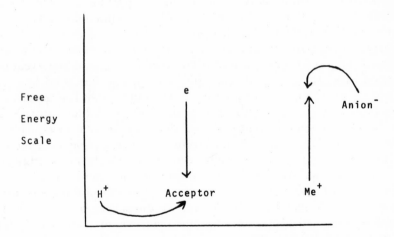

Fig. 3. Coupling of the electron as it moves down its electrochemical gradient to a positively charged species ($Me^+$) moving up its electrochemical gradient. This is formally equivalent to electron–electron coupling. The electron moving up its electrochemical gradient would correspond to anion⁻.

respectively, is a consequence of paired charge separation. Without such pairing the energetics for generating $A^-$ and $H^+$ would be unfavorable.

Electron–$Me^+$ coupling can lead to active transport of a cation as implicit in Fig. 2 or to covalent bond formation as in the synthesis of ATP. To achieve the latter, the chemical potential of a positively charged species that we may represent as (ionophore$-Me^{2+}-Pi^-)^+$ has to be increased to an energy level at which it can react with membrane-bound ADP to form ATP. The chemical potential of a positively charged species can be increased by the electron-driven movement of this species into an increasingly nonpolar region in the membrane.[8] There is thus no basic difference in principle between energy coupling leading to active transport of a cation and energy coupling leading to bond formation.

### Experimental Verification of the Model of Energy Coupling

The first and obvious question is whether the mitochondrial system shows in fact the gross characteristics demanded by the physical model we have proposed for energy coupling. Is the mitochondrial transducing system anchored in a membrane? Is the electron transfer complex housed in a transmembrane complex and does the electron move vectorially across the membrane in the fashion shown in Fig. 1? Is proton ejection an invariant accompaniment of electron flow and is the $H^+/e$ ratio 1:1 for this charge separation? Are protons taken up on the terminus side of the membrane where charge elimination takes place and is the $H^+/e$ ratio 1:1 for this process? Does the movement of each electron through the chain drive the transmembrane transport of one positively charged cationic species? The answer is that all these requirements of the physical model are fully satisfied by the mitochondrial system[1, 7] and thus we can now proceed from the level of gross fit of mitochondrion to model to the level of exact fit (see Addendum).

In this section we shall consider the experimental evidence that bears on the mechanism of uncoupling, on uncoupling as a natural phenomenon, on the mechanism of coupling in cytochrome oxidase, on the role of ionophores and ionophoroproteins in energy coupling, on the control of the energy coupling pattern by control of the ion transfer complexes, and, finally, on the structure of the energy coupling systems and how this structure translates the principle of energy coupling. Limitations of space make it impossible to present the experimental evidence for all the developments listed above, but a sample of the extent of documentation is provided in the treatment of the action of uncouplers.

*Mechanism of Uncoupling*

A considerable number of molecular species of diverse structures have been shown to suppress the coupling between electron flow and ATP synthesis without suppressing electron flow.[9] These species are referred to as "uncouplers" and their action is referred to as "uncoupling". Actually, uncouplers do more than uncouple oxidative phosphorylation. They uncouple the link between ATP hydrolysis and various coupled processes and eliminate protonic and cationic gradients generated by electron flow or hydrolysis of ATP.

The combination of two ionophores (valinomycin and nigericin) in the presence of $K^+$ can duplicate all the known actions of uncouplers and do so at concentrations generally below those of all but the best uncouplers.[10] This combination is known to mediate the cyclical transport of $K^+$ by mitochondria (see Fig. 4). Valinomycin brings the $K^+$ in electrogenically (directly driven by the electron) and nigericin brings the $K^+$ out nonelectrogenically (indirectly driven by the electron). It can be shown that the combination of an electrogenic ionophore in the mitochondrion and the uncoupler acts in exactly the same fashion as the combination of valinomycin and nigericin in mediating coupled cyclical cation transport.[11] But the uncoupler can collaborate with more than one electrogenic ionophore with different cation specificities. Hence

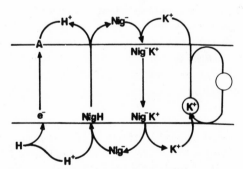

Fig. 4. Cyclical transport of $K^+$ mediated by the combination of valinomycin and nigericin. The two parallel lines represent the inner membrane. Valinomycin is represented as ◯ and the (K) complex of valinomycin as (K). Nigericin is abbreviated as $Nig^-$ in its charged form and as NigH in its protonated form. The movement of the electron ($e^-$) is coupled to the movement of (K). The electron moves through the electron complex to the final acceptor on the other side of the membrane, where the electron and the proton are reunited. (K) moves through the membrane and (K) is released into the aqueous phase on the other side of the membrane. For reasons of simplicity, the charge separation of KA (where A is the anion) corresponding to the charge separation of H into $e^-$ and $H^+$ has been omitted.

uncouplers can mediate cyclical transport of both monovalent and divalent cations.

If the mechanism of action of uncouplers is indeed that of mediating coupled cyclical transport of cations, we would expect the following to be true: uncoupling should require a critical level of the appropriate cation and should be inoperative below this critical level; uncouplers should be ionophores for monovalent and divalent cations; uncouplers should collaborate with electrogenic ionophores in inducing cyclical cation transport; the concentration of the most efficient uncouplers should never be less than the concentration of the electron in an electron transfer complex (it could be more but never less at the point of maximal uncoupling); and, finally, it should be possible to demonstrate by direct means that uncouplers mediate cyclical inflow and outflow of cations. Evidence is available to verify each of these predictions. [11]

The cation requirement for uncoupling was not recognized in previous studies because mitochondria contain a relatively high concentration of bound divalent cations (30 nmoles of $Ca^{2+}$ and 40 nmoles of $Mg^{2+}$ per milligram of protein) and this complement of bound divalent cations is sufficient to sustain uncoupler action. When the level of bound cations is depleted by the use of A23187 (Table I) or EDTA (Table II), then addition of cation is required for uncoupler-mediated release of respiration by uncoupler. The degree of depletion of the bound cation by these two reagents is shown in Table III. [11]

There are two electrogenic ionophores implicated in the release of respiration by uncouplers—one which mobilizes bound divalent cation and is active at relatively low concentrations of divalent metals. The second is active on monovalent cations but its activity is not demonstrable until the concentration of cations is at least 50 mM. This low-affinity monovalent electrogenic ionophore does not come into the picture unless the high-affinity divalent cation electrogenic ionophore is inoperative.

As the data of Tables I and II demonstrate, the degree of uncoupling can be reduced to a small fraction of its original level by reducing the concentration of cation below a critical value. By extrapolation we can infer that there would be no uncoupling in complete absence of suitable cations.

Kessler *et al.* [12] have already demonstrated that all *bona fide* uncouplers are ionophores for both monovalent and divalent cations, although this capability is, with few exceptions, demonstrable only above pH 7. The solution to this dilemma of an unphysiological pH range for the exercise of ionophoric function came with the discovery that the combination of an electrogenic ionophore and uncoupler is highly effective in mediating transmembrane transport of cations at pH 7.0

*Table I. Cation Dependence of Uncoupler action in Beef Heart Mitochondria Depleted of Divalent Metals with A23187 [a]*

| Addition of mitochondria | Rate of oxidation of durohydroquinone (ng atoms O/min/mg protein) | | |
|---|---|---|---|
| | − Uncoupler | + Uncoupler | Δ |
| None | 223 | 345 | 122 |
| KCl (10 mM) | 264 | 416 | 152 |
| KCl (100 mM) | 619 | 1379 | 760 |
| NaCl (100 mM) | 619 | 1480 | 861 |
| Tris-Cl (100 mM) | 219 | 1366 | 1147 |

[a] Heavy beef heart mitochondria were exposed to A23187 (0.5 nmole/mg protein) for 30 min at 30°C in a medium 0.25 M in sucrose, 10 mM in trischloride (pH 7.6), and 1 mM in EDTA. The mitochondria were then washed twice in 0.25 M sucrose, 10 mM in trischloride (pH 7.6), and resuspended in 0.25 M sucrose, 1 mM in trischloride (pH 7.5). In the measurement of respiration by the oxygen electrode method, the substrate was durohydroquinone (1.3 mM) and the uncoupler was mClCCP (2 $\mu$M). The concentration of mitochondrial protein per milliliter was 0.25 mg. The experiment was carried out at 30°C.

(Figs. 5 and 6). Thus it is not the ionophoric capability of the free uncoupler but rather that of the intrinsic electrogenic ionophore–uncoupler combination that is relevant to the mechanism of uncoupling.

The postulate of uncoupler-induced cyclical transport of cations requires that there should be a 1:1 molar relation between the uncoupler and the electron when the cation is monovalent and a 1:2 molar relation when the cation is divalent. [11] We may use as a measure of the electron

*Table II. Restoration of a Normal Uncoupler-Mediated Release of Respiration by Addition of $Mg^{2+}$ or $Ca^{2+}$ to Heavy Beef Heart Mitochondria Depleted of Divalent Metal Cations by Exposure to EDTA [a]*

| Type of mitochondria | Additions | Rate of respiration (ng atoms O/min/mg protein) | | |
|---|---|---|---|---|
| | | Without uncoupler | With uncoupler | Δ |
| Untreated | 0 | 377 | 1640 | 1263 |
| Exposed to EDTA | 0 | 311 | 590 | 279 |
| Exposed to EDTA | 1 mM CaCl₂ | 754 | 3440 | 2686 |
| Exposed to EDTA | 1 mM MgCl₂ | 622 | 2790 | 2168 |

[a] Heavy beef heart mitochondria suspended in a medium 0.25 M in sucrose and 1 mM in trischloride at a final concentration of 10 mg protein/ml were supplemented with the sodium salt of EDTA (3.5 mM) and the pH was adjusted to 7.9. The suspension was incubated for 10 min at 25°C and then cooled on ice before centrifugation. The pellet was resuspended in the original medium (no EDTA added). Durohydroquinone (1.3 mM) was used as substrate and mClCCP (2 $\mu$M) as uncoupler.

*Table III. Bound Ca$^{2+}$ and Mg$^{2+}$ in Different States of Beef Heart Mitochondria$^a$*

| State of mitochondria | Bound divalent metal (nmoles/mg mitochondrial protein) | |
| --- | --- | --- |
| | Ca$^{2+}$ | Mg$^{2+}$ |
| Aggregated | 30 | 40 |
| Orthodox | 0–2 | 3.0 |
| A23187 treated | 0–2 | 4.3 |
| Exposed to EDTA | 19.5 | 5.4 |

$^a$ The mitochondrial suspensions were finally washed in 0.25 M sucrose containing 1 mM trischloride, pH 7.4, before analyzing by atomic absorption by the method of Southard and Green.[28] For details of the preparation of mitochondria in the various states, see Hunter *et al.*[27] on the preparation of orthodox mitochondria.

concentration the concentration of the electron transfer complexes participating in the released respiration in the presence of durohydroquinone as substrate. Since there are two electron transfer complexes involved in this oxidation (complexes III and IV) and each of these complexes is present in beef heart mitochondria at a concentration of 0.21 nmole/mg protein,[13] we would expect that the minimal concentration of the most efficient uncouplers such as SF6847 required for complete release of respiratory control would be 0.42 nmole uncoupler/mg protein when the cation is monovalent and 0.82 nmole/mg protein when the cation is divalent. The data of Fig. 7 are consistent with a 1:2 ratio of electron transfer complex and uncoupler (SF6847), suggesting that in untreated mitochondria, uncouplers induce cyclical transport of divalent cations. The less effective uncouplers such as 2:4-dinitrophenol are required in many times stoichiometric amounts to induce maximal release of respiration.

The mechanism of uncoupler-mediated cyclical transport of cations is a variation of the mechanism formulated in Fig. 4 for cyclical transport of K$^+$ mediated by the valinomycin–nigericin combination. If we represent the neutral electrogenic ionophore as EI and the charged form of the uncoupler as U$^-$, then in cyclical transport of K$^+$ mediated by the combination of EI and U$^-$, K$^+$ is brought into the inner membrane electrogenically as EI-K$^+$ and out of the membrane as the neutral species EI-K$^+$-U$^-$. Similarly, in cyclical transport of Ca$^{2+}$ mediated by the combination of EI and U$^-$, Ca$^{2+}$ is brought into the inner membrane electrogenically as EI-Ca$^{2+}$-U$^-$ and out of the membrane as the neutral species EI-Ca$^{2+}$-(U$^-$)$_2$. Thus uncoupler plays no role in the electrogenic influx of K$^+$, participating only in the nonelectrogenic efflux of K$^+$. However, uncoupler participates in both the electrogenic and nonelectrogenic steps when the cation is divalent. In addition to its role in the cycling of cations, the uncoupler mediates the cycling of protons—entering the inner membrane in its protonated form (UH) and leaving the inner membrane in its charged form (U⁻) with a cation. With each such cycle, a proton is brought in on the electrogenic step and a cation is brought out on the nonelectrogenic step.

The cation requirement established by the data in Tables I and II applies as much to the electrogenic ionophore as to the uncoupler. In fact, it is the electrogenic ionophore that determines which cation can release respiration. Note that the data presented thus far do not prove the presence in mitochondria of electrogenic ionophores that work synergistically with uncouplers. They point to the presence of such ionophores by virtue of the known synergistic action of electrogenic ionophores and uncouplers. In a later section, the direct evidence for electrogenic ionophores in the mitochondrial system will be marshalled.

In cyclical cation transport, cations merely cycle in the membrane phase with no net uptake. Similarly, no gradient of protons is established across the membrane by virtue of the cycling of protons by the uncoupler. Is it possible to demonstrate these cyclical events directly? If the velocities of the electrogenic and nonelectrogenic steps are closely

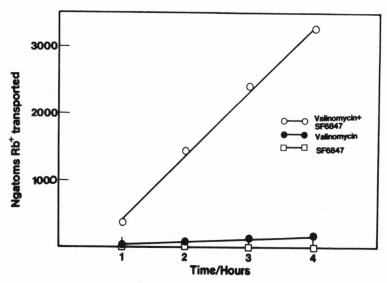

Fig. 5. Synergism of valinomycin and SF6847 in the transport of Rb⁺ in a Pressman cell. The donor aqueous compartment (2 ml) was 5 mM in $^{86}Rb_2SO_4$ and 25 mM in TMA citrate (pH 8.0). The receiver aqueous compartment (2 ml) was 25 mM in TMA citrate (pH 4.0). The organic phase was a 70:30 mixture by volume of chloroform and *n*-butanol. The concentration of valinomycin and SF6847 in the organic phase was 2 mM. The samples taken from the receiver compartment for determination of $^{86}Rb^+$ were 20 μl.

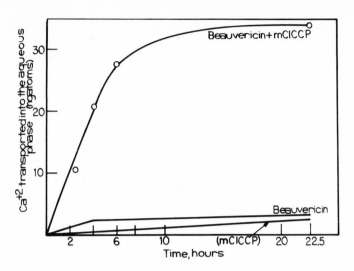

Fig. 6. Cooperative action of beauvericin and uncoupler in mediating transport of $Ca^{2+}$ in a Pressman cell. The donor aqueous compartment (2 ml) was 10 mM in $^{45}CaCl_2$ and 25 mM in TMA Tricine (pH 8.0). The receiver aqueous compartment (2 ml) was 25 mM in TMA citrate (pH 4.0). The organic phase (7 ml) was a 70:30 mixture of chloroform and *n*-butanol and was 2 ml in beauvericin and 0.5 mM in mClCCP when these were added.

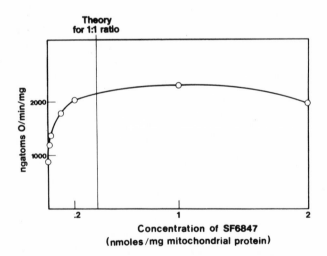

Fig. 7. Molar ratio of uncoupler to electron transfer complex in the release of respiration in A23187-treated mitochondria. Oxygen consumption was measured at 30°C with an oxygen electrode in the presence of durohydroquinone as substrate in a medium 0.25 M in sucrose, 100 mM in tris chloride (pH 7.5), and 1.3 mM in durohydroquinone. The concentration of uncoupler in the mitochondrion is assumed to be equal to the concentration added to the suspension.

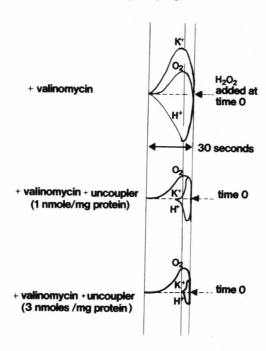

Fig. 8. Protonic and cationic spikes in the presence of uncoupler during pulsing of anaerobic mitochondria with $H_2O_2$. The reaction mixture (8 ml) contained 24 mg beef heart mitochondria, 24 $\mu$g rotenone, 0.4 mg catalase, and 100 nmoles valinomycin. The medium was 0.25 M in sucrose, 1 mM in tris chloride (pH 7.5), and 5 mM in KCl. $H_2O_2$ was added at time 0 and oxygen, pH and $K^+$ changes were recorded simultaneously by the appropriate electrodes. The uncoupler tested was mClCCP at the levels indicated in the figure.

Fig. 9.  Traces showing the ejection of $H^+$ and uptake of $K^+$ by $Ca^{2+}$ state beef heart mitochondria in response to oxygenation in the presence of the substrate durohydroquinone. The medium contained in a final volume of 10 ml: 250 mM sucrose, 4 mM K cacodylate, pH 7.5, 34.2 mg $Ca^{2+}$ state mitochondria, 1.2 mM durohydroquinone, and 0.4 mg catalase. $Ca^{2+}$ state mitochondria were prepared as follows: 2 mg/ml HBHM was incubated at 30°C in 250 mM sucrose–2 mM tris-Cl, pH 7.4, in the presence of 100 nmoles $CaCl_2$/mg

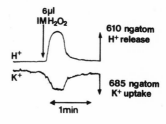

protein. After 15 min, 1 mM EDTA was added and the suspension was cooled for 10 min on ice. Mitochondria were centrifuged for 10 min at 27,000$g$ and resuspended in sucrose–tris-Cl to 2 mg/ml. Centrifugation was repeated and the $Ca^{2+}$ mitochondria were resuspended in sucrose tris-Cl to 50 mg/ml.

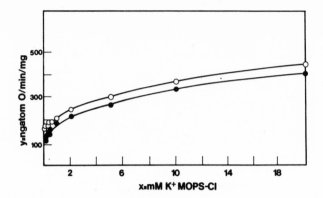

Fig. 10. Dependence of rate of respiration of $Ca^{2+}$ mitochondria on KCl concentration. The medium contained in a final volume of 4 ml: 250 mM sucrose, 0.1 M K MOPS, pH 7.0, KCl as required, 2 mg $Ca^{2+}$ beef heart mitochondria, and 0.75 mM durohydroquinone. 1 $\mu$M mClCCP was added after 25–30% depletion of oxygen. ●, Without uncoupler; ○, with uncoupler.

matched, there would be at best a flicker for the protonic and cationic changes during an oxygen pulse in an anaerobic mitochondrial system containing uncoupler. But if there is an inbalance of the respective velocities of the two steps, we should see a spike for both the protonic and cationic changes, and indeed such a spike is demonstrable (Fig. 8). This has turned out to be a powerful technique for the direct demonstration of uncoupler-induced cyclical transport in mitochondria.

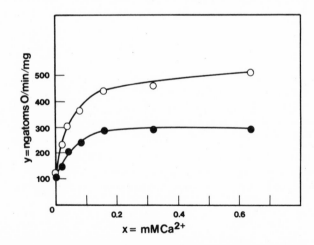

Fig. 11. Dependence of rate of respiration of $Ca^{2+}$ mitochondria on $CaCl_2$ concentration. Details as in the caption of Fig. 10.

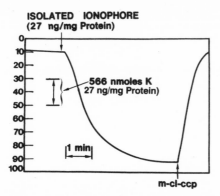

Fig. 12. Uncoupler-mediated efflux of accumulated $K^+$. The uncoupler (mClCCP) was added after $K^+$ had been maximally accumulated by beef heart mitochondria in the presence of the neutral $K^+/Ca^{2+}$ ionophore (the electrogenic ionophore intrinsic to the mitochondrion which has been extracted from the mitochondrion and purified). The suspending medium was 4 mM in potassium acetate, 200 mM in mannitol, 10 mM in tris acetate and tris glutamate. The concentration of mitochondria was 2.14 mg/ml. The temperature was 30°C.

Mitochondria in the $Ca^{2+}$ state show uncoupled respiration that is insensitive to the action of uncoupler since mitochondria in this state contain intrinsic uncoupler. The oxygen pulse technique[14] has been used successfully to demonstrate cyclical cation transport in mitochondria in the $Ca^{2+}$ state (Fig. 9). Moreover, the enhancement of this uncoupled respiration by addition of cations (monovalent of divalent) to the suspending medium is also demonstrable (Figs. 10 and 11).

Another technique that demonstrates not only the cyclical character of uncoupler action but also the synergism of electrogenic ionophore and uncoupler is illustrated in Fig. 12. Mitochondria supplemented with a valinomycin-like ionophore isolated from mitochondria take up $K^+$ under energizing conditions at a given rate. When uncoupler is added, the accumulated $K^+$ is released at about 20 times the inflow rate. The same experiment can be carried out without the ionophore by inducing the intrinsic electrogenic ionophore for $K^+$ by addition of fluorescein mercuric acetate (FMA).[15] When $K^+$ is accumulated by such an FMA-treated mitochondrial suspension, the addition of uncoupler leads to the same dramatically enhanced efflux rate for $K^+$.

Now that we have assembled and exhibited the essential data bearing on the molecular mode of action of uncoupler, let us consider what the implications of these findings are for the mechanism of coupling. These data effectively dispose of the notion that there are uncoupled particles. The hallmark of the so-called uncoupled particle is uncoupled respiration, and this has now been shown to be cyclical cation transport—an eminently coupled process. If coupling is essential for electron flow, then the imperative of pairing in charge flow has to be accepted.

Uncoupling has been shown to depend on a synergistic or collaborative interaction between uncoupler and electrogenic ionophore. Since all coupled processes are affected by uncoupler,[11] it is reasonable to presume that a combination of electrogenic ionophore and uncoupler preempts all coupled processes.

A third important implication of the uncoupler mechanism is that it defines the nature of respiratory control. Electrons will not flow unless this flow is paired to the parallel flow of a positively charged ionophoric species. If pairing is impossible for lack of an ionophore or denied for lack of a suitable cation or is abortive for lack of an uncoupler, then respiratory control is imposed. The phenomenon of the release of respiration via uncoupling is in fact testimony to the imperative of paired moving charges.

It should be mentioned that hydrolysis of ATP can substitute for electron flow in all the phenomena we have examined. We can presume that hydrolysis of ATP generates charged ionophoric species that duplicate the driving action of the electron in coupled reactions. We shall have more to say later about the cation–anion ionophoric species that intervene in the coupled synthesis and hydrolysis of ATP.

## Coupling in Cytochrome Oxidase

Pairing of an exergonic and endergonic center is a precondition for energy coupling according to the PMC model. How would one account, then, for the experimental observation that electron transfer complexes can be isolated in a highly purified state with full retention of electron transfer activity? Presumably these purified complexes are devoid of endergonic centers, and yet they still carry out electron transfer. Is this not direct evidence that electron transfer can take place *without* pairing of centers? It was a prediction of the PMC model that the isolated complexes were duplexes of an electron transfer complex (ETC) and an ionophore-containing, ion transfer complex (ITC); that the duplex should be resolvable into its component complexes; that resolution should lead to the loss of activity of the ETC; and finally that activity should be reconstituted by adding back the ITC. Kessler *et al.*[16] and Vande Zande *et al.*[17] have fully confirmed this prediction. When duplex cytochrome oxidase (*a* heme concentration of 10–12 nmoles/mg protein) is dissolved in a 2:8 mixture of methanol and *t*-butyl alcohol (acidified with 0.05 M HCl), it can be separated into an intensely colored green fraction (*a* heme concentration of 23 nmoles/mg protein) and a colorless fraction. The electron transfer components (*a* heme and copper) are exclusively in the green fraction. The colorless fraction accounts for about 50% of the total protein and contains ion transfer components.

The unresolved ETC shows cyclical cation transport when tested by the oxygen pulse technique [14, 16] as outlined in the legend of Fig. 8. The resolved ETC is inactive in this respect unless supplemented with a suitable ionophore combination such as lipid *c* (Table IV).

The ITC derived from duplex cytochrome oxidase contains a $K^+$/$Ca^{2+}$ ionophoroprotein and an intrinsic uncoupler identified as hydroxyoctadecadienoate. This uncoupler, first described by Blondin, [18] has been shown to be bound to protein in ester link via the hydroxyl group. Uncoupled respiration in cytochrome oxidase involving cyclical cation transport is induced by this combination of ionophore and uncoupler localized in the ITC.

Unresolved cytochrome oxidase can show respiratory control under appropriate conditions. [19, 20] How can this property be reconciled with the thesis that cytochrome oxidase is a duplex of an exergonic and endergonic center? A duplex system should not show respiratory control since all the requirements for coupling are satisfied. Kessler *et al.* [16] have provided the answer. The activity of the ITC can be suppressed by incorporating cytochrome oxidase into liposomes containing mostly acidic phospholipids. Such liposomes form minivesicles with high curvature; the activity of the ITC is suppressed when the curvature of the vesicle exceeds a critical value, and respiratory control is, thereby, imposed. High curvature appears to suppress the coupling capability of the ITC.

The resolution of cytochrome oxidase into ETC and ITC thus provides direct and conclusive evidence that coupling requires the interaction of an exergonic (ETC) and an endergonic (ITC) center; that

*Table IV. Activity of Resolved vs. Unresolved Preparations of Cytochrome Oxidase* [a]

| Assay conditions | | | Activity (nmoles O/min/nm heme *a*) | |
|---|---|---|---|---|
| | | | Resolved | Unresolved |
| Low *c* | High *c* | Asolectin | (15.9 nmoles heme *a*/mg) | (9.1 nmoles heme *a*/mg) |
| + | | + | 140 | 245 |
| | + | + | 295 | 365 |
| + | | — | — | 395 |

[a] Activities measured at 38°C in an oxygen electrode in a medium 2.5 mM in $K^+$ phosphate (pH 7.4), 13 mM in durohydroquinone, 0.2 mM in cytochrome *c* (low) or 20 nM in cytochrome *c* (high), and 25 mM in tetramethyl phenylene diamine. The concentration of cytochrome oxidase was 30 $\mu$g/ml and of asolectin 80 $\mu$g/ml.

the ITC provides the positively charged ionophoric species required for coupling with the electron of the ETC; and finally that an intrinsic uncoupler present in the ITC can mediate cyclical cation transport—the basis of "uncoupled" respiration.

## *Ionophore Strategy in Mitochondrial Energy Coupling*

Mitochondria carry out a variety of coupled processes (coupled ATP synthesis and hydrolysis, energized transhydrogenation, reverse electron flow, energized contraction, and active transport of cations). In addition, mitochondria contain carrier systems for the transport of phosphate (pi), adenosine diphosphate (ADP) and citric cycle intermediates across the inner membrane. The PMC model requires that all these coupled processes, as well as the carrier mechanisms, should be mediated by ionophores. Such mediation would entail a major commitment of the mitochondrion to an ionophore technology. Is there evidence for such a commitment? We have already described the ionophoric equipment of duplex electron transfer complexes such as cytochrome oxidase. About 50% of the total protein is associated with the ion transfer centers intrinsic to such complexes. These proteins include the ionophoroproteins and the structural proteins that form the internal cavity in which ion transfer takes place.

In addition to the ionophoroproteins associated directly with the electron transfer complexes, there is a major set of ionophoroproteins that is intrinsic to the tripartite repeating unit (TRU)—the unit that is responsible for all the coupling functions of the mitochondrion except for cyclical cation transport. In the category of ionophoroproteins, we include both the structural proteins that are required to translate ionophore-mediated transfer reactions and the functional proteins that are ionophores.

The methodology for isolating, purifying, and characterizing ionophoroproteins is now highly developed.[21] The one-by-one isolation of the multiple mitochondrial ionophoroproteins is well under way at the present time. Among the tools and techniques most effective in this program are high-pressure chromatography, the cation- and ionophore-dependent increase in fluorescence of anilinonaphthalene sulfonic acid used in the routine monitoring of electrogenic ionophores, two-phase systems for ionophore-dependent partition of cations, and the three-phase Pressman cell for measuring net ion transport.

The ionophoric capability is detectable by a variety of methods and techniques—mediation of the partition of cations or anions from an aqueous to an organic phase,[21] cation-dependent enhancement of the fluorescence of anilinonaphthalene sulfonate,[16] discharge of a membrane potential established in a black lipid film,[22] and the release of cations from the interior space of liposomal vesicles.[23] Blondin *et al.*[21]

Table V. *Amino Acid Composition of Mitochondrial Electrogenic Ionophore for Monovalent and Divalent Cations*[a]

| Amino acid | Number of residues |
| --- | --- |
| Serine | 1 |
| Proline | 1 |
| Isoleucine | 2 |
| Leucine | 3 |
| Valine | 2 |
| Alanine | 4 |
| Threonine | 1 |
| Glycine | 2 |
| Glutamic Acid | 1 |

[a] The ionophoropeptide has a minimal molecular weight of 1600 and contains 17 amino acid residues. Data of Blondin *et al.*[21]

have isolated a peptidyl ionophore (molecular weight 1600) by proteolysis of a purified ionophoroprotein of beef-heart mitochondria. This ionophore has $K^+$ transport activity comparable to that of valinomycin (see Fig. 12). The amino acid composition of this ionophore is shown in Table V. This ionophore is isolatable from cytochrome oxidase.

Undoubtedly, the most dramatic evidence that mitochondrial ionophoroproteins are intimately tied into energy coupling is provided by the inhibition of ionophoroprotein-mediated processes by reagents that selectively suppress mitochondrial energy coupling. DCDD, a specific inhibitor of oxidative phosphorylation,[24] suppresses the binding of ADP at neutral pH by an ionophoroprotein from beef heart mitochondria.[25] Oligomycin, another specific inhibitor of oxidative phosphorylation,[26] suppresses the transport of $K^+$ and protons mediated by an ionophoroprotein of yeast or heart mitochondria.[22, 27] Ruthenium red, a reagent that specifically inhibits high-affinity active transport of $Ca^{2+}$,[28] suppresses the $Ca^{2+}$ ionophoric activity of an ionophoroprotein isolated from beef heart mitochondria (measured electrometrically in a black lipid film apparatus).[29] Atractylate,[30] a specific inhibitor of the ADP/ATP exchange system, prevents the ionophoroprotein-mediated partition of ADP between an aqueous and an organic phase.[31] These observations leave no room for doubt that energy coupling is an ionophoroprotein-mediated process.

## Mechanism of Oxidative Phosphorylation

The acid test for a model purporting to rationalize mitochondrial energy coupling is that it should lead inexorably to a detailed formulation of the mechanism of oxidative phosphorylation that satisfies all the experimental specifications and predicts the logistical tactics that underlie energy transfer. First a set of key experimental observations had to be selected that could provide the foundation on which to build such a formulation. There were five such observations. (a) Roy and Moudrianakis[32, 33] showed that in chloroplasts ATP is not formed by the direct union of ADP and Pi. Two molecules of ADP interact to form ATP and AMP by transphosphorylation. (b) Criddle *et al.*[34] discovered that one of the subunits of the TRU of yeast mitochondria contains bound pantotheine and that pantotheine can be acylated by oleyl phosphate.[35] (c) The Pi- and ADP-binding ionophoroproteins are $Mg^{2+}$-containing and DCCD-sensitive.[25] (d) Vignais *et al.*[36] implicated the ADP/ATP carrier as an integral component of the ATP synthesizing system in the sense that ADP is delivered directly from the carrier to the enzymic system. (e) Several investigators[37, 38] and notably Meisner[39] have shown that monovalent cations, as well as $Mg^{2+}$, are absolute requirements for oxidative phosphorylation.

Figure 13A, B is a formulation of the mechanism of coupled ATP synthesis by Green and Blondin.[40] This formulation embodies the correct stoichiometry of the reactants and products, accounts for the proton balance in the overall reaction, accounts for the $Pi/H_2O$ and the ADP/ATP exchange reactions,[41] and explains the sensitivity of coupled ATP synthesis in mitochondria to atractylate[30] and N-ethyl maleimide (inhibitor of the Pi carrier).[42] Some important novel features of the formulation are noteworthy: protonation of $Pi^{2-}$ and $ADP^{2-}$ as a means of reducing the activation barrier to bond formation; the carrier strategy by which the synchronization of the incoming anion and the outgoing anion maintain the charge compensation for fixed $Mg^{2+}$; and the matrix strategy by which the anions are taken into the membrane phase only from the matrix side of the inner membrane.

In broad principle the formulation is probably close to the mark, but there are undoubtedly many details yet to be filled in. At least now one can visualize the mechanism in realistic terms. The aura of mystery has finally been dispelled.

The same detailed formulation has been extended to coupled ATP hydrolysis, energized transhydrogenation, energized, contraction, and reverse electron flow. As we anticipated, all coupled processes are variations of one theme.

*Control of the Mitochondrial Coupling Function*

The mitochondrial control system is the first to be studied in depth, and it provides accordingly the first opportunity to relate the strategy of the control mechanism to the principle of energy coupling. Mitochondria of animal cells can exist in two states—the N and $Ca^{2+}$ states.[43-48] The properties in these two states are summarized in Table VI. The conversion of mitochondria from the N to $Ca^{2+}$ state depends on a transition mediated by the control mechanism. [43–45]

The end result of the transition is the substitution of one coupling pattern by another. In the N state all the coupled processes are mediated by the TRU; in the $Ca^{2+}$ state all the coupled processes are mediated by the ion transfer complexes associated with the ETCs. There is thus a shift from coupling of the electron to the positively charged ionophoric species generated in the TRU to coupling of the electron to the positively charged ionophoric species generated in the ion transfer complexes associated with the ETC.

The essence of the control maneuver is the in-and-out movement of the headpiece stalk—into the membrane in the N state and out of the membrane in the $Ca^{2+}$ state. When in the membrane, the ionophoroproteins of the TRU are engaged; when out of the membrane, the ionophoroproteins associated with the electron transfer complexes are engaged. Thus a shift from one coupling pattern to another is accomplished by this in-and-out movement of the headpiece stalk.

The N $\rightarrow$ $Ca^{2+}$ state transition is a two-step process. In the first step (Fig. 14A), the electrostatic link between the matrix system network of proteins and the headpiece stalk is ruptured and the headpiece stalk is now free to be ejected from the membrane. The electrostatic link involves two nucleotides (ADP and NADH) and a divalent cation ($Mg^{2+}$). [45] In the second step (Fig. 14B), the headpiece stalk undergoes a $Ca^{2+}$-induced change in state that compels its explusion from the membrane. [44] The $Ca^{2+}$ state of the headpiece stalk is relatively polar (in this state the headpiece stalk is stabilized by moving into the aqueous phase); the $Mg^{2+}$ state of the headpiece stalk is relatively nonpolar (in this state the headpiece stalk is stabilized by moving into the membrane).

The experimental evidence that underlies the molecular interpretation of the N $\rightarrow$ $Ca^{2+}$ state transition may be summarized as follows. Reagents that induce the conversion of bound ADP to AMP (arsenate) or the conversion of bound NADH to $NAD^+$ (uncouplers, fatty acids, rotenone) destabilize the electrostatic links between the matrix and the headpiece stalk and thereby facilitate the onset of the N $\rightarrow$ $Ca^{2+}$ state transition. [43] In the $Ca^{2+}$ state there is no electrostatic link between the matrix system and the inner membrane. By controlling the level of $Ca^{2+}$ in the suspending medium with EGTA, the mitochondrion can be shifted from the $Ca^{2+}$ state (above 15 $\mu$M $Ca^{2+}$) to the N state (below 15 $\mu$M $Ca^{2+}$) [44]; $Mg^{2+}$ and ADP have comparable action to that of reducing the level of $Ca^{2+}$. [44] Finally, the N $\rightarrow$ $Ca^{2+}$ state transition can be prevented by exposing mitochondria to reagents such as DCCD, *N*-ethylmaleimide, and mercurials. These reagents modify the headpiece stalk to the point that it cannot undergo the transition leading to its expulsion from the membrane. [44] There is a vast body of experimental

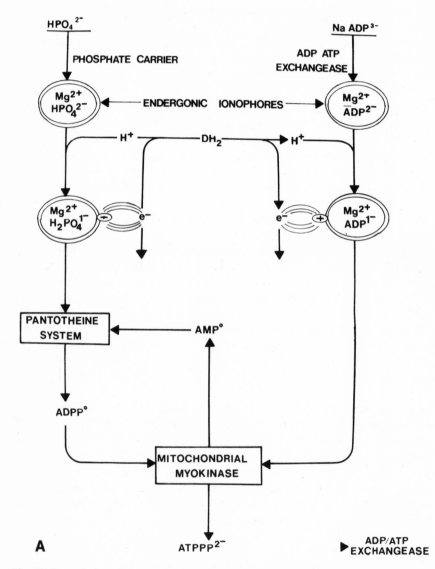

Fig. 13. Formulation of the mechanism of coupled ATP synthesis according to the PMC model. A: Précis of the formulation. B: The detailed formulation. Formulation of Green and Blondin.[40]

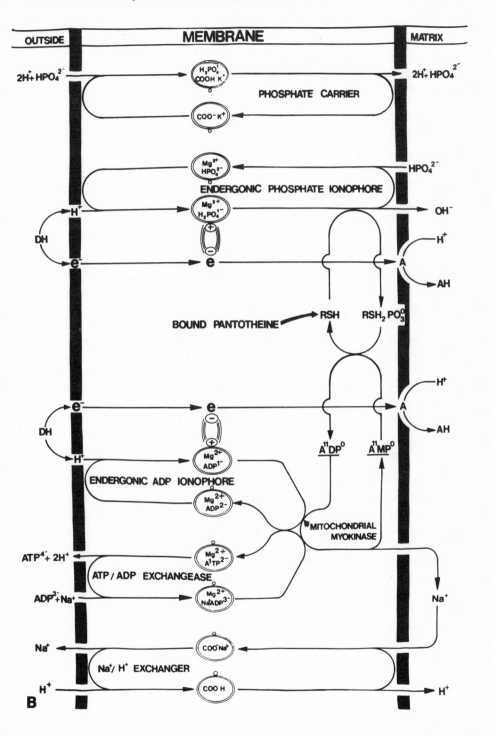

*Table VI. Properties of Mitochondria in the N and Ca²⁺ State*

|                             | N⁺ state   | Ca²⁺ state |
|-----------------------------|------------|------------|
| Oxidative phosphorylation   | +          | 0          |
| Coupled transhydrogenation  | +          | 0          |
| Active transport of cations | +          | 0          |
| Cyclical transport of cations | 0        | +          |
| Respiratory control         | +          | 0          |
| Permeability to sucrose     | 0          | +          |
| Response to uncoupler       | +          | 0[a]       |
| Configuration               | Aggregated | Orthodox   |

[a] No response at levels of Ca²⁺ above 15 μM in the medium. Data from Hunter *et al.*,[45] Hunter and Haworth,[43] and Haworth and Hunter.[44]

observations that can now be rationalized within the framework of the formulation of the control mechanism shown in Fig. 14A and B.

The control mechanism has revealed an entirely new and unexpected dimension of energy coupling. The stalk appears to be the shell of the hollow channel through which all ionophore flow in the TRU is directed. Not only the ionophoroproteins that mediate the electrogenic ion flow, but also the ionophoroproteins that mediate the nonelectrogenic ion flow of ADP, ATP, Pi, etc., circulate in this common channel. It would thus appear that the TRU is a composite system in which exergonic and endergonic centers are merged by virtue of the sharing of a common channel for transmembrane ion flow.

*Structure of the Transducing Assembly*

It is obvious that there has to be an intimate correlation between the principle of energy coupling and the structures in an energy coupling membrane that translate this principle. If the principle is the Mitchellian principle of indirect coupling via a membrane potential and a proton gradient,[49] the arrangement of the transducing units in the membrane would have to reflect and translate that principle. If the principle is that underlying the paired moving charge model, the structure of the transducing units would necessarily have an entirely different character from that appropriate for the Mitchellian principle. The structural realities can thus be highly informative with respect to the operative principle of energy coupling.

The paired moving charge model requires the following structural features. The exergonic and endergonic reaction centers must be vectorial (at right angles to the plane of the membrane), tightly apposed to satisfy the limits of proximity for maximation of electrostatic interac-

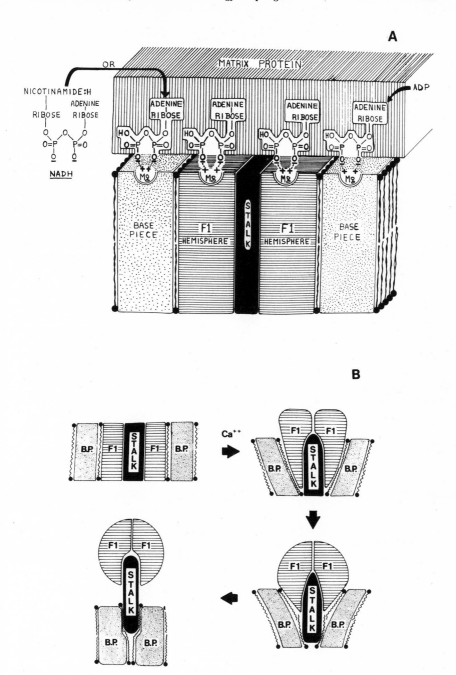

Fig. 14. A molecular interpretation and model of the N → Ca²⁺ state transition. A: Protective system. B: Transition sequence. Formulation of Green and Blondin.[40]

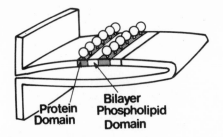

Fig. 15. The inner mitochondrial membrane as a composite of protein and bilayer phospholipid domains. The protein domain would correspond to the ribbon structure.

tion, and so arranged as to make possible the alternative modes of energy coupling inherent in the N and $Ca^{2+}$ states of the inner membrane. These are the three basic structural requirements of the paired moving charge model.

The transducing assembly of the inner mitochondrial membrane has the ribbon structure shown in Fig. 15. The ribbon is a continuum of repeating units—each such unit contains the electron transfer complexes, the ion transfer complexes, and the TRU in the molar proportions and in the arrangement shown in Fig. 16. The experimental justification for this formulation of the transducing assembly has been presented by Haworth *et al.*[50] It will be noted that the structure of the transducing assembly is compatible with the three requirements of the paired moving charge model,—vectorality of the complexes, sufficiently tight association of the exergonic and endergonic centers to satisfy the

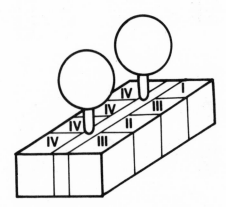

Fig. 16. Molar relations of the four complexes of the electron transfer chain (I, II, III, and IV) to the TRU in a unit segment of the ribbon structure of the mitochondrial inner membrane. Formulation of Haworth *et al.*[50]

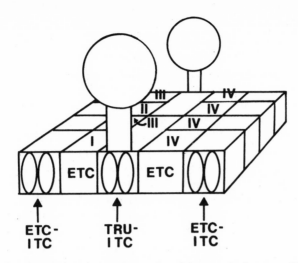

Fig. 17. Alternative coupling modes of the transducing assembly.

limits of proximity, and the possibility of alternative coupling modes (Fig. 17). It should be noted that these structural features of the transducing assembly would have no rationale in terms of the Mitchell model.

The ribbon structure of the mitochondrial transducing system has enormous implications for current theories of membrane structure. The Davson–Danielli model,[53] as well as its latter-day variant, the Singer–Nicholson model,[54] emphasizes the fluid nature of biological membranes and deemphasizes the contribution of intrinsic proteins to the structure of membranes. The necessities of energy coupling require a protein domain in the membrane in which energy coupling can be consummated.[51] The ribbon structure is an expression of the protein domain. The elaborate mechanism by which mitochondrial energy coupling is controlled is yet another manifestation of the protein domain and its relevance to energy coupling. Since energy coupling is an attribute of all biological membranes, it follows that the protein domain has to be a universal attribute of these membranes.

### Enzymic Catalysis and Energy Coupling

The relation between enzymic catalysis and energy coupling has long been an enigma. Until the principle common to enzymic catalysis and energy coupling could be specified, the relation had to remain an enigma. It is now possible to define this relation with some precision, and we can appreciate for the first time the identities and differences between enzymic catalysis and energy coupling. Some very unexpected and dramatic new insights have been gained as a result of exploring this relation.

All chemical reactions in biological systems involving covalent bond formation or rupture are catalyzed by enzymes. There are a few minor exceptions, such as the dissociation of carboxylic acids or the ionization of water, that are not enzyme-mediated. Since covalent bonds are made and broken in every energy coupled reaction, be it electron transfer or synthesis of ATP, it necessarily follows that enzymic catalysis is intrinsic to energy coupling. Then what are the differences between energy coupling and classical enzymic catalysis? Figure 18 diagrammatically defines two of these differences for an electron transfer reaction. A classical enzyme catalyzes the transfer of a hydrogen atom (actually an electron and a proton) from a donor substrate (AH) to an acceptor substrate (B). One enzyme accomplishes the overall catalysis. Energy coupling by contrast would involve the transfer of an electron from AH to B. This transfer would require the enzyme-mediated dehydrogenation of AH and the enzyme-mediated hydrogenation of B. Two enzymes would be needed for this transfer of an electron and for the overall catalysis. In addition, the movement of the electron from A to B would have to be coupled to the movement of a positive charge in an associated endergonic center. The switch from H transfer (transfer of a positive and negative charge simultaneously) to electron transfer (transfer only of a negative charge) is the essential point of difference between classical catalysis and energy coupling. When a single charge is transferred, two enzymes are required, the proton has to be jettisoned at one side of a membrane and taken up at the other, and an electron chain for moving the electron from $S_1$ to $S_2$ has to be provided.

We can conclude from this simple description that in energy coupling, enzymes specialized for transfer of a single charged species have to intervene and that the bond-breaking enzyme is separate and separated from the bond-making enzyme. Does that mean that energy coupling enzymes operate by an entirely different principle than is applicable to classical enzymes? There are in fact several examples of energy coupling enzymes that can be assayed as classical enzymes (succinate dehydrogenase,[54] NADH dehydrogenase,[55] transhydrogenase,[56] and $F_1$ ATPase[57]). In the following section we shall consider soluble energy coupling enzymes, such as xanthine oxidase, and some mixed function oxidases that are unambiguously enzymic and yet are executing energy coupled reactions. We finally perceived that there are two fundamentally different modalities of enzymic catalysis — one mediated by individual classical enzymes, and the other mediated by

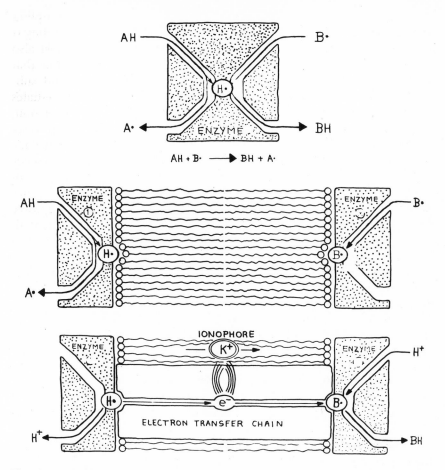

Fig. 18. A dehydrogenation reaction catalyzed by a classical enzyme (top drawing) and an energy coupling enzyme (two lower drawings). Formulation of Green and Blondin.[40]

a *system* of two or more energy coupling enzymes. Classical enzymes induce catalysis via a fixed pair of charges; energy coupling enzymes induce catalysis via a *moving* pair of charges. The technique of mobile charges requires specialized molecular devices (charge and ion transfer chains, internal cavities for ion transfer, ionophoroproteins). The involvement of ionophores provides one of the key diagnostics of an energy coupling system. The principle of paired charge separation applies to both modalities of enzymic catalysis.

As soon as one recognizes that enzymic catalysis is an intrinsic element in energy coupling, then it necessarily follows that the principle of coupling has to be identical for the two processes. Energy coupling and enzymic catalysis are two sides of the same coin. If direct coupling is the characteristic of enzymic catalysis, then direct coupling must also apply to energy coupling. There is now strong evidence available that enzymic catalysis involves direct coupling between enzyme and substrate,[58-60] the token of which is the formation of covalent intermediates between enzyme and groups in the substrate as in transmethylation, transhydrogenation, transamination, and transphosphorylation reactions.[61] It would take us too far afield to pursue this theme, but enzymic catalysis can be formulated in precisely the same terms as energy coupling (Fig. 19). These terms are paired-charge separation, paired-charge substitution, and paired-charge elimination. If it is not necessary to invoke membrane potentials, or their equivalent, as the driving force in enzymic catalysis, then it necessarily follows that membrane potentials cannot be the driving force in energy coupling since the same principle must apply to both processes.

In classical catalysis the molecular instrument of energy transfer is the enzyme protein. In energy coupling, it is the moving ion (the electron in the chain or $Pi^{3-}$ in a $Mg^{2+}$-containing ionophoroprotein) that is the molecular instrument of energy transfer. The enzymes are concerned only with the formation and utilization of these charged species.

We have considered thus far only membrane-centered energy coupling systems. But energy coupling is possible not only in a mem-

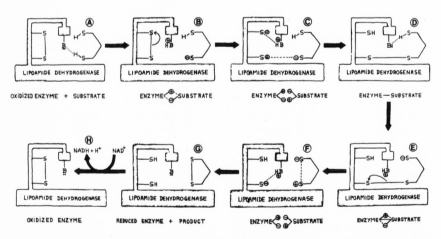

Fig. 19. Formulation of the mechanism of action of lipoamide dehydrogenase. The formulation in terms of the pairing principle corresponds[40] closely to the original formulation of Matthews *et al.*[60]

brane system but also in a filament system, such as actomyosin, and in a soluble enzyme such as xanthine oxidase or DNA polymerase. As long as charges can be separated into two phases (the protein and the aqueous phases) the protein can be part of a membrane or a filament or a soluble complex. Energy coupling is a tactic for energy transfer that is a variation on the theme of classical enzymic catalysis. It is not the exclusive prerogative of membrane systems.

## Unity of Bioenergetics

A strong case has been made for the thesis that the principle underlying energy coupling has to be universal in its application to bioenergetic phenomena to be valid.[62-64] Validation of the principle for mitochondrial energy coupling can only be the starting point for the more stringent test of universality. We have already considered the extension of the pairing principle to enzymic catalysis and concluded that enzymic catalysis and energy coupling are two expressions of the same fundamental principle. But the extension of the pairing principle has actually gone much further. Given an energy coupling system whose properties and specifications are well documented, it should be possible to formulate the mechanism of these systems with the same detail as was introduced in the formulation of coupled ATP synthesis shown in Fig. 13. These formulations must be consistent with a considerable number of experimentally determined specifications, and they must be invariant with respect to the pairing principle. Table VII is a listing of all such formulations that we have made. Three examples of such formulations are given in Fig. 20–22, for muscle contraction (Fig. 20), for the $Na^+/K^+$ ATPase (Fig. 21), and for mixed-function oxidases (Fig. 22). The formulations may be incorrect in some details, but the frameworks for the respective mechanisms must approximate the correct solutions in view of the fidelity with which the experimental specifications are met.

On the basis of these formulations, we are confident that the pairing principle is indeed the principle underlying all bioenergetic phenomena and that an important milestone in the history of bioenergetics has finally been reached.

## Models of Energy Coupling

The chemiosmotic model of energy coupling dominates thinking in bioenergetics at the present time and is accepted by a majority of investigators. If we ask the question—"What contribution has this model made to our understanding of the mechanism of uncoupling, the

Table VII. Energy-Coupled Processes Formulated According to the PMC Model without
Violation of the Experimental Specifications

| System | Coupled process |
|---|---|
| Mitochondria | ATP synthesis and hydrolysis |
| | Reverse electron flow |
| | Active and cyclical transport of cations |
| | Energized contraction and transhydrogenation |
| Sarcoplasmic reticulum | Active transport of $Ca^{2+}$ |
| Plasma membrane | $Na^+/K^+$ ATPase |
| | Adenyl cyclase |
| Actomyosin | Muscle contraction |
| Red blood cell | Spectrin-plasma membrane translation |
| Nerve membrane | Neural transmission[a] |
| Ribosome | Protein synthesis |
| Polymerases | DNA and RNA synthesis |
| Mitochondria, microsomes, soluble enzymes | Mixed-function oxidases |
| Xanthine oxidase | Oxidation of purines, aldehydes, NADH |
| Aldehyde oxidase | Oxidation of aldehydes, quinine |
| Chloroplast | Oxygen evolution |
| | Photosynthetic phosphorylation |
| Nitrogenase | Conversion of $N_2$ to $NH_3$ |
| Hydrogenase | Conversion of $H^+$ to $H_2$ |
| Bacterial membrane | Phosphoenolpyruvate-driven transport of glucose |
| | Active transport of amino acids |

[a] Neural transmission involves a variation of the energy-coupling principle. The formulations listed in the table will appear in a forthcoming monograph of Green and Blondin.[40] Figures 13, 20, 21, and 22 are representative examples of the detail in the formulations.

mechanism of coupling in electron transfer complexes, the structure of the transducing complexes, the mechanism for control of energy coupling, the mechanism of coupled ATP synthesis, the mechanism of enzymic catalysis, and energy coupling systems generally?"—the answer is zero or close to zero. If a model is based on an invalid principle (unpaired-charge separation), it has to be wrong across the board, and we are of the opinion that this is indeed the case. This is not to say that the chemiosmotic model is without merit. We have borrowed two fundamental concepts from this model—vectorial ion flow and the notion of charge separation and charge elimination as primary ingredients in all catalysis. These two concepts were the kingpins in the PMC model. We have only added the pairing principle to the model of energy coupling, and this has made the difference. Peter Mitchells' contribution to bioenergetics dwarfs all other contributions because the vectorial and charge-separating dimensions are the foundation stones for rational

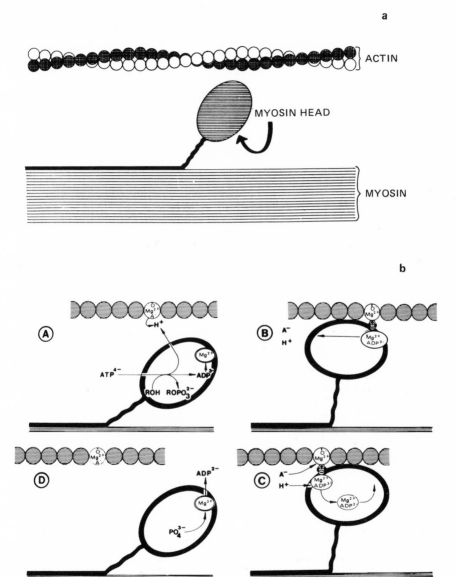

Fig. 20. Formulation of the mechanism of muscle contraction a: The three elements of the muscle contraction system. b: Coupling of the movement of the positively charged actin filament to the movement of $(Mg^{2+} \; ADP^{-3})^{-1}$ and $(Mg^{2+} \; Pi^{-3})^{-1}$ in the myosin head. Formulation of Green and Blondin.[40]

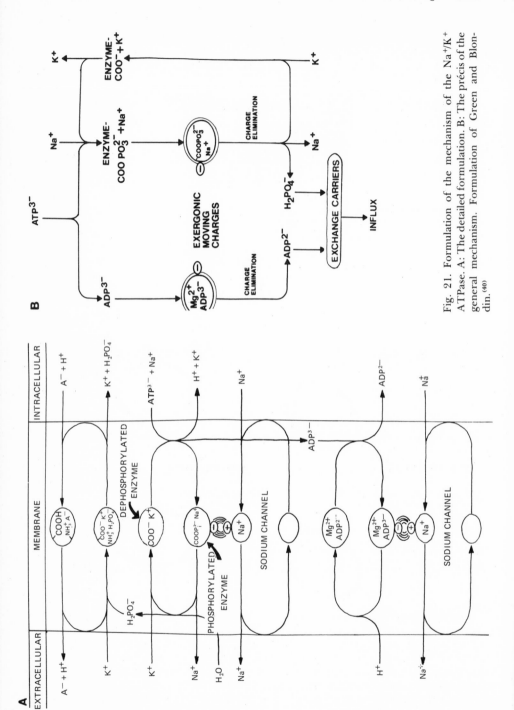

Fig. 21. Formulation of the mechanism of the Na$^+$/K$^+$ ATPase. A: The detailed formulation. B: The précis of the general mechanism. Formulation of Green and Blondin. [40]

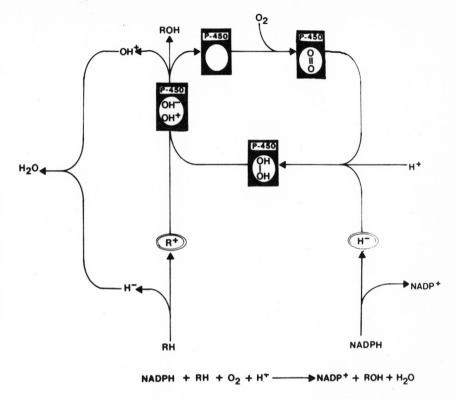

$$NADPH + RH + O_2 + H^+ \longrightarrow NADP^+ + ROH + H_2O$$

Fig. 22. Formulation of the mechanism of a mixed-function oxidase. RH represents the hydroxylatable molecule and ROH the hydroxylated molecule. Formulation of Green and Blondin.[40]

thinking about energy coupling. The reader will undoubtedly ask how, if the chemiosmotic model is based on an invalid principle, this invalidity could have been missed? The answer we think, is that the recognition of the obvious is the most difficult task in science. The recognition of the obviously wrong is equally difficult.

### Addendum

The $H^+$:e ratio cannot be greater than one if H is derived by oxidation of substrate. The reports of ratios greater than one must mean, therefore, that the extra protons are derived by means other than the oxidation of substrate. The ratio that is relevant to our model refers only to H derived from substrate.

## References

1. Blondin, G. A., and Green, D. E. (1975) *Chem. Eng. News Special Report*, Nov. 10, pp. 26–42.
2. Ovichinnikov, Y. A., Ivanov, V. T., and Shkrob, A. M. (1976) in *Membrane Active Complexones*, B. B. A. Library, No. 12, Elsevier, New York.
3. Pressman, B. C. (1973) *Fed. Proc. 32*, 1698–1703.
4. Pressman, B. C. (1970) in *Membranes of Mitochondria and Chloroplasts* (Racker, E., ed.), pp. 213–244, Van Nostrand Reinhold, New York.
5. Lardy, H. A., Graven, S. N., and Estrado, O. S. (1967) *Fed. Proc. 26*, 1355–1360.
6. Eisenman, G., Szabo, G., Ciani, S., McLaughlin, S., and Krasne, S. (1973) in *Progress in Surface and Membrane Science*, Vol. 6, pp. 140–236, Academic Press, New York.
7. Green, D. E., and Reible, S. (1974) *Proc. Natl. Acad. Sci. U.S.A. 71*, 4850–4854.
8. Kassner, R. J. (1972) *Proc Natl. Acad. Sci. USA 69*, 2263–2267; (1973) *J. Am. Chem. Soc. 95*, 2674–2677.
9. Parker, V. H. (1965) *Biochem. J. 97*, 658–662.
10. Montal, M., Chance, T. and Lee, P.-P. (1970) *J. Membr. Biol. 2*, 201–234.
11. Kessler, R. J., Vande Zande, H., Tyson, C., Blondin, G. A., Fairfield, J., Glasser, P., and Green, D. E. (1977) *Proc. Natl. Acad. Sci. USA 74*, 2241–2245.
12. Kessler, R. J., Tyson, C. A., and Green, D. E. (1976) *Proc. Natl. Acad. Sci. USA 73*, 3141–3145.
13. Slater, E. C. (1971) in *Dynamics of Energy Transducing Membranes*, pp. 1–20, B. B. A. Library, Vol. 13, Elsevier, New York.
14. Haworth, R. A., unpublished studies.
15. Southard, J. H., Blondin, G. A., and Green, D. E. (1974) *J. Biol. Chem. 249*, 678–681.
16. Kessler, R. J., Blondin, G. A., Vande Zande, H., Haworth, R. A., and Green, D. E. (1977) *Proc. Natl. Acad. Sci. USA 74*, 3662–3666.
17. Vande Zande, H., Blondin, G. A., Roecker, E., and Green, D. E. (1978) *Proc. Natl. Acad. Sci. (USA)*, in press.
18. Blondin, G. A. (1975) *Ann. N. Y. Acad. Sci. 264*, 98–111.
19. Hunter, D. R., and Capaldi, R. A. (1974) *Biochem. Biophys. Res. Commun. 56*, 623–628.
20. Hinkle, P. C., Kim, J. J., and Racker, E. (1972) *J. Biol. Chem. 247*, 1338–1339.
21. Blondin, G. A., Kessler, R. J., and Green, D. E. (1977) *Proc. Natl. Acad. Sci. USA 74*, 3667–3671.
22. Criddle, R. S., Packer, L., and Shieh, P. (1977) *Proc. Natl. Acad. Sci. USA 74*, 4306–4310.
23. Okamoto, H., Sone, H., Hirata, H., Yoshida, M., and Kagawa, Y. (1977) *J. Biol. Chem. 252*, 6125–6131.
24. Cattell, K. J., Knight, I. G., Lindop, C. R., and Beechey, R. B. (1970) *Biochem. J. 117*, 1011–1015.
25. Blondin, G. A., Kessler, R. J., and Green, D. E. (1978) *Proc. Natl. Acad. Sci. USA*, in press.
26. Lardy, H. A., Connelly, J. L., and Johnson, D. (1964) *Biochemistry 3*, 1961–1968.
27. Shchipakin, V., Chuchlova, E., and Evtodienko, Y. (1976) *Biochem. Biophys. Res. Commun. 69*, 123–127.
28. Moore, C. (1971) *Biochem. Biophys. Res. Commun. 42*, 298–305.
29. Jeng, A. Y., Ryan, T. E., and Shamoo, A. E. (1978) *Proc. Natl. Acad. Sci. USA* in press.
30. Bruni, A., Contessa, A. R., and Luciani, S. (1962) *Biochim. Biophys. Acta 60*, 301–311.
31. Blondin, G. A., and Kessler, R. J., unpublished studies.
32. Roy, H., and Moudrianakis, E. (1971) *Proc. Natl. Acad. Sci. USA 68*, 464–468.

33. Roy, H., and Moudrianakis, E. (1971) *Proc. Natl. Acad. Sci. USA 68*, 2720–2724.
34. Criddle, R., Edwards, T., Partis, M., and Griffiths, D. E. (1977) *FEBS Lett. 84*, 278–282.
35. Griffiths, D. E. (1977) *Biochem. J. 1960*, 809–812.
36. Vignais, P. V., Vignais, P. M., and Doussiere, J. (1975) *Biochim. Biophys. Acta 376*, 219–230.
37. Pressman, B. C., and Lardy, H. A. (1955) *Biochim. Biophys. Acta 18*, 482–487.
38. Amons, R., Van den Bergh, S., and Slater, E. C. (1968) *Biochim. Biophys. Acta 162*, 452–454.
39. Meisner, H. (1971) *Biochemistry 5*, 3919–3925.
40. Green, D. E., and Blondin, G. A. (1978) *Mitochondrial Energy Coupling and the Unity of Bioenergetics*, monograph in preparation.
41. Boyer, P. D. (1967) *Cur. Top. Bioenerg. 2*, 99–149.
42. Hatase, O., Wakabayashi, T., Allmann, D. W., Southard, J. H., and Green, D. E. (1973) *Bioenergetics 5*, 1–15.
43. Hunter, D. R., and Haworth, R. A., unpublished studies.
44. Haworth, R. A., and Hunter, D. R., unpublished studies.
45. Hunter, D. R., Haworth, R. A., and Southard, J. H. (1976) *J. Biol. Chem. 251*, 5069–5077.
46. Hackenbrock, C. R., Rehn, T. G., Gamble, T. L., Jr., Weinbach, E. C., and LeMasters, J. J. (1971) in *Energy Transduction in Respiration and Photosynthesis* (E. Quagliariello, S. Papa, and C. S. Rossi, eds.), pp. 285–305, Adriatica Editrice, Bari.
47. Allman, D. W., Munroe, J., Wakabayashi, T., Harris, R. A., and Green, D. E. (1970) *Bioenergetics 1*, 87–107.
48. Allmann, D. W., Munroe, J., Wakabayashi, T., and Green, D. E. (1970) *Bioenergetics 1*, 331–353.
49. Mitchell, P. (1961) *Biol. Rev. 41*, 445–502.
50. Haworth, R. A., Komai, H., Green, D. E., and Vail, W. J. (1977) *J. Bioenerg. Biomembr. 9*, 141–150.
51. Green, D. E. (1976) *J. Theor. Biol. 62*, 271–285.
52. Green, D. E. (1976) in *The Structural Basis of Membrane Function* (Y. Hatefi and L. Djavadi-Ohaniance, eds.), pp. 241–258, Academic Press, New York.
53. Danielli, J. F., and Davson, H. (1935) *J. Cell Comp. Physiol. 5*, 495–508.
54. Singer, S. J., and Nicholson, G. L. (1972) *Science 175*, 720–724.
55. Hatefi, Y., Stempel, K. E., and Hanstein, W. G. (1969) *J. Biol. Chem. 244*, 2358–2365.
56. Kaplan, N. O. (1967) *Methods Enzymol. 10*, 317–322.
57. Senior, A. E., and Brooks, J. C. (1970) *Arch. Biochem. Biophys. 140*, 257–266.
58. Chipman, D. M., and Sharon, N. (1969) *Science 165*, 454–465.
59. Lipscomb, W. N. (1973) *Proc. Natl. Acad. Sci. USA 70*, 3797–3801.
60. Mathews, R. G., Ballou, D. P., Thorpe, C., and Williams, C. H., Jr. (1977) *J. Biol. Chem. 252*, 3149–3207.
61. Multiple Authors (1973) in *Group Transfer Enzymes* (P. D. Boyer, ed.), Vol. IX, A and B, Academic Press, New York.
62. Green, D. E., and Blondin, G. A. (1978) *Bioscience 28*, 18–24.
63. Green, D. E., and Reible, S. (1975) *Proc. Natl. Acad. Sci. USA 72*, 253–257.
64. Green, D. E. (1976) *J. Theor. Biol. 62*, 271–285.
65. Jackson, J. B., Crofts, A. R., and von Stedingk, L. V. (1968) *Eur. J. Biochem. 6*, 41–54.

*Chapter 11*

# Coupling between the Charge-Separating Devices of the Mitochondrion: Intra- or Extramembrane?

*Harold Baum*

## Introduction

This past year has seen such a plethora of reviews and "mini-reviews" on the mechanism of oxidative phosphorylation (e.g., refs. 1–3) that it would be both tedious and arrogant of me to add to their number. Instead, I propose to offer some personal thoughts on the subject, illustrated by reference to some of our own work, past and present. This selection is not intended to claim any particular priorities; indeed, in some cases the examples are more of pitfalls than of positive contributions. The common thread is that when the contributions have proved to be useful they have been based on the solid conceptual framework of the organization of the inner mitochondrial membrane in general and of the respiratory chain in particular, which is one of the great contributions that David Green has made to science.

The description of the chain as a mosaic of intrinsic complexes, with constant internal stoichiometry, functionally linked through the mobile carriers ubiquinone and cytochrome $c$, has all the hallmarks of a "good" theory.[4] There is not yet, in my opinion, an equally good theory for the overall process of oxidative phosphorylation; but I am optimistic enough to believe that the basis now exists for such a theory. I am also cynical

*Harold Baum* • Department of Biochemistry, Chelsea College, University of London, London SW3 6LX, United Kingdom.

enough to suspect that it would already have been developed but for the siren song of the Nobel Prize. Brillant and powerful hypotheses have been proposed over the past 20 years, but many have been marred by excessive and unjustified elaboration. This detail, grafted onto the dogma of the concept, has led to baroque edifices of multiorder hypotheses that have obscured the original, central ideas. Such edifices also tend to overshadow more modest constructions, or incorporate them so that their identity and usefulness are lost. Even more confusing has been the tendency to adopt an existing system (possibly with some change in terminology) and then modify it in an apparently arbitrary manner, to give it a new, personal hallmark. Much of this has been possible because there are so many chemically feasible mechanisms of oxidative phosphorylation,[5] and because the intact mitochondrion (or chloroplast, or bacterium) is so complex that almost any observation can be rationalized in terms of almost any model.

The time has surely come to dismantle all currently fashionable systems, set aside (without prejudice) all components that are not essential (or that have required so much modification during their evolution as to render suspect their inherent validity), and attempt a reconstruction from all the core ideas and well-tested observations that survive this scrutiny. The remaining pieces may all fit. If they do not, then at least the surviving anomalies and contradictions ought to provoke fresh creative thought and experiment.

## The Respiratory Chain

### Charge Separation

There is now little doubt that the intact respiratory chain is a charge-separating device. The prediction and demonstration of this fundamental principle are a major achievement of Peter Mitchell's chemiosmotic hypothesis.[6] (As with all good theories, it had a prior "footprint," in R. J. P. Williams's concept that respiration generated localized protons.[7])

What, however, are the charged species that initially separate? What is their stoichiometry with respect to electron flow? How far can they separate without compensatory ion movements? Above all, what is the mechanism of charge separation?

### Proton Stoichiometry

When cations are permitted to move across the membrane in electrophoretic response to the primary charge separation, protons are released in exchange, so that it is likely that the positively charged species of the separating pair is $H^+$ or $RH^+$.

The stoichiometry of proton release (i.e., proton–cation exchange) to electron flow has become a major issue of current controversy. For example, Rottenberg[8] and Wiechman *et al.*[9] have calculated, from phosphate potential measurements, that around $3H^+$ are released per pair of electrons per "coupling site." But once stoichiometries come into question are we sure any more what a coupling site is? That is just the kind of terminology that can become a straitjacket to our thinking. Moreover, even if the concept of coupling sites is valid, they are not necessarily equivalent.

We have shown[10, 11] that the steady-state degree of swelling of mitochondria, associated with the respiration-driven uptake of potassium acetate, potassium phosphate, or calcium acetate, depends on the respiratory span utilized. Even at very high respiratory rates, the span succinate–ferricyanide could maintain only a fraction of the swelling achieved by the span succinate–oxygen or ascorbate–oxygen.

Of course, when the whole chain is operative the oxidized/reduced ratios of all the components will poise to be in "equilibrium" with the steady-state "energy pressure." But there is no reason in principle (apart from single-site $P/2e^-$ measurements) why each site should be producing the same number of protons (per pair of electrons) at that pressure.

A more convincing demonstration that the stoichiometry of proton–cation exchange per coupling site is greater than 2 (the number enshrined in the chemiosmotic hypothesis) is the demonstration, by actual pH measurement, that the number may be as great as 4, if precautions are taken to prevent compensatory ion movements, particularly phosphate–hydroxyl exchange.[12]

Proponents of the "classical" stoichiometry point to "cleaner" systems, of purified respiratory complexes reconstituted into vesicles, where anion-exchange carriers are absent and where "ideal" proton stoichiometries are obtained.[13] However, the lipid/protein ratio in these systems is very high, and we have evidence that, in the case of cytochrome oxidase vesicles, extensive exchange of, say, $Cl^-$ for $OH^-$ may be taking place across the bulk lipid.[14]

This controversy is by no means settled, but it is a healthy one; indeed, it is only unfortunate that—as with the Emperor's new clothes—it has taken so long for the "classical" stoichiometry to be questioned.

*Respiratory Control*

The question of how far charges separate before respiratory control ensues, or compensatory ion movements occur, is fundamental to the mechanism of coupling between charge-separating devices, as will be discussed later. Certainly, the precise nature of the "energized state" under conditions of respiratory control is far from clear. The existence of endogenous ionophores[15] might permit some electrogenic cation transport in intact mitochondria, and the bulk lipid in reconstituted vesicles is not entirely impermeable to cations.[16] Nevertheless, it does seem that the osmotic component in respiratory control may be very small, and it is conventional to refer to the major component as a "membrane potential." The exact meaning of this term in a complex assembly like the inner mitochondrial membrane is ambiguous,[17] and the use of probes to report on the nature and precise localization of the fields involved has not yielded definitive information.[18]

*Mechanism of Charge Separation*

The original formulation of the chemiosmotic hypothesis was quite precise in attributing the generation of a proton motive force to a "looping" of the respiratory chain, with alternate hydrogen- and electron-carrying prosthetic groups. This is open to a number of serious objections.

First, no reasonable sequence of carriers can be written that is compatible both with what is clearly established concerning the complexes of the respiratory chain and with the currently accepted coupling spans. Second, the pyridine nucleotide transhydrogenase is reversibly coupled to ATP synthesis/hydrolysis, so that the span $NADPH \rightarrow NAD^+$ must be considered as a coupling site of an extended respiratory chain. It is well known that this transhydrogenase mediates a stereospecific transfer of hydride that does not exchange with the bulk medium, and no reasonable "loop" has been devised to accommodate this. As a further challenge to classical "looping," it has recently been reported[19] that the oxidation of ferrocyanide by cytochrome oxidase is associated with a release of protons, even though the redox reaction itself involves only electron transfer.

A more potent criticism still would be if the proton stoichiometry per coupling site were unequivocally established as greater than the number of electrons involved (as discussed above), since a simple loop mechanism could not then apply.

Attempts are now being made to formulate schemes that overcome some of these objections, and that also account for other characteristics of the respiratory chain. These are the so-called Q cycles.[20] Some of these are more feasible than others, but at present they are merely highly ingenious paper chemistry, inspired by a commitment to hydrogen carriers as the source of ejected protons.

What alternatives are there? For many years the concept of the "transmembrane Bohr effect" has been mooted (and there are elegant experiments to demonstrate pH dependences of oxidation-reduction potentials of cytochromes[21]). In some formulations these are seen as functioning additionally to loops. (Certainly one cannot ignore the fact that the reduction of oxygen concomitant with the oxidation of reduced cytochrome $c$ involves transmembrane electron transfer, which is at least half a loop).

In my opinion it would be better to reserve judgment on the mechanism of charge separation until further studies have been made of the mechanism of electron transfer within individual complexes of the chain. In our early studies on complex III, for example, we demonstrated a major structural rearrangement when a particular component became reduced.[22] We also obtained evidence suggesting protonic involvement at this, or some closely related, step.[23] Much more is now known about this complex, and in particular about the cytochromes $b$.[24] It ought therefore to be possible now to devise unambiguous experiments to identify the charge-separating events associated with electron flow. Perhaps the cyclical changes in the tightness of association between cytochrome $c_1$ and the cytochromes $b$ are manifestations of a real Q cycle, or its analogue—but unprejudiced chemistry is the only route to a useful answer.

Other very useful systems for further probing into the mechanism of charge separation will be bacteria, first because some have relatively simple respiratory chains, readily amenable to genetic manipulation, and second because in the bacteriorhodopsins there exists a machinery for light-induced charge separation that can serve as a paradigm for membrane energisation without redox loops.[25]

## Relative Stoichiometry of Complexes

There is one other aspect of the respiratory chain that ought to be emphasized when considering any formulation for the organization of the total machinery for oxidative phosphorylation. Apparently, there are no completely independent, parallel, respiratory chains within a given membrane, but a network of complexes functionally linked through the mobile carriers. There may or may not be domains of complexes in relatively closer association within which there is preferential electron flow, but efficient coupled respiration can take place in mitochondria with variable ratios of content of the respiratory complexes relative to each other and to the ATP synthetase. Such variations occur, for example, during development or tissue regeneration, or can be induced by simple nutritional means.[26] Models accounting for energy transfer between complexes (as in respiration-driven reversed electron transfer) must incorporate this fact into their architecture.

## The ATP Synthetase

Analytical, reconstitutional, and genetic techniques, as well as the extensive use of specific inhibitors and probes, have provided a great deal of information concerning the subunit organization of the ATP synthetase complex. ATP is abortively hydrolyzed by the detachable segment ($F_1$). When $F_1$ is attached to that segment of the complex intrinsic to the membrane, then ATP hydrolysis appears to be obligatorily and reversibly coupled to an energization that can be expressed in terms of a separation of charge, and (if countertransport of cations is permitted) the release of protons. Similarly, if a sufficiently large osmotic gradient of protons is imposed across the membrane, ATP synthesis can be achieved. It also seems that removal of $F_1$ exposes, in the membrane component of the complex, a specific, protonophorous channel.[27]

### Proton Stoichiometry

In the chloroplast, where the ATPase responsible for photophosphorylation has an outward orientation, it seems likely that the stoichiometry of proton uptake per ATP hydrolyzed (or proton efflux per ATP synthesized) is about $3:1$.[28] In mitochondria, with the opposite orientation, the most generally accepted number is around $2H^+:ATP$.[29] The controversy regarding the proton stoichiometries for respiration, and the electrogenicity of the adenine nucleotide exchange-diffusion carrier (see below), now requires for each system studied an impartial reconciliation of $P/2e^-$ ratios with the proton stoichiometries of all components of the overall machinery for oxidative phosphorylation. If the numbers do not reconcile, then it can no longer be valid to assume that the energy transmitted between the component systems can be entirely expressed in terms of proton–cation exchange.

### Mechanism of Charge Separation

Yet another of the important conceptual contributions of Peter Mitchell was the suggestion that the ATP synthetase is a reversible, electrogenic proton pump. Certainly it can function in that manner, and clearly an understanding of how this is achieved is crucial to the full description of oxidative phosphorylation.

Competing views have included the asymmetrical access of substrates and products to the hydrolytic center,[30] and a variety of analogies to other cation pumps.[31] However, the very convincing oxygen exchange studies of Boyer *et al.*[32] suggest that the hydration-dehydration step itself involves little change in free energy, and that the major exergonic step in ATP hydrolysis is ATP binding. Conversely, the major energy-requiring step in ATP synthesis is product release. In other words, energy transduction involves conformational energy of the enzyme complex itself. So again we have a divergence of views, analogous to the question of charge separation in the respiratory chain. Are the substrates the direct source of the charged species (loops, asymmetrical access to hydrolytic center); or are the complexes themselves the source (transmembrane Bohr effect, conformational energy of substrate binding)? Again, the paradigm may come from another system, the binding of ATP to myosin, which generates a new conformational species[33] and may be associated with the release of a proton.

A timely warning that even the above "objective" statement of alternatives can close our minds to other possibilities comes from the recent studies of Griffiths.[34] He has adduced an intriguing body of evidence to support a mechanism of ATP synthesis that involves a sequence of group transfer reactions via an acyl lipoate intermediate (analogous to the terminal reactions of GTP synthesis via succinyl lipoate in the TCA cycle). It is very tempting to write models accounting for charge separation by this system, or to graft it on to the Boyer model (with, say, the conformational changes favoring the cyclical formation of an initial thioester). At this stage, that would be fruitless speculation. The possibility of acyl lipoate involvement must first be rigorously tested. If validated, then the characteristics of this new mechanism must be incorporated in any new overall model of oxidative phosphorylation.

## Control of the ATPase

Control of the ATPase by the "membrane potential" (plus osmotic gradients) that it is capable of generating is (as with the nature of respiratory control) central to the problem of coupling between transducing devices. The ATPase is, however, also subject to other possible control mechanisms, in particular those associated with the inhibitor protein,[35] with anions,[36] with $Mg^{2+}$,[37] and with tightly bound adenine nucleotides.[38] In some cases, particularly the last, the question of control may be an inextricable aspect of the central mechanism of action of the enzyme complex.[39] In other cases, the action of effectors may confound the interpretation of apparently straightforward experiments. For example, certain uncouplers stimulate the ATPase by a mechanism apparently independent of their action on the energized state.[40] Similarly, Griffiths has shown that certain uncouplers will prevent ATP synthesis from oleyl lipoate, whereas others will not.[34, 34a] This may account for (although not explain) Hatefi's findings of specific uncoupler-binding sites on the ATPase complex, and of uncoupling by membrane-impermeant reagents.[41]

A particular aspect of the control of the ATPase complex that has received a great deal of attention relates to its potential modes of functioning as an ATP synthetase on the one hand and as an ATP hydrolase on the other. The endogenous inhibitor protein may function to prevent futile cycling via these dual activities, by virtue of the various factors that control the binding and release of the protein.[35] This, however, may not be the whole story; for example, David Green has proposed an elegant scheme accounting for the reciprocal modulation of potential coupling modes of the complex.[42]

In connection with these modalities, there has been much speculation as to whether the catalytic site on the complex responsible for ATP hydrolysis is different from that involved in ATP synthesis. The basis for much of this discussion (and also for much otherwise unnecessary detail in various proposed mechanisms for oxidative phosphorylation) is the paradoxical inhibitory characteristics of the aurovertins. These apparently bind very specifically to the $F_1$-ATPase and, in terms of nmoles of inhibitor per milligram of mitochondrial protein, they are virtually stoichiometric inhibitors of oxidative phosphorylation.[43] Yet they are apparently relatively ineffective as inhibitors of soluble or membrane-bound ATPase and exhibit a variable degree of inhibition of ATP-driven processes.

However, we have shown that these apparent anomalies are explicable in relatively simple terms.[44] Soluble ATPase is inhibited by relatively high concentrations of aurovertin. If it is assumed that the inhibitor binds reversibly (and noncooperatively) to the enzyme, and when bound decreases the $V_{max}$ of the enzyme to about 15% of its initial rate, then, at 50% of the maximal attainable inhibition (i.e., at about 43% actual inhibition), half of the available enzyme sites will be occupied by inhibitor. Knowing the molar concentrations of inhibitor and enzyme in the assay medium, it is then possible to calculate a dissociation constant for the enzyme–inhibitor complex. We were able to demonstrate that, over a tenfold range of concentration of enzyme the same constant applied, although, as expected for any simple binary dissociating system of this kind, the sensitivity to inhibitor *in terms of nmoles per milligram of protein* increased with increasing protein concentration. It then transpired that essentially the same dissociation constant could be obtained for membrane-bound ATPase *and for oxidative phosphorylation*. In other words, the apparent discrepancy in sensitivities simply arose from the fact that oxidative phosphorylation is routinely assayed at a much higher protein concentration than is ATPase.

This explanation, however, did not fully account for the variable sensitivity to aurovertin of ATP-driven processes, some of which are assayed at quite high concentrations of mitochondrial protein. But if we assume that, with all sites saturated, aurovertin only ever inhibits the total ATPase capacity by 85%, then even this difficulty disappears. For example, the total capacity of the inner membrane for coupled ATP hydrolysis far exceeds that for the transhydrogenase reaction, so that an 85% decrease in ATPase activity resulting from complete titration with aurovertin need not give an equivalent inhibition of ATP-driven trans-hydrogenase. The actual sensitivity of the coupled process would then depend on factors such as membrane "leakiness," and hence account for variabilities in published data.

A nice demonstration of this concept is that ATP-driven swelling of fresh mitochondria associated with the uptake of potassium phosphate was found to be insensitive to levels of aurovertin sufficient to saturate the ATPase. Addition of subthreshold levels of uncoupler (i.e., sufficient uncoupler for the full ATPase capacity to be necessary to maintain the steady state of swelling) rendered the system extremely sensitive to contraction on addition of aurovertin (Fig. 1).

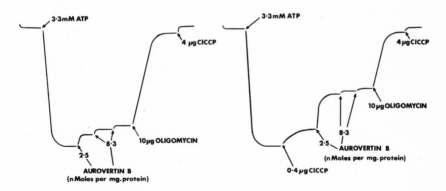

Fig. 1. Effect of low levels of uncoupler on the sensitivity to aurovertin of ATP-dependent accumulation of potassium phosphate by rat liver mitochondria. Uptake was monitored by swelling-induced changes in absorbance at 540 nm. The reaction mixture (3 ml, 20°C) was 125 mM in sucrose, 10 mM in tris-HCl, 0.1 $\mu$M in rotenone, and 15 mM in potassium phosphate, pH 7.4. Mitochondria (3 mg protein) and valinomycin (0.5 $\mu$g) were added prior to initiation of swelling by ATP.

## The Adenine Nucleotide Carrier

Oxidative phosphorylation can clearly take place without the intervention of the ADP:ATP exchange-diffusion carrier, as in bacteria or submitochondrial particles. There are, nevertheless, two cogent reasons for considering the carrier as part of the machinery of oxidative phosphorylation, at least in the case of the intact mitochondrion.

First, there is evidence, albeit controversial, that ADP entering on the carrier is phosphorylated preferentially to the endogenous pool of free ADP and that the ATP thus formed is exported without equilibrating with free internal ATP.[45, 46] If validated, this would suggest a tight association between the carrier and the ATP synthetase complex, to facilitate such substrate channeling. (We have demonstrated[47] that such channeling is not obligatory in the case of phosphate; the internal phosphate pool is readily available for ATP synthesis, even when the phosphate carrier is completely inhibited.)

An even more relevant aspect of the adenine nucleotide carrier is its apparent electrogenicity.[48] At neutral pH the approximate ionizations of the species exchanged are, respectively, $ADP^{3-}$ and $ATP^{4-}$. (By contrast, the phosphate–hydroxyl exchange is essentially electroneutral.) It therefore follows that to maintain the flux of oxidative phosphorylation one proton (or its equivalent) has to be ejected for every ATP synthesized. In other words, if the proton stoichiometry for ATP synthesis itself is 2 (see above), then the proton stoichiometry for a coupling span in intact mitochondria ought to be at least 3 to allow for ATP export.

Direct evidence that the adenine nucleotide carrier is indeed electrogenic is the demonstration of an energized, osmotic swelling mediated by the carrier. We have shown that mitochondria, treated with fluorescein mercuric acetate (FMA) and suspended in a sucrose medium, will swell rapidly on addition of ATP plus a potassium salt.[49] The swelling is sensitive to oligomycin and atractylate, and is reversed by ADP plus a respiratory substrate. Very significantly the swelling is insensitive to (or stimulated by) uncouplers.

Apparently the overall reaction, mediated by the carrier and the ATPase, is

$$\underset{\text{(out)}}{ATP^{4-}} + \underset{\text{(out)}}{K^+} \rightarrow \underset{\text{(out)}}{ADP^{3-}} + \underset{\text{(in)}}{K^+} + \underset{\text{(in)}}{Pi^-}$$

(This ignores any internal acidification due to subsequent ionization of $Pi^-$.) The demonstration of this phenomenon requires pretreatment with the mercurial, because FMA both inhibits the phosphate carrier and renders the membrane permeable to $K^+$.

## Oxidative Phosphorylation

In what form and by what route is energy transmitted between the respiratory chain and the ATP synthetase complex? Certainly, each of these transducers, quite independently of even the existence of the other in the same membrane, is capable of the reversible separation of charge. As charge separation is the earliest form of conserved energy, common to both systems, that has yet to be unambiguously demonstrated and that is potentially capable of transmission, it is very tempting to attempt to describe coupling in these terms. It is true that some formulations implicate conformational energy states as precursors of charge separation in each system, in which case the most direct coupling link might be by the propagation of conformational energy. It should also not be forgotten that in substrate level phosphorylation oxidation-reduction reactions are obligatorily linked to bond formation. As ATP hydrolysis can give rise to charge separation, so in principle could the cleavage of a bond generated through the functioning of the respiratory chain. Hence bond formation could precede charge separation in each transducer, and coupling between respiratory complexes and the ATPase could conceivably occur by a sequence of group transfer reactions with the conservation of bond energy (Fig.2).

There are two routes to exploring these possibilities. One is by gaining further insight into the characteristics of the transducers themselves. The other is by studying the characteristics of the overall coupled process. Any acceptable theory must be compatible with information gleaned from each approach.

When we come to consider what is actually known about oxidative phosphorylation as an overall process, most of the information is self-evidently pertinent to the characteristics of the individual transducers and is not very helpful in identifying the nature of the normal functional link between them. For example, in the presence of appropriate ionophores or permeant ions each system can reversibly generate an osmotic gradient. Therefore, in model systems we would expect to be able to achieve some kind of coupling via such gradients. In intact mitochondria performing oxidative phosphorylation, no such bulk gradients have been demonstrated, so the model systems only reinforce what was already known about the transducing capabilities of the individual complexes.

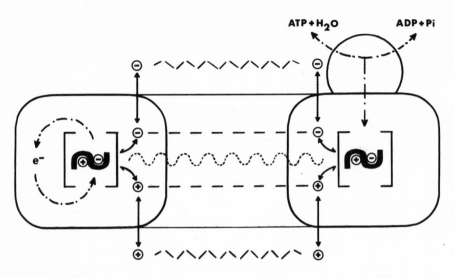

Fig. 2. Possible modes of coupling between a respiratory span (left-hand side) and ATP synthetase (right-hand side) in a given membrane. [ 🔵🔵 ], initial energized state; ∿∿, propagation of conformational energy *or* sequence of group transfer reactions; − − −, intramembrane transmission of charge; ∿∿∿, coupling via bulk membrane potential.

A crucial feature of oxidative phosphorylation is that it is essentially a stoichiometric process, even under conditions that are far from equilibrium. (The P/O ratio for succinate oxidation when [succinate]:[fumarate] is very high and [ATP]:[ADP] [Pi] is very low is close to that observed when the system is close to respiratory control; the excess free energy in the former case is released as heat.) Whatever is transferred between the systems therefore is not just "energy" but entities. The entities that at present are most favored are not "conformons,"[50] or chemical bonds, but protons. This is an eminently tenable hypothesis, for reasons that have already been stressed. But are the other characteristics of physiological oxidative phosphorylation compatible with this concept?

*Uncoupling and Charge Separation*

It is claimed that since all true uncouplers are protonophores, the link in oxidative phosphorylation must be protons ejected electrogenically into the bulk aqueous phase. (Since bulk pH gradients are not a prerequisite for oxidative phosphorylation, the conserved free energy resides in the "membrane potential" generated by uncompensated proton ejection.) However, uncouplers have lipophilic anions, so that their very presence as counterions might facilitate proton release into the bulk phase from an intramembrane charge-separated ion pair. In other words, uncouplers might permit and then abort a proton ejection that normally cannot occur.

It is also pertinent to the question of whether uncouplers divert normal energy transduction into a pathway which is then rendered cyclic that the Madison group has demonstrated a cation requirement for uncoupler action.[51] By a similar token, the uncoupling action of nigericin plus valinomycin demonstrates only that the initial separation of charge (in either the ATP synthetase complex or the complexes of the respiratory chain) can be translated into a bulk $H^+:K^+$ antiport when the membrane is rendered permeable to $K^+$.

The question raised by these considerations is the one referred to previously in relation to the individual transducers. If countertransport does not take place, i.e., if coupling is not via osmotic gradients, can charges be separated across the entire thickness of the membrane?

We have attempted to study this question using two different systems. We looked at the degree of stimulation by uncouplers of ATPase of well-coupled submitochondrial particles as a function of inhibition of the enzyme by oligomycin[52] or by dicyclohexylcarbodiimide (DCCD) or the inhibitor protein.[53] The idea was that, if charge was separated across the entire thickness of the membrane, then each ATPase would be influenced by the "membrane potential" generated by the entire population of ATPases in that membrane. If membrane conductance remained constant, then the steady-state magnitude of this potential (and hence the degree to which residual ATPases were under potential-dependent control) would decrease as the population of functional ATPases in the membrane was titrated with irreversible inhibitors. Various uncoupling systems were used, to eliminate artifacts arising from direct stimulation of the enzyme itself, yet it was routinely found that the degree of stimulation of residual ATPase, on release of control by any means, was independent of the fraction of the total ATPase that had been titrated out by inhibitor. In other words, the system behaved as if each ATPase were a microscopic transducer, primarily controlled by local, intramembrane events (Fig. 3).

In retrospect, these experiments are not so definitive as they seemed. First, any residual endogenous inhibitor protein might have introduced an additional variable, since inhibitor binding is affected by energization. Second, Nicholls[54] has provided evidence that, for some mitochondria at least, membrane conductance is not constant but varies sharply with "membrane potential." This latter argument also counters the otherwise "unambiguous" observation that the remarkably high acceptor control of vesicles of *Paracoccus denitrificans* oxidizing NADH is

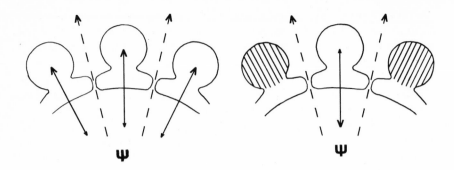

Fig. 3. Expected effect of inhibition of a fraction of the transducers in a given membrane in decreasing the extent of control of the remaining transducers by the residual membrane potential (Ψ). This model assumes that all transducers contribute to a macroscopic separation of charge and that there is a constant potential leak (— — →). (Inhibition is represented by hatching.) Constancy of control of residual transducers, at all levels of inhibition, might be taken to imply that control is not a macroscopic phenomenon.

relatively constant, over a limited range of inhibition of respiration by rotenone.[55]

The other system that we have used (to overcome artifacts due to the anomalous conductance characteristics of the inner mitochondrial membrane) is cytochrome oxidase reconstituted into phospholipid vesicles. This system exhibits respiratory control when oxidizing external reduced cytochrome $c$, and the extent of control can be titrated as a function of inhibition by, say, azide, cyanide, or carbon monoxide. If we corrected for effects due to preferential binding of cyanide to the oxidized complex, and to the energized accumulation of azide inside the vesicle, then it did seem as if control were independent of the number of functional complexes within the same vesicle.[56] Again, it was as if each complex controlled itself only.

Even this result is not as clear-cut as it seemed. Full release of control by uncoupler alone was always very slow, taking up to 5 min, whereas the same final steady state of respiration could be achieved almost instantaneously if valinomycin (itself without effect on the controlled respiration) was also present.[16] On the one hand this reinforces Green's view of the involvement of cations in uncoupling; on the other hand it reveals how little we actually know of the characteristics of these model systems and of the precise nature of uncoupling.

*Macroscopic vs. Microscopic Coupling*

Another way of formulating the question of the distance that charges can be separated without compensatory ion flow relates to coupling between systems. If charge can be separated into the bulk phase, then all transducers within the same membrane are coupled indiscriminantly via a common, macroscopic potential gradient. If charge separation remains intramembranous, then coupling becomes microscopic, within domains whose extent depends on the nature of the intramembrane channels for the specific lateral conductance of, say, uncompensated protons.

It is of course possible that both forms of coupling might operate, the latter being kinetically favored, the former occurring under more artificial conditions. This would be analogous to what might be the case for the relationship between ADP transport and ATP synthesis, as discussed previously.

The macroscopic vs. microscopic controversy has been debated on theoretical grounds.[17] We have attempted to discriminate between these alternative forms of coupling by direct kinetic experiments. (Other workers have used complementary approaches and reached similar conclusions,[57, 58] which are open to similar criticisms.) The general strategy underlying all of these experiments was essentially the same— to titrate a coupled function with an irreversible inhibitor of one of the two transducers involved and then repeat the titration at various levels of inhibition of the other transducer. For strictly microscopic coupling the family of titration curves were expected to be identical in shape. For macroscopic coupling, once one transducer had become rate limiting, it should be possible initially to inhibit the other without further effect on the rate of the overall coupled reaction.[52]

In fact, we observed identical titration curves in the following sets of "double titration" experiments: ATP-dependent $NAD^+$ reduction by succinate (titrated with oligomycin or DCCD ± rotenone, or *vice versa*) and aerobic transhydrogenase (titrated with cyanide ± tryptic digestion of the transhydrogenase).[52, 53] As a control, to demonstrate that mobile intermediates between domains could be detected in such a kinetic system, we studied the titration curves with oligomycin or rotenone of ATP-driven $NAD^+$ reduction by succinate, at various levels of inhibition by malonate (and *vice versa*). We routinely observed that the residual activity, following partial inhibition by malonate, was relatively insensitive to oligomycin or rotenone (and *vice versa*). The rate-limiting step had become the reduction of the ubiquinone pool, and there was a kinetic excess of transducers available to utilize this limited supply of substrate (Fig. 4).

These apparently definitive experiments are open to a major criticism. Inevitably, the particles used in these experiments were "leaky," so that, in chemiosmotic terminology, oligomycin would decrease the steady-state membrane potential, under conditions where the

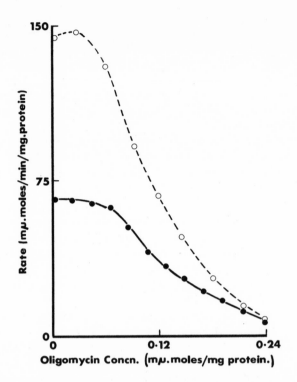

Fig. 4. Relative insensitivity to oligomycin of ATP-dependent reduction of $NAD^+$ by succinate following partial inhibition by malonate. Beef heart submitochondrial particles were used at a concentration of 0.14 mg protein/ml. The concentration of succinate was 10 mM. ○, Control; ●, plus malonate (0.33 mM).

potential (treated as a substrate term) was not saturating. Hence the coupled rate would be independently sensitive to oligomycin (variation in substrate concentration) and rotenone (variation in enzyme concentration).

Now, these criticisms are not watertight when applied to the whole body of our data; and we have devised other, more complex experiments that throw further doubts on this precise formulation of the counterarguments. Nevertheless, it does seem that the leakiness of the particles (and the peculiar conductance characteristics of the membrane) introduce so many variables that, after all, this particular approach is probably not very helpful.

We are currently pursuing a related kinetic approach that appears to be internally controlled against all criticisms of the above kind. The strategy here is to compare the effects on the overall rate of ATP-driven transhydrogenase of different modes of inhibition of the ATPase. If coupling is entirely macroscopic, then a given rate of ATP hydrolysis should always (with a given preparation of particles) give rise to a particular rate of $NADP^+$ reduction. If coupling is constrained within restricted domains, then decreasing the turnover rate of all ATPases (e.g., by lowering the concentration of substrate) ought to have a different effect than decreasing the number of functional ATPases with irreversible inhibitors.

Clear-cut differences would be strong evidence in favor of microscopic coupling, but the converse would not distinguish between coupling via the bulk phase and intramembrane coupling over relatively extensive distances within the membrane. Indeed, it is difficult to envisage a kinetic distinction between these two mechanisms. Furthermore, once the possibility of lateral transfer of protons (or other coupling entity) is accepted, there are fewer difficulties in envisaging the organization of a membrane where there is a variable stoichiometry between individual classes of complex. An experimental approach involving physiological modifications of these stoichiometries might therefore not be fruitful.

*Alternative Approaches*

Intramembrane coupling does not, in principle, require an intact membrane. If counterion movement could be completely abolished and if membrane ultrastructure could be preserved, then high rates of oxidative phosphorylation ought to be demonstrable in unsealed membrane preparations. Such systems have been described,[59] and deserve the closest attention. The critical test is not electron microscopy, which is always potentially artifactual, but sensitivity to any electrogenic ion movement. As both sides of the membrane are now exposed to the same compartment, any such transport is effectively cyclic and would hence uncouple the system. (The prescription for such an experiment ought therefore to include an inhibitor for the adenine nucleotide carrier.)

Another approach would be to seek a specific inhibitor of the conducting channel, such that each transducer could still generate an osmotic gradient in the presence of permeant ions, but direct coupling between transducers was blocked. (It did seem to us that perhaps FMA was such a reagent,[60] but recent work has shown that the mercurial is extremely multipotent; e.g., it facilitates the lytic effects of an endogenous $Ca^{2+}$-dependent lipase,[61] so that such a simple interpretation is not valid.)

What is more difficult to devise is a strategy to discriminate *against* intramembrane coupling, since failures of the above "discriminatory" approaches are effectively neutral in implication. Whatever the

shortcomings of the original formulations of the details of the chemiosmotic theory, and whatever the theoretical objections, the overall concept of "proticity" has been incredibly fruitful. We should therefore not
abandon it except on the basis of very cogent experimental evidence.
Throughout the history of science, too many healthy babies have already
been thrown out with the bath water. On the other hand, we must not be
so charmed by the beauty of a particular offspring that we ignore the
claims of less elegant, but perhaps healthier siblings!

ACKNOWLEDGMENTS

I am grateful to my colleagues, particularly Dr. R. B. Beechey, for
helpful discussion and for permission to cite unpublished observations. I
am also grateful to the organizers of this meeting both for inviting me to
participate and for making it possible for me to accept.

Above all, I am grateful to Dr. David E. Green, in whose honor this
meeting is held, for his encouragement and inspiration over the past 23
years. I have given only cursory citation in this chapter to his monumental contributions to bioenergetics. His own recent work is reviewed
elsewhere in the present volume, and his early work is so well established
as to be taken for granted.

## *References*

1. Mitchell, P. (1976) *Biochem. Soc. Trans. 4*, 399–430.
2. Ernster, L. (1977) *Ann. Rev. Biochem. 46*, 981–995.
3. Ferguson, S. J. (1977) *Biochem. Soc. Trans. 5*, 582–588.
4. Baum, H. (1974) *Ann. N.Y. Acad. Sci. USA 227*, 675–680.
5. Wang, J. H. (1976) *J. Bioenergetics 8*, 209–220.
6. Mitchell, P. (1966) *Biol. Rev. 41*, 445–502.
7. Williams, R. J. P. (1961) *J. Theor. Biol. 1*, 1–17.
8. Rottenberg, H. (1975) *J. Bioenergetics 7*, 61–74.
9. Wiechman, A. H. C. A., Beam, E. P., and VanDam, K. (1975) in *Electron Transfer Chains and Oxidative Phosphorylation* (Quagliariello, E., *et al.*, eds.), North-Holland, Amsterdam.
10. Wrigglesworth, J. M., and Baum, H. (1973) *Biochem. Soc. Trans. 1*, 413–416.
11. Fonyo, A., Baum, H., and Elsden, J., unpublished observations.
12. Brand, M. D., Reynafarge, B., and Lehninger, A. L. (1976) *Proc. Natl. Acad. Sci USA 73*, 437–441.
13. Leung, K. H., and Hinkle, P. C. (1975) *J. Biol. Chem. 250*, 8467–8471.
14. Wrigglesworth, J. M., Nichols, P., and Baum, H., unpublished observations.
15. Blondin, G. A., DeCastro, A. F., and Senior, A. E. (1971) *Biochem. Biophys. Res. Commun. 43*, 28–35.
16. Wrigglesworth, J. M., Elsden, J., and Baum, H. (1976) *10th Int. Congr. Biochem.*, Hamburg, p. 346.

17. Williams, R. J. P. (1975) in *Electron Transfer Chains and Oxidative Phosphorylation* (Quagliariello, E., *et al.*, eds.), North-Holland, Amsterdam.
18. Azzi, A., and Montecucco, C. (1976) *J. Bioenergetics 8*, 257–269.
19. Wikstrom, M. K. F. (1977) *Nature 266*, 271–273.
20. Mitchell, P. (1975) *FEBS Lett. 59*, 137–139.
21. Papa, S. (1976) *Biochim. Biophys. Acta 456*, 38–84.
22. Baum, H., Rieske, J. S., Silman, H. I., and Lipton, S. H. (1967) *Proc. Natl. Acad. Sci. USA 57*, 798–805.
23. Baum, H., and Rieske, J. S. (1966) *Biochem. Biophys. Res. Commun. 24*, 1–9.
24. Rieske, J. S. (1976) *Biochim. Biophys. Acta 456*, 195–247.
25. Kayushin, L. P., and Skulachev, V. P. (1973) *FEBS Lett. 39*, 39–42.
26. Baum, H., and Pollak, J. K. (1977) *CIBA Symp. 51*, 79–90.
27. Lee, C. P., and Ernster, L. (1965) *Biochem. Biophys. Res. Commun. 18*, 523–529.
28. Boeck, M., and Witt, H. T. (1972) *Proc. 2nd Int. Congr. Photosynthesis Res. Stresa* (Forti, G., Avron, M., and Melandri, A. eds.), pp. 903–911, Dr. W. Junk, Publ., The Hague, The Netherlands.
29. Moyle, J., and Mitchell, P. (1973) *FEBS Lett. 30*, 317–320.
30. Mitchell, P. (1974) *FEBS Lett. 43*, 189–194.
31. Skulachev, V. P. (1975) *Energy Transducing Mechanisms*, Vol. 3 (Kornberg, H. L., and Phillips, D. C., eds.), pp. 31–73, MTP International Review of Science, Butterworths, London.
32. Boyer, P. D., Cross, R. L., and Momsen, W. (1973) *Proc. Natl. Acad. Sci. USA 70*, 2837–2839.
33. Tonomura, Y., and Inoue, A. (1975) *Energy Transducing Mechanisms*, Vol. 3 (Kornberg, H. L., and Phillips, D. C. eds.), pp. 121–161, MTP International Review of Science, Butterworths, London.
34. Griffiths, D. E. (1976) *Biochem. J. 160*, 809–812.
34a. Griffiths, D. E., personal communication.
35. Van de Stadt, R. J., de Boer, B. L., and Van Dam, K. (1973) *Biochim. Biophys. Acta 292*, 338–349.
36. Mitchell, P., and Moyle, J. (1971) *J. Bioenergetics 2*, 1–11.
37. Moyle, J., and Mitchell, P. (1975) *FEBS Lett. 56*, 55–61.
38. Harris, D. A., and Slater, E. C. (1975) *Biochim. Biophys. Acta 387*, 335–348.
39. Kayalar, C., Rosing, J., and Boyer, P. D. (1976) *Fed. Proc. 35*, 1601.
40. Selwyn, M. J. (1967) *Biochem. J. 105*, 279–288.
41. Hanstein, W. G., and Hatefi, Y. (1974) *Proc. Natl. Acad. Sci. USA 71*, 288–292.
42. Hatase, O., Wakabayashi, T., Hayashi, H., and Green, D. E. (1972) *J. Bioenergetics 3*, 509–514.
43. Roberton, A. M., Holloway, C. T., Knight, I. G., and Beechey, R. B. (1968) *Biochem. J. 108*, 445–456.
44. Linnett, P. E., Mitchell, A. D., Beechey, R. B., and Baum, H. (1977) *Biochem. Soc. Trans. 5*, 1510–1511.
45. Bertagnolli, B. L., and Hanson, J. B. (1973) *Plant Physiol. 52*, 431–435.
46. Vignais, P. V., Vignais, P. M., and Doussiere, J. (1975) *Biochim. Biophys. Acta. 376*, 219–230.
47. Fonyo, A., Ligeti, E., and Baum, H. (1977) *J. Bioenerg. Biomembr. 9*, 213–221.
48. Vignais, P. V. (1976) *J. Bioenergetics 8*, 9–17.
49. Al-Shaikhaly, M. M., and Baum, H., unpublished observations.
50. Ji, S. (1974) *Ann. N.Y. Acad. Sci. USA 227*, 211–226.
51. Kessler, R. J., Van de Zande, H., Tyson, C., Blondin, G. A., Fairfield, J., Glasser, P., and Green D. E., in preparation.

52. Baum, H., Hall, G. S., and Nalder, J. (1971) *Energy Transduction in Respiration and Photosynthesis*, pp. 747–755, Adriatica Editrice, Bari.
53. Nalder, J., and Baum, H., unpublished observations.
54. Nicholls, D. G. (1974) *Eur. J. Biochem. 50*, 305–315.
55. Ferguson, S. J., and John, P., unpublished observations.
56. Baum, H., Elsden, J., and Wrigglesworth, J. (1975) unpublished observations.
57. Ernster, L., Junni, K., and Asami, K. (1973) *J. Bioenergetics 4*, 149–161.
58. Ferguson, S. J., Lloyd, W. J., and Radda, G. K. (1976) *Biochim. Biophys. Acta 423*, 174–188.
59. Komai, H., Hunter, D. R., Southard, J. H., Haworth, R. A., and Green, D. E. (1976) *Biochem. Biophys. Res. Commun. 69*, 695–704.
60. Southard, J., Nitisewojo, P., and Green, D. E. (1974) *Fed. Proc. 33*, 2147–2153.
61. Al-Shaikhaly, M. M., and Baum, H. (1977) *Biochem. Soc. Trans. 5*, 1093–1095.

*Chapter 12*

# Coupling Factors in Oxidative Phosphorylation

## D. Rao Sanadi, Saroj Joshi, and Fariyal M. Shaikh

### Introduction

A classical approach to the study of the mechanism of multienzyme systems involves resolution of the constituent parts and reconstitution of the system from the purified components. Application of this approach to the study of the mechanism of oxidative phosphorylation was first initiated in David Green's laboratory. He devised the procedures for the large-scale isolation of mitochondria from bovine heart.[1] Even now this is the source for much of the fractionation studies. In 1958 the preparation of phosphorylating submitochondrial particles, $ETP_H$, and the first coupling factor[2] was reported. General use of these pioneering techniques led to the isolation and characterization of four generally accepted coupling factors, shown in Table I.

Since the organizers had asked me to review the work of my laboratory in this area, I shall confine my remarks primarily to coupling factors A and B.

### Factor A

Much against criticism at the time, we chose to use the convenient ATP-driven reversed electron flow from succinate to NAD[3] as the routine assay in our search for coupling factors. Coupling factor A was

*D. Rao Sanadi, Saroj Joshi, and Fariyal M. Shaikh* • Department of Cell Physiology, Boston Biomedical Research Institute, Boston, Massachusetts 02114, and Department of Biological Chemistry, Harvard Medical School, Boston, Massachusetts 02115.

*Table I. Coupling Factors and Their Properties*

| Factor | Assay particle | Intrinsic activity | Properties |
|---|---|---|---|
| $F_1$ (factor A) | A or urea particle | ATPase (overt or latent) | 360,000 dalton oligomer of five subunits, cold labile; ATPase decreases on binding to membrane |
| Factor B | AE particle | None | 29,200 dalton (dimer) with two —SH groups; poor staining by dyes, monomer and oligomers are known |
| OSCP | Membrane proteins from OS-ATPase | None | In its presence, the membrane-bound ATPase becomes sensitive to oligomycin; 18,000 dalton protein |
| $F_6$ | STA particle | None | Heat stable, trypsin labile, hydrophobic 8000 dalton protein; promotes binding of ATPase to membrane |

extracted by sonication of mitochondria and purified extensively. It was cold stable and had low ATPase activity, but closer examination of the ATPase activity showed that it could be enhanced by exposing the preparation to higher temperatures (a step which is used in the preparation of $F_1$-ATPase). The heat-activated ATPase activity declined on cooling but did not lose its ability to stimulate energy-linked reactions, which led us to postulate that Factor A was a different form of $F_1$.[4] Experiments to determine whether the ATPase activity of Factor A was low because of bound ATPase inhibitor[5] showed that the inhibitor content was the same in both heated Factor A with high ATPase activity and unheated factor with low ATPase activity.[6] The bound inhibitor was less than 0.5 mole per mole of Factor A. Furthermore, it could not be removed from heated Factor A by ammonium sulfate fractionation, although exogenously added inhibitor is removed under these conditions. Based on these data, we proposed tentatively that the difference in ATPase activity between Factor A and $F_1$ may be attributed to differences in conformational structure of the coupling factor.[5] Pullman and Schatz,[7] Penefsky,[8] and Beechey[9] have maintained that the low ATPase activity is due to bound inhibitor.

We have recently analyzed more highly purified preparations of Factor A which had undergone gel filtration for inhibitor content by two methods: SDS-PAGE using purified inhibitor as a standard, and inhibitor activity assay. The first method is illustrated in Fig. 1, showing the scans for Factor A, heated extracts of Factor A containing the heat-stable inhibitor,[6] and a purified preparation of inhibitor prepared according to Pullman and Monroy.[5] The inhibitor content was computed from the area under the respective peaks.

Table II shows the inhibitor analysis of Factor A and two preparations of $F_1$ made according to the procedure of Horstman and Racker[10]— a preparation which is low in inhibitor content. There is good agreement between the two methods, and on a molar basis there is less inhibitor in Factor A with an ATPase activity of 5 units/mg than in two samples of $F_1$ with 70–75 units/mg.

It is clear from our results that the low ATPase activity of coupling Factor A cannot be attributed to bound Pullman inhibitor, and other

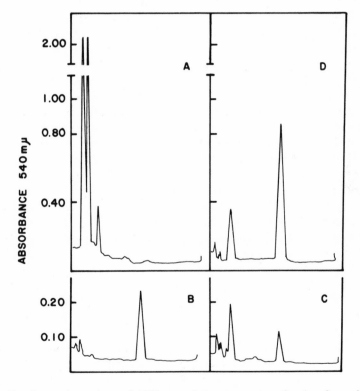

Fig. 1. Densitometric tracings of ATPase and the supernatant fraction from the heat treatment of ATPase at 70°C. SDS-PAGE was in 12% acrylamide gels. (A) 50 μg Factor A. (B) Supernatant fraction after heating 150 μg $F_1$ according to Warshaw *et al.*[6] (C) Supernatant fraction after heating 150 μg Factor A. (D) 1 μg Pullman inhibitor and supernatant from Factor A.

Table II. ATPase Inhibitor Analysis of Factor A and $F_1$[a]

| Factor | ATPase (units/mg) | Inhibitor (units/mg) | | Inhibitor (mole/mole factor) | |
|--------|------------------|----------------|--------------|----------------|--------------|
|        |                  | Enzymic assay | SDS analysis | Enzymic assay | SDS analysis |
| Factor A | 5.0 | 1.8 | 3.2 | 0.025 | 0.045 |
| $F_1$ | 70.0 | 3.6 | 7.6 | 0.05 | 0.10 |
| $F_1$ | 75.0 | 10.2 | 22.0 | 0.14 | 0.30 |

[a] Factor A used in the above experiments was purified by filtration through Sepharose 4B after DEAE-cellulose chromatography.[13] $F_1$ was made according to the procedure of Horstman and Racker.[10] The enzymic assays for inhibitor analyses were carried out according to the procedure of Pullman and Monroy.[5] For estimation of inhibitor from SDS gels, the areas under the respective peaks were computed using Pullman inhibitor as a standard.

explanations have to be invoked. These findings do not conflict with the observation that the Pullman inhibitor blocks ATPase activity without affecting coupling activity; this may occur by a different mechanism unrelated to the difference between $F_1$ and Factor A. The role of the inhibitor in oxidative phosphorylation appears to be more complex than was thought originally, since its interaction with the adenine nucleotide translocase is a newly described phenomenon.[11]

Recent work on the mitochondrial ATPase in the laboratory of Moudrianakis[12] has revealed three electrophoretically distinct forms of $F_1$, all with six subunits, including presumably the 10,000 dalton Pullman inhibitor protein. Since Factor A and $F_1$ of Horstman and Racker have insignificant amounts of the inhibitor, the preparation of Moudrianakis may be yet another variant. Polymorphism in the isolated membrane-associated proteins and protein complexes seems to be a fairly common phenomenon, and Factor A and $F_1$ represent still another example of it.

### Factor B

Comparison of the activities of crude and purified Factor A preparations in stimulating ATP-driven NAD reduction by succinate showed consistently higher maximal activity with the former. This led to the working hypothesis that the crude extracts contained a second factor. Further examination of this phenomenon using the AE particle, which appears to be selectively responsive to the second factor, led to the isolation of coupling factor B.[13] The factor stimulated the activity of the

AE particle in all of the energy-linked reactions that were tested. It appeared homogeneous by the criteria of sedimentation in the analytical ultracentrifuge, disk gel electrophoresis, and immunoelectrophoresis.[14 15] The antiserum to Factor B showed a single band in immunoelectrophoresis and inhibited Factor B stimulated activities. The sedimentation constant in 0.1 M tris sulfate, pH 8.0, in the presence of 1 mM DTT was 3.11–3.39 S with four independent preparations. A value of 3.38 S was obtained with a sample that had been treated with iodoacetamide. The molecular weight of the preparation from these values and by sedimentation in a sucrose density gradient using hemoglobin and cytochrome $c$ markers was estimated to be 32,000. The amino acid analysis gave a minimum molecular weight of 14,600, which suggested that Lam's Factor B was a 29,200 dalton dimer. Factor B activity is inhibited also by mercurials and reactivated by —SH compounds.

Factor B has been obtained in other forms by procedures which are essentially similar to those for Lam's preparation. The 46,000 dalton protein[16] is relatively less stable than the 30,000 dalton preparation but has distinctly higher specific activity in the stimulation of succinate-linked NAD reduction in AE particles. A 15,000 dalton preparation with even higher specific activity has been reported recently[17] and has also been obtained in our laboratory (unpublished results).

Since Factor B has no intrinsic activity, its role in oxidative phosphorylation has to be determined by indirect means. The simpler the reaction system for these experiments, the easier would the task be. In our present work we have chosen the Pi-ATP exchange system for detailed studies.

## Pi-ATP Exchange

Several types of oligomycin-sensitive ATPase complexes have recently been reported; these have Pi-ATP exchange activity (Table III). The oligomycin-sensitive ATPase[18] and the Pi-ATP exchangease of Sadler *et al.*[19] have exchange activities which are 2 to 3 times higher than that of ETP$_H$. They are relatively low in electron transport components. The OS-ATPase, on supplementation with phospholipid and F$_1$, shows exchange activity of roughly 200 nmoles $\times$ min$^{-1}$ $\times$ mg$^{-1}$, and is further stimulated nearly twofold by the addition of Factor B. These activities, obtained by the cholate dilution procedure,[22] are higher than our earlier data[23] obtained by the cholate dialysis procedure. The exchangease shows no increase in activity with phospholipids and only 20% increase on addition of Factor B.

*Table III. Properties of the Different Mitochondrial Energy-Transducing ATPase Complexes* [a]

| Preparation | ATPase ($\mu$moles $\times$ min$^{-1}$ $\times$ mg$^{-1}$) | Pi-ATP Exchange (nmoles $\times$ min$^{-1}$ $\times$ mg$^{-1}$) | Phospholipid (mg/mg protein) | Reference |
|---|---|---|---|---|
| ETPH | 1.0[b] | 200[b](38°C) | 0.43[b] | |
| OS-ATPase | 8.3[b] | 0 | 0.43 | (18) |
| | | 400[b, c] (38°C) (+ P-lipid) | | |
| Exchangease | 3.5 | 186 | — | (19) |
| | | 600[b, c] (38°C) | 1.0[b] | |
| Complex V | 3.5 | 110 | Depleted | (20) |
| H$^{+}$-translocating ATPase | 9.3 | 91 | <0.06 | (21) |
| Reconstituted membrane protein | 6.0[b] | 200[b, c] (38°C) | 2.1[b] | |

[a] Activities were determined at 30°C except as indicated
[b] Values from this laboratory.
[c] Factor B added.

Tzagoloff *et al.*[18] had described a procedure for further extraction of ATPase complex with 3.5 M NaBr which caused dissociation of $F_1$ and yielded a membrane protein fraction which was capable of rebinding $F_1$ with restoration of oligomycin-sensitive ATPase activity. Extraction of Sadler's ATPase complex in a similar manner also yields a membrane pellet deficient in $F_1$. The SDS-PAGE scans of the ATPase complex and the membrane protein fraction are seen in Fig. 2. The latter shows mainly six bands compared to the 13-banded profile of the ATPase complex. The twin $F_1$ peaks around 55,000 daltons are eliminated from the membrane pellet. The membrane protein fraction has negligble ATPase and Pi-ATP exchange activities, but these could be induced by addition of purified $F_1$, and the Pi-ATP exchange activity increased further twofold by Factor B (Fig. 3). The amount of $F_1$ required to saturate the system is 400–600 $\mu$g/mg membrane protein, as seen by Tzagoloff *et al.*[24] for the ATPase activity. Approximately 1.5–2 units of Factor B is required per milligram of protein for maximal effect. No exchange activity was detected in the absence of $F_1$, even with Factor B added.

The exchange activity of the reconstituted system is 90% inhibited by energy transfer inhibitors (2 $\mu$g oligomycin, 5 $\mu$g rutamycin, 10 $\mu$M DCCD) and uncouplers (50 $\mu$M ClCCP, 2 mM dinitrophenol). The

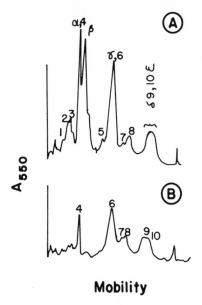

**Mobility**

Fig. 2. Scans from SDS-PAGE of Pi-ATP exchangease (A) and membrane proteins obtained after NaBr treatment (B).

activity is 25% inhibited by 2 $\mu$g of valinomycin plus K$^+$ or 4 $\mu$g nigericin separately but is totally inhibited by the ionophoric combination (1 $\mu$g each) of valinomycin (plus K$^+$) and nigericin. In all respects, the reconstituted system behaves much like mitochondrial particles.

### ATPase Inhibition

We expected that since Factor B was capable of stimulating the Pi-ATP exchange, it might also inhibit ATPase activity. Under the normal assay conditions shown in Fig. 4, using saturation levels of F$_1$, Factor B had no inhibitory effect on the ATPase. Considering the fact that the ATPase activity of the system is at least an order of magnitude higher than its exchange activity, it seemed desirable to carry out the experiments at low levels of F$_1$ where the ATPase activity would be in the range of the exchange activity. Figure 4 shows a titration of exchange and ATPase activities with different levels of F$_1$ per milligram of membrane protein. These experiments are carried out by incubating membrane protein fraction with different levels of F$_1$, centrifuging off unbound F$_1$, and then carrying out the assay with or without added

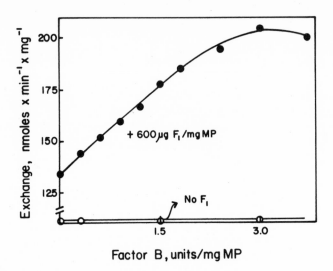

Fig. 3. Effect of Factor B on the Pi-ATP exchange activity of the membrane proteins. Membrane proteins were obtained by extracting the exchangease twice with 3.5 M NaBr.[24] Factor B was incubated with 250 $\mu$g membrane proteins, 120 $\mu$g $F_1$, and 40 $\mu$g OSCP for 10 min at 23 °C in the presence of 100 $\mu$moles Tricine KOH (pH 8.0), 0.5 $\mu$moles MgCl$_2$, 1 $\mu$mole DTT, and 2.5 mg BSA in 0.5 ml of volume. Then 0.1 ml solution containing 10 $\mu$moles of $^{32}$Pi (2–3 × 10$^5$ cpm) was added, and the reaction was started by adding 0.4 ml of solution containing 15 $\mu$moles ATP, 5 $\mu$moles ADP, and 20 $\mu$moles MgCl$_2$. Samples were incubated for 15 min at 38°C.

Factor B. In the absence of Factor B, the stimulation of the exchange with increasing $F_1$ is sigmoidal, and the curve becomes linear in the presence of Factor B. At saturation levels of $F_1$, the stimulatory effect of Factor B is still seen, although in this particular experiment it was considerably lower than usual. The ATPase activity shows the inverse pattern. The curve of ATPase vs. $F_1$ is linear in the absence of Factor B and becomes sigmoidal when Factor B is added. Thus at low levels of $F_1$ the stimulation of exchange and the inhibition of the ATPase are quite striking.

From the linear part of the curves in Fig. 5 (i.e., 25–50 $\mu$g levels of $F_1$) the catalytic activity of bound $F_1$ may be calculated. Independent experiments have shown that at these levels practically all of the added $F_1$ is bound by the membrane pellet. The ATPase activity of free $F_1$ decreases to less than one-third on binding to the membrane protein. It is further reduced to less than one-half that level in the presence of Factor B (Table IV). At the same time, the activity of $F_1$ in the exchange is enhanced by an order of magnitude.

## AE Particle Stimulation

Early in our research on Factor B, we had drawn attention to the fact that it stimulated AE particle activity in energy-linked reactions in a manner similar to that obtained with oligomycin at low levels, but the maximum stimulation provided by the factor was generally slightly higher (Fig. 5 of Adolfsen *et al.*[12]). It was also noted that relatively high levels of Factor B gave small but significant inhibition of the ATPase activity of the AE particle (Fig. 6 of Adolfsen *et al.*[12]).

These results prompted a restudy of the phenomenon described earlier, *viz.*, the stimulation of AE particle activity in ATP-driven NAD reduction by succinate by both Factor B[12] and oligomycin.[12, 25] Figure 5 shows a titration with oligomycin in the presence of different levels of Factor B. The oligomycin level for peak activity shifts to lower and lower concentrations with increasing amounts of Factor B. The data might be interpreted as reflective of a competition between oligomycin and Factor

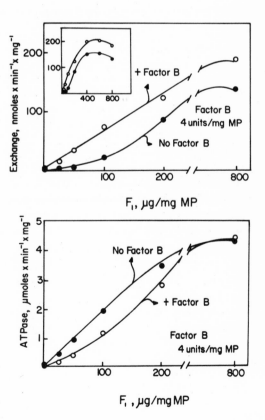

Fig. 4. Effect of Factor B on the Pi-ATP exchange (top) or ATPase (bottom) activities of membrane proteins. In the Pi-ATP exchange experiment, membrane proteins (250 $\mu$g) were incubated with/without Factor B (4 units/mg MP) and OSCP (160 $\mu$g/mg MP) and indicated levels of $F_1$ for 10 min at 23°C. Activity was then started as described in the caption of Fig. 3. For ATPase activity 1 mg membrane protein was incubated with the indicated levels of $F_1$ and OSCP (160 $\mu$g) for 10 min at 23°C in 0.2 ml of 0.25 M sucrose, 10 mM tris acetate (pH 7.5), 2 mM ATP, and 1 mM EDTA. Sucrose-tris (1 ml) was then added to the solution and the particles were collected by centrifugation. The pellet was washed twice with sucrose-tris and resuspended in 1 ml sucrose-tris containing 4 units of Factor B where indicated. After 10 min incubation at 23°C, aliquots were drawn for ATPase activity.

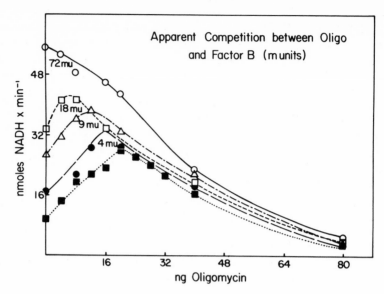

Fig. 5. Stimulation of the AE particle activity by Factor B and oligomycin using ATP-driven NAD reduction by succinate as a measure of activity.[13]

Table IV. Catalytic Activities of $F_1(\mu moles \times min^{-1} \times mg^{-1}F_1)^a$

| | ATPase | | Exchange | |
|---|---|---|---|---|
| | −Factor B | +Factor B | −Factor B | +Factor B |
| $F_1$ (no MP) | 70 | 70 | | |
| $F_1$ + MP | 19.2 | 8.4 | 0.05 | 0.68 |

[a] The catalytic activities of $F_1$ + membrane proteins were calculated from the slopes of the initial rates from experiments described in Fig. 4.

B, although they do not have to act at the same site. A more likely explanation is that oligomycin promotes the stimulation by facilitating the action of Factor B that may be present in an inactive form. This possibility is supported by the fact that our experiments looking for similar apparent competition between oligomycin and Factor B in the membrane protein fraction have not been successful.

## Conclusion

Several explanations are possible for the sigmoidal shape of the $F_1$ vs. exchange activity curve and differences in initial slopes in Fig. 5 with and without Factor B present. Our working hypothesis assumes that two types of binding sites of $F_1$ are present in the membrane protein fraction—some that are coupled and some uncoupled. These sites may exist in different assemblies of the ATPase complex. Melnick *et al.*[26] considered the possibility of $F_1$ existing in two different states—one energy coupled and another energy uncoupled—in submitochondrial particles to explain the differential effect of AMP-PNP in whole mitochondria and in particles. The preferred binding of Factor B may be at the coupled sites. Factor B converts some of the uncoupled sites to coupled sites, although the remaining sites are uncoupled even in the presence of excess Factor B. These studies are beginning to give us a glimmer of how Factor B acts in oxidative phosphorylation, but much work will be needed to arrive at a precise description of its role.

## References

1. Crane, F. L., Glenn, J. L., and Green, D. E. (1956) *Biochim. Biophys. Acta 22*, 475.
2. Linnane, A. W. (1958) *Biochim. Biophys. Acta 30*, 221.
3. Sanadi, D. R., and Fluharty, A. L. (1963) *Biochemistry 2*, 523.
4. Andreoli, T. E., Lam, K. W., and Sanadi, D. R. (1965) *J. Biol. Chem. 240*, 2644.
5. Pullman, M. E., and Monroy, G. C. (1963) *J. Biol. Chem. 238*. 3762.
6. Warshaw, J. B., Lam, K. W., Nagy, B., and Sanadi, D. R. (1968) *Arch. Biochem. Biophys. 123*, 385.
7. Pullman, M. E., and Schatz, G. (1967) *Ann. Rev. Biochem. 36*, 539.
8. Penefsky, M. S., (1967) *J. Biol. Chem. 242*, 5789.
9. Beechey, R. B. (1974) in *Membrane ATPases and Transport Processes* (Bronk, J. R., ed.), p. 41, Biochemical Society Special Publication No. 4, The Biochemical Society, London.
10. Horstman, L. L., and Racker, E. (1970) *J. Biol. Chem. 245*, 1336.
11. Chan, S. H. P., and Barbour, R. L. (1976) *Biochem. Biophys. Res. Commun. 72*, 499.
12. Adolfsen, R. A., McClung, J. A., and Moudrianakis, E. N. (1975) *Biochemistry 14*, 1727.
13. Lam, K. W., Warshaw, J. B., and Sanadi, D. R. (1967) *Arch. Biochem. Biophys. 119*, 477.
14. Lam, K. W., Swann, D., and Elzinga, M. (1969) *Arch. Biochem. Biophys. 130*, 175.
15. Lam, K. W., and Yang, S. S. (1969) *Arch. Biochem. Biophys. 133*, 366.
16. Shankaran, R., Sani, B. P., and Sanadi, D. R. (1975) *Arch. Biochem. Biophys. 168*, 394.
17. You, K., and Hatefi, Y. (1976) *Biochim. Biophys. Acta 423*, 398.
18. Tzagoloff, A., Byington, K. H., and MacLennan, D. H. (1968) *J. Biol. Chem. 243*, 2405.
19. Sadler, M. H., Hunter, D. R., and Haworth, R. A. (1974) *Biochem. Biophys. Res. Commun. 59*, 804.
20. Hatefi, Y., Stiggall, D. L., Galante, Y., and Hanstein, W. G. (1974) *Biochem. Biophys. Res. Commun. 61*, 313.
21. Serrano, R., Kanner, B. I., and Racker, E. (1976) *J. Biol. Chem. 251*, 2453.
22. Racker, E., Chien, T. F., and Kandrach, A. (1975) *FEBS Lett. 57*, 14.
23. Joshi, S., Shaikh, F. M., and Sanadi, D. R. (1975) *Biochem. Biophys. Res. Commun. 65*, 1371.
24. Tzagoloff, A., MacLennan, D. H., and Byington, K. H. (1968) *Biochemistry 7*, 1596.
25. Lee, C. P., and Ernster, L. (1965) *Biochem. Biophys. Res. Commun. 18*, 523.
26. Melnick, R. L., DeSousa, J. T., Maguire, J., and Packer, L. (1975) *Arch. Biochem. Biophys. 166*, 139.

*Chapter 13*

# Studies of Energy-Linked Reactions: A Biochemical Genetic Approach to the Mechanism of Oxidative Phosphorylation

## David E. Griffiths

### Introduction

Current theories of the mechanism of oxidative phosphorylation can be divided into three different concepts: (a) the chemical coupling hypothesis based on the mechanism of substrate-linked phosphorylation,[1, 2] (b) the conformational coupling hypothesis,[3, 4] and (c) the chemiosmotic (electrochemical) hypothesis.[5] Many variants of these hypotheses have been proposed, and the evidence in favor and the limitations of each hypothesis have been summarized recently.[6] There is general agreement that energy made available in respiratory chain oxidations is primarily conserved in a form other than ATP, and this form (termed "$\sim$" or "X $\sim$ 1") can be utilized to drive a number of endergonic processes such as ATP synthesis, ion transport, energy-linked transhydrogenation and energy linked reversal of electron transport (Fig. 1) (*cf.* ref. 7). Such reactions can also be driven by ATP hydrolysis and under these conditions are inhibited by oligomycin, triethyl tin, and DCCD as well as uncoupling agents, and, in addition, it can be demonstrated that the intermediate energy form ($\sim$ or X $\sim$ 1) is common to all three coupling sites.[8]

***David E. Griffiths*** • Department of Molecular Sciences, University of Warwick, Coventry, United Kingdom.

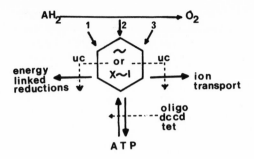

Fig. 1. The mitochondrial energy conservation system.

The versatility and ubiquitous nature of the hypothetical inter-mediate energy state ($\sim$ or X $\sim$ 1) have been puzzling features of the chemical hypothesis, as is the lack of a mechanism for the mode of action of uncoupling agents, and any component or intermediate isolated should provide an adequate explanation for the large body of information which has led to the development of the concept of the intermediate energy state ($\sim$ or X $\sim$ 1). The variety of reactions de-manding the use of $\sim$ or X $\sim$ 1 would argue that it serves a common group transfer function and that reaction specificity is achieved by interaction of $\sim$ or X $\sim$ 1 with specific enzyme species, hence the variety of coupling factors isolated which have a function in energy-linked reactions and oxidative phosphorylation. The acyl carrier protein in the fatty acid synthetase complex may provide a useful analogy for the structure and function of the intermediate energy state ($\sim$ or X $\sim$ 1).

The differential sensitivity of oligomycin and DCCD as compared to uncouplers of energy-linked reactions and the demonstration of ex-change reactions such as Pi-ATP exchange provide the strongest avail-able evidence in favor of the chemical coupling hypothesis, particularly as the conformational and chemiosmotic hypotheses do not provide an adequate mechanistic interpretation. A knowledge of the mode of action of these inhibitors and their interaction sites will thus provide important evidence on components of the oxidative phosphorylation system as well as their functional organization. However, no firm evidence for energy-rich intermediates (X $\sim$ 1 or X $\sim$ P) has been obtained, despite extensive investigation, as emphasized repeatedly by adherents of the chemios-motic hypothesis.

### Organization of Components of the Inner Membrane

The components of the electron transport chain have been frac-tionated into four complexes, I, II, III, and IV,[9-12] and their properties have since been extensively investigated by many workers, from the standpoint of electron transport activity, protein composition, membrane location and organization, and biogenesis.[13-15] All of the complexes examined contain protein subunits in excess of those required minimally for electron transport activity.[14, 15] It is possible that these additional subunits serve a structural or organizational function, but their prime function may be to serve as interaction points with the ATP synthetase complex (complex V), the other operational complex of the oxidative phosphorylation system, or with a component which links the respiratory chain and the ATP synthetase complex. Demonstration of a proton or H-carrier function of these components would be particularly relevant, but no evidence is available to date.

The ATP synthetase has been obtained as a defined operational entity in various degrees of complexity, e.g., oligomycin-sensitive ATPase,[16] OS-ATPase with Pi-ATP exchange activity,[17] complex V of Hatefi,[13] and the lysolecithin preparation of Hunter *et al.*[18] These preparations contain numbers of subunits varying from ten to fourteen, including six which are associated with $F_1$-ATPase and the oligomycin-sensitivity-conferring protein. The remaining four to eight components contain the interaction sites of the inhibitors of oxidative phosphorylation, oligomycin, venturicidin, DCCD, and triethyl tin. The best-defined preparation is the yeast OS-ATPase preparation of Tzagoloff,[16] which contains ten subunits and has been shown to contain the sensitivity sites for oligomycin, venturicidin, DCCD, and triethyl tin.[19, 20] Other preparations such as complex V,[13] proton-translocating ATPase,[17] and the lysolecithin-extracted preparation[18] catalyze the Pi-ATP exchange reaction, and evidence is presented that in the lysolecithin preparation[21] the exchange activity is catalyzed by a membrane-bound vesicle-free system. This claim is hotly contested by the adherents of the chemiosmotic hypothesis, who claim that a closed vesicle system is a basic requirement for generation of a proton gradient and membrane potential which provides the driving force for ATP synthesis.

Many experiments on the reconstitution of oxidative phosphorylation have been reported[22, 23] which demonstrate reconstitution of overall oxidative phosphorylation and Pi-ATP exchange activity from purified mitochondrial fractions such as cytochrome oxidase, purified ATPase, and hydrophobic protein fraction. Despite this significant advance in methodology, no information on any intermediate energy states has been obtained. Major emphasis is laid on the necessity for vesicle formation and the proper orientation of components in membranes, and the experiments have been construed by Racker as evidence for the chemiosmotic hypothesis.[24] Other interpretations such as the interaction of membrane components resulting in a link between the respiration chain complexes and the ATP synthetase complex have not been considered. It is thus pertinent to summarize at this stage the current evidence as to the structure and function of the components used, particularly the ATPase complex. The organization of the OS-ATPase (ATP synthetase) is assumed to be the familiar $F_1$ headpiece, OSCP stalk, and $F_0$ basepiece described in many publications.[25] While there is good evidence that the $F_1$-ATPase is located on the inner side of the mitochondrial membrane, it is implicitly assumed that the $F_0$ components are located in the membrane and extend across to the outer side of the inner membrane and form a protonophore which spans the membrane. There is no firm evidence that this is so, and there is no evidence for assuming that there is a stoichiometric relationship between $F_1$ and $F_0$ components.

Various estimates of the protein composition of the mitochondrial inner membrane ascribe 60–70% of the protein components to complexes I–IV, the ATP synthetase complex (complex V), and ADP translocase: the major portion of the remainder can be ascribed to an insoluble protein fraction termed "structural protein" by Green and "hydrophobic protein" by Racker. These preparations contain several components, the majority being low molecular weight components in various stages of aggregation (or denaturation ?). Some of these components may include hydrophobic components of the ATP synthetase complex and also subunits of unknown composition. This hydrophobic protein fraction in one form or another has been utilized in all reconstitution experiments with cytochrome oxidase or other electron transport complexes and purified OS-ATPase. It is significant that no reconstitution has been achieved with purified OS-ATPase and a purified respiratory chain complex. A hydrophobic protein is required in all cases, even in bacteriorhodopsin-driven reactions.[26] The possibility that an energy coupling component is operative in these reconstituted systems present in the ATP synthetase or the hydrophobic protein fraction, or both fractions, has not been considered.

There is no evidence that ATP synthetase complexes are specifically associated in the membrane with any of the electron transport complexes, and Green and Ji[27] claim that the electron transport and phosphorylation centers are at least 50 Å apart. Energy transfer by direct conformational interactions of electron transport complexes and ATP synthetase[28] is unlikely. Studies with fungal and bacterial systems where there is rapid turnover of energy conservation components provide no evidence for a tight association of electron transport and ATP synthetase complexes, and are consistent with extensive evidence for mobility of protein components in a variety of biological membranes.[29] Recent studies in this laboratory[30] also indicate that the microenvironment of the ATP synthetase complex in yeast mitochondria can be differentiated from that of the electron transport components. There is thus no evidence for a highly organized integrated multienzyme complex such as is found in fatty acid synthesis or $\alpha$-ketoacid oxidation, although the degree of organization in heart mitochondria is significantly greater as compared with that in yeast and bacterial systems.

The major outstanding question and the major source of controversy in oxidative phosphorylation is how the link between the respiratory chain complexes and the ATP synthetase complex is achieved. The chemiosmotic hypothesis states that the link between the proton-translocating ATPase and the proton-translocating respiratory chain is a transmembrane electrochemical potential of hydrogen ions $(\Delta\mu H^+)$[31] and emphasizes the lack of evidence for chemical intermediates. There is also well-documented evidence in mitochondria, chloroplasts, and bacteria for proton-translocating electron transport and ATPase systems that are specifically oriented in bioenergetic membranes[31] and have a function in ion transport.

The driving force of the proton gradient postulated in the chemiosmotic hypothesis is a novel principle in biochemical energy transfer mechanisms, but no evidence as to the mechanism is available. As to date all biochemical reactions leading to the synthesis of ATP follow chemical reaction mechanisms as expressed in the mechanisms of enzyme catalysis and ligand–protein and protein–protein interactions, the novel principles of the chemiosmotic hypothesis as applied to ATP synthesis are a curious anomaly. However, if one accepts the proposition that proton translocation events involve defined chemical functional groups (proton acceptors and donors) on membrane components, then any apparent differences in the chemical and chemiosmotic hypotheses are a matter of semantics, as the same (or equivalent) proton donors and acceptors could be "intermediates" in the chemical hypothesis. That proton acceptor/donor species involved in proton translocation are present in membranes is accepted by some adherents of the chemiosmotic hypothesis to explain the mode of action of inhibitors such as oligomycin and DCCD,[32] but the implications are ignored as they are at variance with a basic tenet of the chemiosmotic hypothesis, i.e., there are no "intermediates."

There is ample evidence from studies of the $F_1$-ATPase, OS-ATPase, and membrane-bound ATPase in many laboratories that the ATPase complex undergoes major conformational changes during ligand (substrate) binding and various energized states of the membrane. That the mitochondrial ATPase complex is maintained in a mode for ATP synthesis has been demonstrated recently.[33] It is possible that maintenance of the complex in the ATP synthesis mode is achieved via a cyclic series of interactions with a ligand generated in mitochondria by the respiratory chain; i.e., a link component with the properties of an *energy coupling factor* (ECF) is required which serves the function of X $\sim$ 1 in the chemical hypothesis.

In view of the structural and mechanistic constraints outlined above, particularly those imposed by the structural organization of the inner membrane, any link component or energy coupling factor should satisfy the following requirements:

a. ECF should be a relatively small lipophilic molecule, possibly a mobile species capable of diffusion in mitochondrial membranes and interaction with different mitochondrial components.

b. ECF should have one or more specific functional groups and could serve as a proton donor/acceptor system.

c. ECF may be present in excess as compared with the electron transport chain complex and the $F_1$-ATPase to avoid kinetic restraints in a membrane system containing mobile protein components. It may serve a "pool" function.

d. ECF may react directly with the electron transfer complexes. If so, specific interaction sites with isolated electron complexes should be demonstrable. Alternatively, interaction of electron transfer complexes with ECF may be mediated by another proton donor/acceptor system.

e. ECF should react specifically with the ATP synthetase complex and form a dissociable ligand of the ATP synthetase complex Activation or "energization" of the ATP synthetase complex occurs during association/dissociation reactions with the ECF.

f. Uncoupling agents interact with ECF or its precursor either by converting to an active form or by preventing the interaction of ECF or its precursor with the complexes of the respiratory chain or the ATP synthetase complex.

This chapter will summarize the results of experiments that indicate the existence of components which satisfy the requirements of an energy coupling factor (ECF) in the mitochondrial inner membrane. The properties of two specific lipid cofactors of oxidative phosphorylation are also described.

### Experimental

Investigations of the mechanism of oxidative phosphorylation over the past decade have given several indications of the presence of intermediary reactions (intermediates) in oxidative phosphorylation: the properties of the energy-linked transhydrogenase reaction are consistent with the participation of a third component ($\sim$ or X $\sim$ 1) in the reaction mechanism[34] but no evidence as to its nature has been obtained. The demonstration that the reaction involves a direct hydrogen transfer[35, 36] even in the presence of an uncoupling agent[36] is clearly inconsistent with any chemiosmotic interpretation of this reaction. In addition, solvent extraction experiments[37, 38] with heart mitochondria indicated a function for quinone and other lipid components in the ATP-driven transhydrogenase reaction and the ATP-driven reduction of NAD by succinate. Similar results were obtained with *Escherichia coli* membrane vesicles.[38] These experiments and related solvent extraction experiments on chromatophores by Horio *et al.*[39] indicated that energy-conserving systems contained lipophilic moieties which were in equilibrium with ATP via the ATPase or ATP synthetase system. The identification and functional characterization of lipophilic components (hydrophobic subunits) of the ATP synthetase complex thus became a prime objective, especially as these components were associated with the site of action of the known inhibitors of energy transfer and oxidative phosphorylation, oligomycin and DCCD. The limitations of solvent extraction experiments and the poor state of the art with respect to fractionation of hydrophobic proteins dictated another experimental approach to the problem—the application of biochemical genetic techniques,[40, 41] in particular the mitochondrial genetics of baker's yeast, *Saccharomyces cerevisiae*.

## Biochemical Genetics of Drug-Resistant Mutants

The isolation of oxidative phosphorylation mutants and inhibitor-, uncoupler-, and ionophore-resistant mutants has been described elsewhere,[40–44] and the properties of relevant antibiotic-resistant mutants are summarized in Table I.

The rationale for these investigations was the premise that drugs (inhibitors, uncouplers, ionophores) that affect mitochondrial energy conservation reactions have specific inhibitor (receptor) sites associated with specific protein subunits of the oxidative phosphorylation complex.[40, 41] Genetic analysis should then reveal the number of components (mitochondrial gene products) involved. Different mitochondrial gene products have been shown to be involved in the binding site for oligomycin, venturicidin, and triethyl tin, this evidence not being previously available from biochemical studies. These findings correlate well with biochemical studies[19] and studies of mitochondrial biogenesis carried out by Tzagoloff *et al.*[14, 15] In addition, the demonstration in our laboratory of new mitochondrial loci (*OLI, OLII, OLIII*) has proved to be of great value in mitochondrial genetics in "mapping" of the mitochon-

Table I. Mutants of S. cerevisiae Resistant to Inhibitors, Uncouplers, and Ionophores Which Affect Oxidative Phosphorylation

| Group | Inhibitor | Resistant mutant isolated[a] |
|---|---|---|
| Electron transport | Antimycin A | Yes (M) |
| Uncouplers | TTFB (CCCP, DNP) | Yes (C) |
| | 1799 | Yes (C and M) |
| ATP synthetase inhibitors | Oligomycin | Yes (M) |
| | Venturicidin | Yes (M) |
| | Triethyl tin | Yes (C and M) |
| | DCCD | Yes (C and M) |
| $F_1$-ATPase inhibitors | Aurovertin | Yes (N) |
| ADP translocase | Bongkrekic | Yes (N and C) |
| | Rhodamine 6G | Yes (C and M) |
| Ionophores | Valinomycin | Yes (M?) |
| | Nigericin | Yes (C?) |

[a] N, Nuclear; M, mitochondrial; C, "cytoplasmic."

drial genome. (These aspects and related aspects of mitochondrial biogenesis are discussed in detail in contributions to this volume by Tzagoloff and by Linnane and Hall.) Of particular significance, in terms of mitochondrial function, are the studies on uncoupler- and ionophore-resistant mutants[40, 41] that indicate that mitochondrial membranes contain *protein* components involved in the mode of action of uncouplers and ionophores. The genetic evidence thus supports the concept that the various inhibitors of the ATP synthetase complex such as oligomycin, venturicidin, and triethyl tin react at different loci in a multistep reaction sequence.

Mutants resistant to triethyl tin, a known inhibitor of oxidative phosphorylation, Pi-ATP exchange, and OS-ATPase, in contrast to specific oligomycin- and venturicidin-resistant mutants, show cross-resistance to venturicidin, the uncoupling agent 1799, rhodamine 6G, and bongkrekic acid.[44, 46] Triethyl tin resistance may thus reflect modification of a component which serves as a link between the ATP synthetase complex and interacting systems. To identify this component a specific-affinity label derivative for the trialkl tin binding site in ATP synthetase was developed, dibutyl chloromethyl tin chloride (DBCT). DBCT was shown to have the properties of a site-directed inhibitor which competes for the triethyl tin binding site and which reacts covalently with a mitochondrial component[47, 48] and has all the inhibitory properties of its parent compounds, triethyl tin and tributyl tin. Excess DBCT can be removed by washing, but 8–9 nmoles DBCT/mg protein remains bound to the mitochondrial inner membrane, despite extensive washing and treatment with trichloroacetic acid. Studies of [³H]-DBCT-labeling of yeast submitochondrial particles and analysis by gel electrophoresis of the OS-ATPase complex showed two radioactive bands of molecular weight 46,000 and 8000. These correspond to the aggregated and monomeric forms of subunit 9, a component of OS-ATPase and a major product of mitochondrial protein synthesis in yeast[49] which appears to be present in excess as compared with ATP synthetase. DBCT thus appears to label subunit 9, a component of the OS-ATPase complex. However, examination of the labeled proteolipid by thin-layer chromatography showed that the radioactivity was associated with a nonprotein lipophilic component which is separable from subunit 9. Further studies showed that DBCT labeled one mitochondrial species, a nonprotein lipophilic component which was completely extractable by chloroform–methanol mixtures. Thus DBCT inhibition is due to reaction with a nonprotein lipophilic component of the mitochondrial inner membrane which is present in high concentration, 8–9 nmoles/mg protein; i.e., DBCT inhibits a lipid cofactor of oxidative phosphorylation.[48, 50]

## Evidence for a Role for Lipoic Acid in Oxidative Phosphorylation

a.   The lipophilic component which reacts with DBCT has been shown to contain lipoic acid.[48, 50] It has been tentatively identified as a lipoic acid conjugate containing an unidentified nonprotein lipophilic component.

b.   Lipoic acid (as a conjugate) has been shown to be present in high concentration in all bioenergetic membranes examined (heart SMP, yeast SMP, *E. coli* vesicles, chloroplasts, and chromatophores) by [³H]-DBCT binding and lipoic acid bioassay.[48, 50] Purified preparations of ATP synthetase, yeast OS-ATPase,[16] complex V,[13] and proton-translocating ATPase[17] contain stoichiometric amounts of lipoic acid (4–6 moles/mole complex).[50, 51]

c.   Oxidative phosphorylation, ATP-driven energy-linked reactions, and ATP synthetase (OS-ATPase) are inhibited by titration of the lipoic acid component present in the inner membrane by DBCT.[52] Inhibition by DBCT is specifically reversed by dihydrolipoic acid.[50, 52]

d.   8-Methyl lipoic acid, a known growth inhibitor[53] and inhibitor of lipoic acid requiring reactions,[54, 55] is a potent energy transfer inhibitor in the reduced form, and its mode of action is analogous to that of the classical energy transfer inhibitors of oxidative phosphorylation.[52]

    e.   Dibutyl tin dichloride, a known inhibitor of $\alpha$-ketoacid oxidation and a dithiol reagent with a mode of action similar to that of arsenite,[56] has been shown to be a potent inhibitor of oxidative phosphorylation, ATP-driven energy-linked reactions, and ATP synthetase (OS-ATPase).[57] In contrast to inhibition by DBCT, inhibition by dibutyl tin dichloride is reversible by a variety of thiols and dithiols, including dihydrolipoic acid.

    f.   Studies with a lipoic acid requiring mutant of *E. coli*[58] have demonstrated a specific lipoic acid requirement for oxidative phosphorylation.[59] Membrane vesicles prepared from lipoic-deficient cells have a normal electron transport chain but are unable to catalyze net ATP synthesis. ATP synthesis is restored simply by addition of cofactor amounts of lipoic acid to the incubation mixture (Table II).

    g.   Dihydrolipoate-dependent ATP synthesis has been demonstrated in mitochondria and submitochondrial particles where the respiratory chain has been rendered nonfunctional by inhibition by rotenone and antimycin A.[50, 51] The reaction is sensitive to uncoupling agents, oligomycin, and $F_1$-ATPase inhibitors but insensitive to ionophores.

    h.   Dihydrolipoate-dependent ATP synthesis has been demonstrated in yeast promitochondria[60] which do not contain the cytochrome components of the respiratory chain, again indicating that a functional respiratory chain is not required.

    i.   Therefore, the results indicate that lipoic acid residues have a specific role in the terminal reactions of oxidative phosphorylation and that all of the components required for ATP synthesis are present in the

*Table II. Restoration of Oxidative Phosphorylation in Membrane Vesicles from Lipoic Deficient E. coli (W1485 lip 2)[a]*

| Additions | ATP synthesis (nmoles/min) | | Oxygen uptake (ngatom/min) |
|---|---|---|---|
| | $\Delta$Pi | $\Delta$G6P | |
| None | 0 | 0 | <10 |
| Lipoate (1 $\mu$mole) | 0 | 0 | <10 |
| Succinate (10 $\mu$moles) | 0 | 0 | 170 |
| Succinate + lipoate (1 $\mu$mole) | 155 | 145 | 170 |
| Succinate + lipoate (100 nmoles) | 135 | 130 | 170 |
| Succinate + lipoate (10 nmoles) | 100 | 95 | 165 |
| Succinate + lipoate (5 nmoles) | 50 | 50 | 170 |

[a] ATP synthesis was assayed in a glucose-hexokinase trap system.[50, 51] Cofactors and inhibitors were added to the reaction medium 5 min before initiation of the reaction with succinate (10 $\mu$moles). *E. coli* vesicles from lipoic-deficient cells,[58] 1 mg protein, 20 min incubation at 30°C.

ATP synthetase complex. The lipoic acid component involved has not yet been identified but has been shown to be a nonprotein lipoic acid conjugate.

## Evidence for a Pool Function for Lipoic Acid

The high concentration of lipoic acid residues in the mitochondrial inner membrane (twice the concentration of ubiquinone) indicates a "pool" function for a component which is capable of linking the ATP synthetase complex with the respiratory chain. The specific titration of this component by dibutyl[$^3$H]chloromethyl tin chloride provides a method for examination of its distribution in the inner membrane.[50, 52] Fractionation of DBCT-labeled yeast mitochondrial inner membrane (8–9 nmoles DBCT/mg protein) shows that only 60% of the labeled component is extracted by 0.5% Triton X-100, a procedure which extracts all of the ATP synthetase (OS-ATPase). Only 60–70% of the component in the Triton X-100 extract is precipitated by an antibody to yeast OS-ATPase, and 50% of the component separates from the OS-ATPase on density gradient centrifugation of the Triton X-100 extract.[16] Thus only 30% of the DBCT-binding component (lipoate residues) in the inner membrane is specifically associated with the OS-ATPase complex. These residues appear to be associated with the subunit 9 component of the OS-ATPase complex. The location of the remainder of the DBCT binding component (6 nmoles/mg protein) has not been established and no specific association with the complexes of the respiratory chain has been demonstrated, as yet. A similar distribution was obtained during fractionation of the beef heart mitochondria using bile salts.

The question arises as to whether this pool of lipoate residues is a "free" mobile component of the inner membrane or is specifically associated with a proteolipid such as subunit 9, which also may be a mobile membrane component. Subunit 9 is the major product of protein synthesis in yeast mitochondria[14, 49] and appears to be present in excess as compared with ATP synthetase, and its role is of increasing interest as it is also the site of action of DCCD, whose mode of action appears to involve the other cofactor of oxidative phosphorylation, oleic acid.[60, 61] Studies of the mode of action of DBCT in heart mitochondria and the specific reversal of DBCT inhibition by dihydrolipoate[50, 52] indicate a specific site of action of DBCT on the ATP synthetase complex which is reversible either by added dihydrolipoate or by dihydrolipoate generated by the action of the respiratory chain which is part of the membrane-bound "pool" of lipoate residues not specifically associated with the ATP synthetase complex.[50, 52] The results are consistent with "cycling" of a mobile lipophilic membrane cofactor through specific interaction sites in the ATP synthetase complex.

Further evidence for this conclusion comes from experiments in which reversal of DBCT inhibition is achieved by mixing inhibited particles with noninhibited particles (Fig. 2A) and mixing DBCT-inhibited particles with particles inhibited by another inhibitor which reacts at a different reaction site, e.g., efrapeptin and DCCD (Fig. 2B). The results can be explained by protein subunit exchange reactions occurring after membrane fusion or by exchange of a lipophilic cofactor of ATPase. While the former possibility has not been excluded, interchange and reincorporation of a low molecular weight lipophilic cofactor are indicated. The process is enhanced by conditions under which oxidative phosphorylation occurs and is inhibited by uncouplers. The results also provide further evidence that the DBCT interaction site is different from the DCCD interaction site.

### *Evidence for a Cofactor Role for an Unsaturated Fatty Acid–Oleic Acid*

Investigation of ATP synthesis by isolated ATP synthetase preparations showed no ATP synthesis in the presence of dihydrolipoate alone, which suggested a deficiency in ATP synthetase preparations of a component normally present in the mitochondrial inner membrane necessary for generation of an active intermediate.[50, 51] Attention has then focused on the possibility that there was another lipophilic cofactor in the membrane, possibly a carboxylic acid by analogy with substrate-level phosphorylation. Experiments on the mode of action of DCCD similar to those described for DBCT in Fig. 2A,B indicated that the mode of action of DCCD also involved a mobile component of the mitochondrial inner membrane and that unsaturated fatty acids could reverse the DCCD inhibition of membrane-bound ATPase. These findings and the extensive literature on a requirement for unsaturated fatty acids for aerobic growth of yeast UFA auxotrophs and effects of UFA deficiency on oxidative phosphorylation[62-64] pointed to a role for unsaturated fatty acids. Table III shows that net synthesis of ATP can be achieved in a detergent-dispersed preparation of ATP synthetase (complex V) which is dependent on dihydrolipoate and cofactor amounts of oleoyl CoA and oleic acid.

Similar results have been obtained with heart proton-translocating ATPase, yeast OS-ATPase, and partially purified *E. coli* DCCD-sensitive ATPase complex. A specific requirement for unsaturated fatty acids is seen in all preparations and saturated fatty acids and equivalent acyl

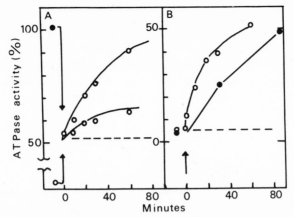

Fig. 2. Recovery of ATPase activity after DBCT inhibition. ATPase activity was assayed at 30°C as described previously.[50, 51] Experiment A: Particles 98% inhibited with 0.5 nmole DBCT/mg protein were mixed with an equal volume of noninhibited particles (1 mg/ml, specific activity 2.5 $\mu$moles/min/mg). The mixtures were then incubated at 4°C or 30°C, and 0.1-ml samples were taken for ATPase activity at times shown. Experiment B: Particles (1 mg/ml) 95% inhibited with 0.4 nmole DBCT/mg protein were mixed with an equal volume of particles (1 mg/ml) 95% inhibited with 0.2 $\mu$g/mg protein, efrapeptin (O), or mixed with particles 95% inhibited with 0.2 $\mu$g/mg protein DCCD (●). The dashed line represents the theoretical maximal activity.

CoA derivatives are inhibitory in situations where a requirement for added oleate and oleoyl CoA has been demonstrated. It appears that the unsaturated fatty acid cofactor is removed during the bile salt fractionation procedures or Triton X-100 fractionation procedures used to prepare isolated ATP synthetase. In membrane preparations such as mitochondria and SMP, the cofactor is still present in the membrane and there is no requirement for added fatty acid cofactors, and added

*Table III. Net Synthesis of ATP by ATP Synthetase[a]*

| Additions | Pi uptake ($\mu$moles/20 min) |
|---|---|
| Oleate (10 nmoles) | 0.0 |
| Oleoyl CoA (10 nmoles) | 0.0 |
| Oleoyl CoA + oleate[b] | 2.2 |
| Oleoyl CoA + palmitate (10 nmoles) | 0.0 |
| Palmitoyl CoA + oleate | 0.0 |

[a] A glucose-hexokinase trap system was used as described previously.[50] 0.5 mg complex V; reaction was started by addition of 2.85 $\mu$moles dihydrolipate.
[b] The reaction is sensitive to oligomycin, DCCD, and uncoupling agents such as 1799 and CCCP and to $F_1$ inhibitors such as efrapeptin.

saturated fatty acids such as palmitate do not inhibit dehydrolipoate-dependent ATP synthesis. However, if the mitochondrial membrane is depleted of UFA by manipulating the nutrition or growth conditions of the cell, then a requirement for oleate and oleoyl CoA is observed in the mitochondrial preparation. Under these conditions, saturated fatty acids such as palmitate are inhibitory, as in the case of isolated ATP synthetase preparations.

These observations have been made with yeast promitochondria grown under conditions of UFA depletion[60] and with the yeast UFA auxotroph KD 115, where depletion of mitochondrial UFA from 75% to 35% causes loss of oxidative phosphorylation[61] and loss of dihydrolipoate-dependent ATP synthesis.[50, 51] This reaction can be restored by addition of cofactor amounts of oleate and oleoyl CoA.[50, 61] These experiments provide further evidence that oleic acid has a cofactor role and that the mitochondrial inner membrane contains a "free" cofactor pool of unsaturated fatty acid which is distinct from the UFA pool forming the structural components of membrane phospholipids. This cofactor pool is also responsive to the physiological and nutritional status of the cell and may be an important regulatory point in cellular energy metabolism.

Studies of the specificity of the requirement for unsaturated fatty acids indicate a high degree of specificity for $cis$-$\Delta^9$-monoenoic acids[65]; the *trans* isomers are inactive, and in many cases are potent inhibitors. A similar specificity has been reported by Walenga and Lands[63] in their studies of the growth yield of an UFA auxotroph on a variety of unsaturated fatty acids. A previous report from our laboratory[51] that arachidonic acid is active in this assay has not been confirmed with other samples of arachidonic acid and must be assumed to have been due to a contamination.

These studies provide the first demonstration of a biochemical function for unsaturated fatty acids in intermediary metabolism and provide an explanation for the longstanding association of UFA deficiency with reduced efficiency of oxidative phosphorylation[66] and for the lack of ability of *E. coli* and yeast UFA auxotrophs to grow on oxidizable substrates.[67–69]

### Studies of the Mechanism of ATP Synthesis

The demonstration of two specific cofactors of oxidative phosphorylation, lipoate and oleate, and the generation of ATP in dihydrolipoate-, oleoyl CoA-, and oleate-dependent reactions indicate a series of reactions analogous to those of substrate-level phosphorylation. These reactions, involving the generation (by oxidative decarboxylation of a ketoacid) of succinyl lipoate, transacylation to CoA, and generation of GTP (ATP) via a succinyl phosphate intermediate, form the basic model system for the mechanism of oxidative phosphorylation[1,2] and provide the only logical mechanistic basis for the Pi-ATP exchange reaction characteristic of oxidative phosphorylation. The analogous series of components in oxidative phosphorylation would then be:

$$\text{Respiratory chain} \rightarrow \text{oleoyl} \sim S\text{-lipoate} \nrightarrow \text{oleoyl} \sim \text{CoA (ACP)} \rightarrow$$
$$\text{Pi} \searrow \quad \text{CoA?}$$
$$\text{oleoyl CoA(ACP)} \rightarrow \text{oleoyl} \sim P \rightarrow \text{ATP}$$

The following experiments indicate that the above reactions are relevant to the mechanism of oxidative phosphorylation:

a. Oleoyl-$S$-lipoate has been shown to generate ATP in the absence of added CoA in an uncoupler- and inhibitor-sensitive reaction catalyzed by mitochondrial membranes and ATP synthetase preparations from heart, yeast, and *E. coli*.[50,51] Elaidoyl-$S$-lipoate is inactive in these reactions. In addition, oleoyl-$S$-lipoate has been shown to generate ATP in chloroplast preparations, indicating that a similar series of reactions are involved in photophosphorylation.[70]

b. Transphosphorylation from oleoyl phosphate to ADP has been demonstrated, catalyzed by mitochondria, submitochondrial particles, chloroplasts, and ATP synthetase preparations from heart, yeast, and *E. coli*.[50,51] The reaction catalyzed by heart preparations is not inhibited by uncouplers or by oligomycin, but is sensitive to DCCD and to efrapeptin, and $F_1$-ATPase inhibitor.[51]

c. ATP-driven energy-linked reactions such as the energy-linked transhydrogenase and energy-linked reduction of NAD by succinate are inhibited by dibutylchloromethyl tin chloride and dibutyl tin dichloride, reagents which react with lipoic residues in the mitochondrial inner membrane, indicating a role for a lipoate derivative as an intermediate in these reactions. Dihydrolipoate has previously been reported to be a coupling factor for the energy-linked transhydrogenase reaction in *Rhodospirillum*,[71] and recent studies have shown that dihydrolipoate drives the energy-linked transhydrogenase reaction.[71a]

The available evidence is consistent with a role for acyl lipoate and acyl phosphate intermediates in the mechanism of oxidative phosphorylation. There is no direct evidence at present for a role for oleoyl CoA, and the available evidence indicates that a bound form of CoA is present in the ATP synthetase as one of the ATP synthetase subunits has been shown to contain bound pantothenate.

A major unresolved mechanistic problem is the synthesis of oleoyl-*S*-lipoate, a possible intermediate. Two general hypotheses are being tested on the basis of the available experimental evidence. Any system that continuously reduces the lipoate pool will facilitate the reductive acylation of a dithiol residue in the ATP synthetase complex, which, by transacylation, phosphorolysis, and transphosphorylation, leads to the synthesis of ATP in a reaction sequence analogous to the reaction catalyzed by succinic thiokinase in substrate-level phosphorylation (Fig. 3). Such a reaction sequence is implied in the demonstration of dihydrolipoate-dependent ATP synthesis by SMP and dihydrolipoate-oleoyl CoA-, and oleate-dependent synthesis of ATP by ATP synthetase preparations. Supporting evidence comes from experiments with *Halobacterium* purple membrane preparations mixed with rotenone- and antimycin A-inhibited submitochondrial particles or ATP synthetase preparations. Light-dependent ATP synthesis has been demonstrated by simply mixing the preparations together (Table IV) and irradiating at 560 nm. Similar results are obtained if SMP is replaced by ATP synthetase preparations.

These results are unlikely to be the result of the generation of a proton gradient or membrane potential as claimed in the Racker–Stoeckenius experiment.[26] The reaction rate is enhanced by addition of oxidized lipoate, and the simplest explanation is that lipoate or a component in redox equilibrium with it is photoreduced by *Halobacterium* purple membrane. Direct photoreduction of lipoate by purple membrane has been demonstrated.[70] The capacity of purified respiratory chain complexes to catalyze disulfide reduction and lipoate reduction is under investigation, as is the possible location of DBCT-binding sites (dithiol residues) in respiratory chain complexes.

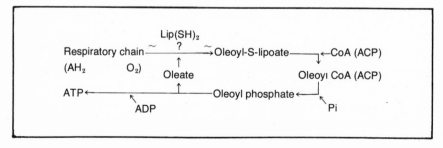

Fig. 3. The "oleoyl cycle" of oxidative phosphorylation.

Table IV. ATP Synthesis with Purple Membrane:
Light (560 nm) → PM + SMP → ATP[a]

| Additions | Light/dark | ΔPi (μmoles) 20 min | 40 min |
|---|---|---|---|
| Reduced lipoate (2 μmoles) | D | 0.67 | 1.08 |
| Oxidized lipoate (100 nmoles) | D | 0 | 0 |
| PM (0.4 mg) + oxidized lipoate | D | 0 | 0 |
| PM (0.4 mg) | L | 0.27 | 0.67 |
| PM + oxidized lipoate | L | 1.08 | 1.48 |

[a] Glucose-hexokinase trap system, 30°C, 0.4 mg *Halobacterium* purple membrane Heart SMP inhibited with antimycin A + rotenone, 2 mg. Light source 500 W projector. Filter 500–600 nm, cutoff 560 nm max.

Alternatively, a high-energy intermediate may be generated at the level of the respiratory chain and oleoyl lipoate generated by a transacylation reaction. A similar general mechanism of oxidative phosphorylation involving a carboxylic acid has been proposed by Falcone[72] and can be investigated by appropriate oleoyl exchange reactions and $^{18}O$ exchange reactions. Some evidence for an intermediate step prior to the involvement of lipoate residues has been obtained in studies of the energy-linked transhydrogenase reaction as the succinate-driven transhydrogenase reaction is not inhibited by levels of DBCT which are assumed to completely titrate the lipoate pool in the mitochondrial inner membrane.[50] The generation of an intermediate at the respiratory chain level prior to the involvement of lipoate residues is thus indicated. It should be possible to examine these possibilities by development of oleoyl exchange assays for testing of respiratory chain complexes and testing of oleoyl ~ carrier derivatives such as oleoyl ubiquinol.

The information obtained to date provides strong evidence for a chemical mechanism for oxidative phosphorylation with respect to the energy transfer steps and phosphorylation steps, but definitive information as to the link with the respiratory chain has not yet been obtained. Biochemical genetic techniques can provide a major insight into the problem as a large number of yeast mutants[73-75] and *E. coli* mutants[76-78] are available with defects in oxidative phosphorylation. These mutants are now amenable for intensive study of the biochemical defects involved as a result of the developments described in this chapter.

ACKNOWLEDGMENTS

I was particularly pleased to receive an invitation to contribute to this symposium volume in honor of Professor David Green. It was my privilege to work in his laboratory at the Institute for Enzyme Research from 1958 to 1961 and to participate in the research into the enzymology of ubiquinone, the fractionation of the respiratory chain, the isolation of the respiratory chain complexes, and the role of lipids in energy conservation reactions. They were exciting, pioneering days and the general principles which evolved have been the basic guidelines for work in this field and form the framework for the studies of the mechanism of oxidative phosphorylation summarized in this chapter.

I wish to thank my research students for their support over the past 8 years and for their hard work and efforts which are the necessary biological requirements to overcome a major gradient. This work was supported by grants from the Science Research Council and the Medical Research Council and by NATO Grant 572.

## References

1. Slater, E. C. (1966) *Compr. Biochem. 14*, 327–396.
2. Reed, L. J. (1966) *Compr. Biochem. 14*, 99–126.
3. Boyer, P. D. (1965) in *Oxidase and Related Redox Systems* (King, T. E., *et al.*, eds.), pp. 994–1008, Wiley, New York.
4. Green, D. E. (1974) *Biochim. Biophys. Acta 346*, 27–78.
5. Mitchell, P. (1966) *Biol. Rev. 41*, 445.
6. Klingenberg, M. (1976) in *Energy Transformation in Biological Systems*, Ciba Foundation Symposium 31, pp. 22–40, Associated Scientific Publishers, Amsterdam.
7. Van Dam, K., and Meyer, A. L. (1971) *Ann. Rev. Biochem. 40*, 115–160.
8. Lee, C. P., and Ernster, L., (1968) *Eur. J. Biochem. 3*, 385–390.
9. Hatefi, Y., Haavik, A. G., and Griffiths, D. E. (1962) *J. Biol. Chem. 237*, 1676–1680.
10. Ziegler, D. M., and Doeg, K. A. (1959) *Biochim. Biophys. Res. Commun. 1*, 344–349.
11. Hatefi, Y., Haavik, A. G., and Griffiths, D. E. (1962) *J. Biol. Chem. 237*, 1681–1685.
12. Griffiths, D. E., and Wharton, D. C. (1961) *J.Biol. Chem. 236*, 1850–1856.
13. Hatefi, Y., Hanstein, W. G., Galante, Y., and Stiggall, D. L. (1975) *Fed. Proc. 34*, 1699–1706.
14. Tzagoloff, A., Rubin, M. S., and Sierra, M. F. (1973) *Biochim. Biophys. Acta 301*, 71–104.
15. Schatz, G., and Mason, T. (1974) *Ann. Rev. Biochem. 43*, 51–87.
16. Tzagoloff, A., and Meagher, P. (1971) *J. Biol. Chem. 246*, 7328–7336.
17. Serrano, R., Kanner, B. I., and Racker, E. (1976), *J. Biol. Chem. 251*, 2453–2461.
18. Hunter, D. R., Komai, H., and Howarth, R. A. (1974) *Biochem. Biophys. Res. Commun. 56*, 647–653.
19. Griffiths, D. E., and Houghton, R. L. (1974) *Eur. J. Biochem. 46*, 157–167.
20. Cain, K., and Griffiths, D. E. (1977) *Biochem. J. 162*, 575–580.
21. Komai, H., Hunter, D. R., Southard, J. H., Howarth, R. A., and Green, D. E. (1976) *Biochem. Biophys. Res. Commun. 69*, 695–704.
22. Ragan, C. I., and Racker, E. (1973) *J. Biol. Chem. 248*, 6876–6882.
23. Racker, E., and Kandrach, A. (1973) *J. Biol. Chem. 248*, 5841–5847.
24. Racker, E. (1975) *Biochem. Soc. Trans. 3*, 785–802.
25. Senior, A. E. (1973) *Biochim. Biophys. Acta 301*, 249–277.

26. Racker, E., and Stoeckenius, W. (1974) *J. Biol. Chem. 248*, 662–663.
27. Green, D. E., and Ji, S. (1973) *Proc. Natl. Acad. Sci USA 56*, 904–908.
28. Ernster, L., Nordenbrand, K., Chude, O., and Juntti, K. (1974) in *Membrane Proteins in Transport and Phosphorylation* (Azzone, G. F., *et al.*, eds.), pp. 29–41, North-Holland, Amsterdam.
29. Singer, S. J. (1974) *Ann. Rev. Biochem. 43*, 805–833.
30. Bertoli, E., Finean, J. B., and Griffiths, D. E. (1976) *FEBS Lett. 61*, 163–165.
31. Mitchell, P. (1976) *Biochem. Soc. Trans. 4*, 399–430.
32. Skulachev, V. P. (1971) *Curr. Top. Bioenergetics 4*, 127–190.
33. Ferguson, S. J., John, P., Lloyd, W. J., Radda, G. K., and Whatley, F. R. (1976) *FEBS Lett. 62*, 272–275.
34. Ernster, L., and Lee, C. P. (1964) *Ann. Rev. Biochem. 33*, 729–788.
35. Lee, C. P., Simard-Duquesne, N., Ernster, L., and Hoberman, H. D. (1965) *Biochim. Biophys. Acta 105*, 397–409.
36. Griffiths, D. E., and Roberton., A. M. (1966) *Biochim. Biophys. Acta 118*, 453–464.
37. Sweetman, A. J., and Griffiths, D. E. (1970) *FEBS Lett. 10*, 92–96.
38. Sweetman, A. J. (1970) Ph.D. thesis, University of Warwick.
39. Horio, T., Nishikawa, K., Akayama, S., Horuiti, Y., Yamamota, Y., and Katutani, Y. (1968) *Biochim. Biophys. Acta 153*, 913–916.
40. Griffiths, D. E. (1972) in *Mitochondria: Biogenesis and Bioenergetics* (Van den Bergh *et al.*, eds.), pp. 95–104, North-Holland, Amsterdam.
41. Griffiths, D. E. (1976) in *The Structural Basis of Membrane Function* (Hatefi, Y., and Djavadi-Ohaniance, L., eds.), pp. 205–214, Academic Press, New York.
42. Avner, P. R., and Griffiths, D. E. (1973) *Eur. J. Biochem. 32*, 312–321.
43. Griffiths, D. E., Houghton, R. L., Lancashire, W. E., and Meadows, P. A. (1975) *Eur. J. Biochem. 51*, 392–402.
44. Lancashire, W. E., and Griffiths, D. E. (1975) *Eur. J. Biochem. 51*, 377–392.
45. Tzagoloff, A. (1976) in *Genetics and Biogenesis of Chloroplasts and Mitochondria* (Bucher, T. *et al.*, eds.), pp. 419–426, North-Holland, Amsterdam.
46. Carignani, G., Lancashire, W. E., and Griffiths, D. E. (1977) *Molec. Gen. Genet. 151*, 49–56.
47. Cain, K., Partis, M. D., and Griffiths, D. E. (1975) in *International Symposium on Electron-Transfer Chains and Oxidative Phosphorylation*, abst. vol., p. 47, University of Bari, Bari.
48. Cain, K., Partis, M. D., and Griffiths, D. E. (1977) *Biochem. J., 166*, 593–602.
49. Tzagoloff, A., Akai, A., and Foury, F. (1976) *FEBS Lett. 65*, 391–396.
50. Griffiths, D. E. (1976) in *Genetics and Biogenesis of Chloroplasts and Mitochondria* (Bucher, T., *et al.*, eds.), pp. 175–185, North-Holland, Amsterdam.
51. Griffiths, D. E. (1976) *Biochem. J. 160*, 809–812.
52. Cain, K., and Griffiths, D. E. (1977) *Biochem. Soc. Trans. 5*, 205–207.
53. Stokstad, E. L. R. (1954) *Fed. Proc. 13*, 712–714.
54. Reed, L. J. (1957) *Adv. Enzymol. 18*, 319–347.
55. Schmidt, U., Graften, P., Atland, K., and Goedde, H. W. (1969) *Adv. Enzymol. 32*, 423–469.
56. Aldridge, N., and Cremer, J. (1955) *Biochem. J. 61*, 406–418.
57. Cain, K., and Griffiths, D. E. (1977) *FEBS Lett. 82*, 23–28.
58. Herbert, A. A., and Guest, J. R. (1970) *Methods Enzymol. 18A*, 269–272.
59. Partis, M. D., Hyams, R. L., and Griffiths, D. E. (1977) *FEBS Lett., 75*, 47–51.
60. Griffiths, D. E., Hyams, R. L., and Bertoli, E. (1977) *FEBS Lett., 74*, 38–42.
61. Griffiths, D. E., Hyams, R. L., and Bertoli, E. (1977), *Biochem. Biophys. Res. Commun., 75*, 449–456.
62. Haslam, J. M., Proudlock, J. W., and Linnane, A. W. (1971), *J. Bioenergetics 2*, 351–370.

63. Walenga, R. W., and Lands, W. E. M. (1975) *J. Biol. Chem. 250*, 9130–9136.
64. Marzuki, S., Cobon, G. S., Haslam, J. M., and Linnane, A. W. (1975) *Arch. Biochem. Biophys. 169*, 577–590.
65. Griffiths, D. E., Hyams, R. L., and Carver, M. (1977) unpublished work.
66. Klein, P. D., and Johnson, R. M. (1954) *J. Biol. Chem. 211*, 103–110.
67. Silbert, D. F., and Vagelos, R. (1967) *Proc. Natl. Acad. Sci. USA 58*, 1579–1586.
68. Esfahani, M., Barnes, E., and Wakil, S. J. (1969) *Proc. Natl. Acad. Sci. USA 64*, 1057–1064.
69. Resnick, M. A., and Mortimer, R. K. (1966) *J. Bacteriol. 92*, 597–600.
70. Hyams, R. L., and Griffiths, D. E. (1977), unpublished results.
71a. Cain, K., Carver, M., and Griffiths, D. E., unpublished work.
71. Orlando, J. A., (1970) *Arch. Biochem. Biophys. 141*, 111–120
72. Falcone, A. B., (1966) *Proc. Natl. Acad. Sci. USA 56*, 1043–1046.
73. Kovac, L., Lachowicz, T., and Slonimski, P. (1967) *Science 158*, 1564–1567.
74. Tzagoloff, A., and Foury, F. (1976) *Mol. Gen. Genet. 149*, 43–50.
75. Lancashire, W. E., Darlison, M., and Griffiths, D. E., unpublished work.
76. Cox, G. B., and Gibson, F. (1974) *Biochim. Biophys. Acta 346*, 1–26.
77. Schairer, H. U., Friedl, P., Schmid, B. I., and Vogel, G. (1976) *Eur. J. Biochem. 66*, 257–268.
78. Simoni, R. D., and Postma, P. W. (1975) *Ann. Rev. Biochem. 44*, 523–554.

## Chapter 14

# Ion Transport by Mitochondria

## Gerald P. Brierley

### Introduction

The inner membrane of the mitochondrion fulfills several crucial roles in cell metabolism. It segregates the enzymes and metabolites of the major energy-yielding metabolic pathways from alternative reactions in the cytosol and, at the same time, provides pathways for regulated interchange of metabolites between the matrix and nonmitochondrial compartments.[1, 2] In addition, a large body of evidence suggests that the primary mode of energy coupling in the mitochondrion consists of pH and electrical gradients across[3] or within[4] the inner membrane of the mitochondrion. It has become apparent that the movement of protons, anions, and cations across the inner membrane can profoundly affect both energy coupling and the regulation of metabolism. It is also clear that we have only begun to understand the pathways that are available to these components in the mitochondrial membrane, and we can expect that future work will provide a complete explanation of the mechanism and control features of these processes.

My own studies of mitochondrial ion transport were begun in 1961 in collaboration with David Green at the Enzyme Institute. With Drs. Erik Murer and Elizabeth Bachmann we found that heart mitochondria took up massive amounts of $Mg^{2+}$ and Pi in an energy-dependent reaction which was an alternative to oxidative phosphorylation, rather than an ATP-dependent transport.[5-7] This accumulation, like that of $Ca^{2+}$ [8] and $Mn^{2+}$,[9] attained high levels of ion uptake as a result of

*Gerald P. Brierley* • Department of Physiological Chemistry, College of Medicine, Ohio State University, Columbus, Ohio 43210.

deposition of the insoluble salts of the divalent cations and was also accompanied by a near stoichiometric extrusion of $H^+$. In these early studies our emphasis was on the transport of Pi in connection with the phosphorylation reaction,[5] but the demonstration of energy-dependent membrane loading of divalent cations in the absence of $Pi$[9, 8] shifted attention to the transport of cations. In the years that have followed, the concept has emerged that an electrogenic extrusion of $H^+$ is the primary event in mitochondrial ion transport reactions and that cation and anion uptake and extrusion are direct reflections of the pH gradient ($\Delta pH$)* and the negative interior potential ($\Delta\Psi$) which are produced by electron transport or the proton-translocating ATPase.[3] This view is presently held by a large number of workers in the field of bioenergetics, but is by no means universal. A number of recent reviews have summarized the experimental evidence and various alternative interpretations.[3, 8, 10-23] In the present chapter I wish to present a brief overview of the ion movements which occur in mitochondria and to summarize some recent experiments from our laboratory on monovalent cation uptake and extrusion which I feel are best interpreted in terms of the chemiosmotic coupling model.[3]

## Overview of Mitochondrial Ion Movements

The various pathways available for protons, anions, and cations across the mitochondrial membrane are summarized in Fig. 1. The primary energy-dependent event appears to be the electrogenic separation of $H^+$ and $OH^-$ across the membrane so that the matrix becomes more alkaline.[3] Kinetic analyses of the formation and decay of $\Delta pH$ in intact mitochondria[3, 17] and SMP[15] indicate that the proton ejection is a direct result of the redox reactions of the electron transport system or the action of the proton-translocating ATPase. The chemiosmotic coupling concept is also supported by a wide variety of reconstitution and model membrane studies (e.g., see refs. 13, 16). The production of $\Delta pH$ by electron transport and membrane ATPase is not unique to mitochondria, but appears to be the basis of substrate uptake, energy coupling, and osmotic volume control in many microorganisms and in chloroplasts.[3, 13, 24]

Most permissible anion movements in mitochondria occur via a

---

* Abbreviations: SMP, submitochondrial particles, CCP, $m$-chlorocarbonylcyanidephenyl-hydrazone; $DDA^+$, dimethyldibenzylammonium; $\Delta pH$, proton differential across the inner membrane of the mitochondrion; $\Delta\psi$, electrical potential across the inner membrane.

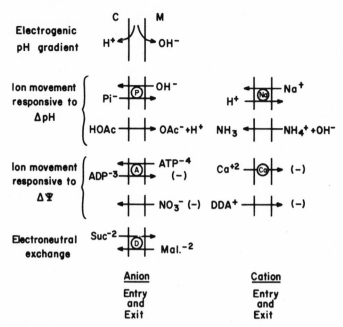

Fig. 1. Pathways for $H^+$, anions, and cations to cross the inner membrane of the mitochondrion. The electrogenic ejection of $H^+$ on the cytosol (C) side of the membrane is brought about by either the electron transport system or the proton-translocating ATPase.[3] See the text for explanation and details.

complex series of exchangers present in the membrane.[1, 2, 19, 25] Some of these exchangers involve $H^+$ or $OH^-$, such as the $Pi^-$ for $OH^-$ exchange shown in Fig. 1 for the phosphae transporter or the equivalent mediated symport of $H^+$ and $Pi^-$.[26] Such exchange reactions provide a means of converting $\Delta pH$ to anion gradients or a membrane potential with the interior negative ($\Delta \Psi$). Penetration of acetate and other weak acid anions appears to occur via the free acid as shown (Fig. 1) and will consequently reflect the $\Delta pH$. The entry of acetate into the matrix does not appear to be mediated by a transporter, and the extent to which this type of anion uptake may occur physiologically is uncertain.

A number of electroneutral anion exchanges (Fig. 1) involving phosphate and malate can occur via the dicarboxylate and ketoglutarate transporters, and these exchanges seem to be of great importance in the communication of metabolites between mitochondrial and extramitochondrial reaction pathways.[1, 2] At least two anion transporters, the adenine nucleotide and the glutamate-aspartate translocaters, promote an exchange which is not balanced electrically (Fig. 1) and, consequently, will be responsive to the membrane potential. Thus the interior negative charge will promote $ATP^{4-}$ extrusion and $ADP^{3-}$ uptake on the adenine nucleotide transporter.[19]

A wide variety of inorganic (nitrate, $SCN^-$) and organic ($PCB^-$, phenyldicarbaundecaborane; see ref. 16) anions penetrate natural and artificial membranes at significant rates. Since these components appear to move as the charged species, the extent and direction of their mobility will be determined by the membrane potential (Fig. 1, nitrate). This type of anion movement does not appear to occur physiologically, however.

The options available for cation movement are more limited than the pathways for anions. Weak bases, such as $NH_3$, can cross the mitochondrial membrane and generate a cation ($NH_4^+$) by ionization (Fig. 1). Obviously, this type of cation movement will be responsive to $\Delta pH$, and numerous examples are available in which respiration produces extrusion of $NH_4^+$ salts.[27, 28] Considerable evidence (to be discussed below) indicates that a cation$^+$/$H^+$ exchanger is present in the membrane which will respond to $\Delta pH$ as shown in Fig. 1. The exchanger shows cation discrimination with greater activity for $Na^+$ than $K^+$ (for review, see ref. 10).

The avid uptake of $Ca^{2+}$ by mitochondria appears to be a mediated process[29-30] and is generally regarded as an electrophoretic movement of cation in response to $\Delta\Psi$ as shown in Fig. 1.[29-31] Lipid-soluble synthetic cations, such as $DDA^+$,[16] appear to distribute across the membrane as a function of the membrane potential. Thus such lipid-soluble cations uncouple intact mitochondria, in which the matrix potential is negative, but have little effect on SMP, in which the interior is positive.[16]

### The Uptake and Extrusion of Monovalent Cations by Mitochondria

Chappell and Crofts[32] provided a rationale for mitochondrial swelling and contraction reactions when they pointed out that osmotic swelling of the organelles will occur in salt media only when both the anion and cation can enter the matrix. For example, the rapid swelling observed when mitochondria are suspended in $NH_4^+$ acetate[32] is readily explained in terms of entrance of $NH_3$ and free acetic acid into the matrix (*cf.* Fig. 1), neutralization of the acetic acid by $NH_3$ to produce an increase in matrix salt concentration, and water movement to equate solvent activity on the two sides of the membrane and bring about osmotic swelling. The light-scattering changes which accompany such osmotic swelling and contraction reactions provide a convenient way to

examine ion movements in mitochondria and have been used extensively in our laboratory to follow $K^+$ and $Na^+$ uptake and extrusion.[10, 33]

Nonenergized mitochondria do not seem to be permeable to $K^+$ or
to be able to exchange $K^+$ for $H^+$ at rates sufficient to support passive
swelling in $K^+$ nitrate or $K^+$ acetate.[28, 33] Since rapid swelling occurs
under these conditions when transmembrane permeability to $K^+$ is
increased by addition of valinomycin in the nitrate medium and when
$K^+/H^+$ exchange is enhanced by addition of an exogenous exchanger,
such as nigericin, in the acetate medium, it is clear that the low rate of $K^+$
influx is the limiting factor in the absence of these ionophores.[33] The
rapid swelling seen when mitochondria are suspended in $Na^+$ acetate
has been taken as one of the pieces of evidence for the presence of a
$Na^+/H^+$ exchanger in the mitochondrial membrane.[10, 33] In this case,
penetration of acetic acid would provide a source of $H^+$ in the matrix,
and the exchange of $Na^+$ (in) for $H^+$ (out) would result in osmotic
swelling due to accumulation of $Na^+$ acetate.

In contrast to these results with nonenergized mitochondria, respiring mitochondria swell rapidly in $K^+$ as well as $Na^+$ acetate (Fig. 2A). A
rationale for this energy-dependent, uncoupler-sensitive net uptake of
$K^+$ and acetate is provided in the diagram below the swelling record in
Fig. 2A. In 100 mM $K^+$ acetate the neutralization of the respiration-
dependent pH gradient by free acetic acid penetration is rapid and
complete to the extent that $\Delta pH$ is virtually eliminated as a component
of proton motive force. What results can really be considered to be an
acetate "pump," with acetate accumulation producing a large increase in
membrane potential ($\Delta\Psi$). Under the conditions of Fig. 2A this $\Delta\Psi$
apparently becomes so large that the normal permeability barrier to $K^+$
entrance is overcome and $K^+$ entry and osmotic swelling result. Studies
with $^{42}K^+$ indicate that, when matrix $K^+$ is maintained in a steady state,
$^{42}K^+$ influx shows saturation kinetics,[34–38] a result which suggests that a
specific $K^+$ entry mechanism may be present. However, at higher
concentrations of $K^+$ (above 40 mM), there is some question as to
whether $K^+$ influx can be saturated,[37] and respiration-dependent
swelling in 100 mM acetate salts occurs with almost all cations tested.[39]
The reaction is more efficient with $K^+$ than with tetramethylam-
monium$^+$, for example,[39] but there appears to be a real question as to
whether a specific $K^+$ uniport is functional under these conditions.
Whereas the $K^+$ concentration (100 mM) in the net influx experiment
(Fig. 2A) is close to physiological levels, the high level of acetate appears
to provide an unphysiological conversion of $\Delta pH$ to an acetate gradient
which is free of the constraints imposed on the mediated exchange

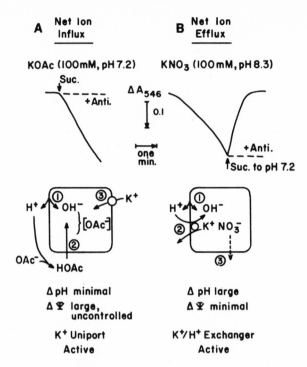

Fig. 2. Comparison of net ion influx with net ion efflux in mitochondria. For influx, the accumulation of $K^+$ acetate is followed as decreased absorbance at 546 nm in an Eppendorf photometer. Beef heart mitochondria (0.5 mg/ml) were added to a medium of $K^+$ acetate (100 mM) containing tris acetate (2 mM, pH 7.2) and rotenone (1 μg/ml) at 25°C. Swelling was initiated by addition of $K^+$ succinate (4 mM). The effect of antimycin is shown in the dashed trace. To demonstrate net ion efflux, the mitochondria are swollen at 35°C and pH 8.3 in a medium of $K^+$ nitrate (100 mM) containing tris nitrate (2 mM) and rotenone (1 μg/ml). After an appropriate period of passive swelling, tris succinate (4 mM) is added to initiate respiration and bring the pH to 7.2. Contraction and ion extrusion are seen as an increase in absorbance at 546 nm. The dashed trace shows the effect of antimycin. The two diagrams summarize the salient features of respiration-dependent ion accumulation (swelling) as opposed to respiration-dependent ion extrusion (contraction).

reactions. It appears quite likely that if a $K^+$-specific uniport is present in the mitochondrion it may be "voltage-gated"[40] and inoperative under conditions of low $\Delta\Psi$.

Evidence in favor of the model shown for net $K^+$ acetate influx (Fig. 2A) includes the following points: (a) Valinomycin and gramicidin increase transmembrane permeability to $K^+$ and strongly activate the net ion influx in acetate media.[10, 33] Such increases in transmembrane permeability would be expected to promote ion uptake and swelling by reducing the barrier to $K^+$ influx in the model shown in Fig. 2A, but would be expected to undermine the efficiency of models which require $K^+$ pumping by electrogenic[41] or electroneutral[14] mechanisms (for a review and more detailed exposition, see ref. 10). (b) Increased efflux of $^{42}K^+$ during net $K^+$ influx in the presence of valinomycin has been demonstrated.[42] This result is clearly incompatible with suggestions that valinomycin operates in series with an otherwise cryptic $K^+$ pump.

The low passive permeability of the mitochondrial membrane to $K^+$ (and to $Na^+$, when the $Na^+/H^+$ exchange pathway is not operative) can be overcome by increasing the pH to 8.3 at 35°C in a nitrate medium.[43-45] Mitochondria swollen under these conditions provide an ideal demonstration of the ability of the organelle to extrude monovalent cations in an energy-dependent reaction (Fig. 2B). The respiration-dependent, uncoupler-sensitive, net efflux of $K^+$ and nitrate can be rationalized as shown in the diagram below the contraction record (Fig. 2B). Under these conditions, when respiration is initiated the proton motive force would be expressed largely as $\Delta pH$, since the permeant nitrate anion would minimize $\Delta\Psi$. The high $\Delta pH$, would bring the endogenous cation$^+/H^+$ exchanger into play to extrude internal $K^+$ at the expense of the pH gradient, and nitrate efflux would follow to produce osmotic contraction.

Evidence in favor of the participation of the cation$^+/H^+$ exchanger in net $K^+$ and $Na^+$ efflux (Fig. 2B) includes the following: (a) The reaction is inhibited by reagents which increase transmembrane permeability to cations, such as gramicidin,[43] valinomycin,[44] and elevated pH,[43] but the efficiency is increased by $Mg^{2+}$, which decreases transmembrane permeability to $Na^+$ and $K^+$.[43] (b) The reaction is stimulated by nigericin[44] and closely resembles the ion uptake reaction seen in SMP.[46] Since the particles have the opposite orientation to intact mitochondria, ion uptake in SMP should correspond to ion extrusion in mitochondria. (c) Ion extrusion shows the cation selectivity sequence $Na^+ > Li^+ > K^+ >$ choline$^+$, the identical sequence to that of passive swelling in the corresponding acetate salts.[44] As discussed above, the passive swelling in acetate has been explained by the action of a cation$^+/H^+$ exchanger. (d) Ion extrusion (contraction) shows the expected competition between external cation and external $H^+$ (Fig. 3). Mitochondria swollen in increasing concentrations of $NaNO_3$ show decreasing rates of ion extrusion as the external $Na^+$ concentration increases (Fig. 3A). At an unfavorable $Na^+$ concentration (180 mM), extrusion is increased by addition of either (i) nigericin, which provides an exogenous, parallel pathway for $Na^+$ (out) and $H^+$ (in), or (ii) mersalyl, which blocks alternative pathways for utilization of $\Delta pH$ (Fig. 3B), or by addition of (iii) nitric acid to decrease external pH and increase the ease of protonating the exchanger in the high salt exterior

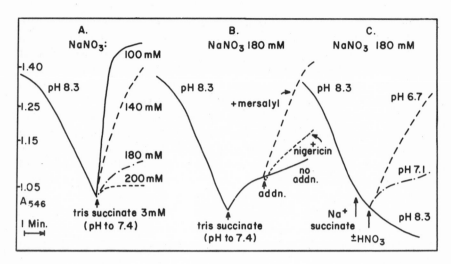

Fig. 3. Contraction of heart mitochondria swollen in NaNO₃. A: Heart mitochondria (0.5 mg/ml) were swollen at 35°C in a medium of NaNO₃ at either 100, 140, 180, or 200 mM as indicated, tris nitrate (2 mM, pH 8.3), and rotenone (1 μg/ml). Contraction was initiated by addition of tris succinate (3 mM), which brought the pH to 7.4. Absorbance at 546 nm was recorded using an Eppendorf photometer. B: Identical to A but NaNO₃ at 180 mM, showing the effect of mersalyl (20 nmoles/mg) and nigericin (1 × 10⁻⁷ M) on the rate of contraction at pH 7.4. C: Identical to B, but showing the effect of adding Na⁺ succinate (at pH 8.3) and of further adjustment of pH to 6.7 and 7.1 with small volumes of 0.1 M nitric acid.

medium (Fig. 3C). (e) Decreased extrusion of $H^+$ ($H^+$/O) is detected by a glass electrode when succinate is added to previously nonrespiring mitochondria (*cf.* ref. 47) under conditions of net contraction as compared to the $H^+$ extruded when no volume charge is occurring.[45] (f) Contraction is inhibited by low concentrations of acetate, and, as discussed above, acetate converts $\Delta pH$ to $\Delta\Psi$.[45]

The models shown in Fig. 2 provide an explanation for net ion influx and net efflux in terms of the nature of the anion present and the relative preponderance of $\Delta pH$ and $\Delta\Psi$ which results from respiration. Studies of influx and efflux of $^{42}K^+$ under conditions in which a near steady-state level of mitochondrial $K^+$ is maintained are also compatible with the operation of both an inwardly directed uniport for $K^+$ which responds to $\Delta\Psi$ and a $K^+/H^+$ exchanger which responds to $\Delta pH$.[35-38] Measurements of $^{42}K^+$ influx into heart mitochondria respiring in KCl-sucrose media have been carried out in our laboratory[37] and agree with previous reports by Diwan *et al.*[34, 48-50] which have established that the influx requires respiration, is sensitive to uncouplers, and shows saturation kinetics. The reaction in heart mitochondria shows a rather high $K_m$ (12 mM) in the concentration range from 5 to 40 mM K$^+$ and there are indications of a second, even lower-affinity uptake of $^{42}$K$^+$ at higher concentrations of KCl.[37] Influx is inhibited by Mg$^{2+}$ [37] and activated by Pi, thiol-group reagents such as mersalyl, and elevated pH.[37, 48–50] Influx in the presence of Pi is strongly inhibited by ADP, a result which indicates that $^{42}$K$^+$ influx competes poorly with ATP formation.[37]

Steady-state influx of $^{42}$K$^+$ has been estimated by centrifuging the mitochondria through a layer of silicone oil,[50] by centrifugation through a layer of cold sucrose containing Mg$^{2+}$,[37] and by a Millipore filtration procedure.[37] Whereas there is reasonably good qualitative agreement among the results obtained by the three methods, some quantitative difficulties are apparent and all three procedures result in considerable scatter of influx data at concentrations of KCl near the physiological range (140 mM). We have found it more convenient to follow the efflux of $^{42}$K$^+$ from previously labeled mitochondria when turnover is to be studied under conditions of steady-state K$^+$. In addition, the study of the efflux of labeled matrix K$^+$ reveals much about the influx process.[36, 38]

In the experiment shown in Fig. 4, the loss of $^{42}$K$^+$ from labeled heart mitochondria is compared with a simultaneous record of absorbance at 546 nm to measure volume changes (and hence net uptake or loss of matrix ions). In the absence of respiration, mitochondria lose only about 14% of matrix $^{42}$K$^+$ to a medium of KCl (100 mM). The swelling record shows no volume change, so the loss of labeled K$^+$ must be countered by influx of unlabeled K$^+$ from the medium. With respiration, this loss of matrix $^{42}$K$^+$ increases to 34%, again with no net change in matrix K$^+$ (Fig. 4b). Addition of Pi (2 mM) to respiring mitochondria produces some swelling under these conditions (Fig. 4b) and results in an accelerated loss of $^{42}$K$^+$. Studies of the time course of efflux by Millipore filtration (data not shown) have established that immediately following the addition of Pi the efflux is decreased to a very low level and that a new and higher steady-state turnover of K$^+$ is established after the period of net influx, which is indicated by the swelling record. Mersalyl induces a rapid efflux of $^{42}$K$^+$ with no net change in matrix K$^+$ (Fig. 4c) at levels of the mercurial which are sufficient to block the phosphate and dicarboxylate exchangers but not high enough to induce passive swelling. Addition of low concentrations of acetate (5 mM) to the KCl medium results in a slight net uptake (swelling) but increased retention of internal $^{42}$K$^+$ (Fig. 4d), and the combination of mersalyl and acetate produces an extensive net influx of K$^+$ without the $^{42}$K$^+$ efflux seen when mersalyl is added to respiring mitochondria in the absence of acetate.

These effects can be explained in terms of the model shown in Fig.

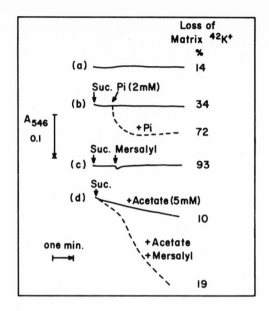

Fig. 4. Retention and loss of matrix $^{42}K^+$ under conditions of net $K^+$ influx and of steady-state $K^+$. Beef heart mitochondria were equilibrated with $^{42}K^+$ as described by Chávez *et al.*,[36] washed in cold 0.25 M sucrose, and suspended (0.5 mg/ml) in a medium of unlabeled KCl (100 mM) containing tris-Cl (2 mM, pH 7.2) and rotenone (3 μg/ml). Swelling was recorded, and after 4 min the mitochondria were separated by rapid centrifugation and the amount of $^{42}K^+$ lost was determined by scintillation counting. The numbers given represent the percentage of the original radioactivity which is lost during the course of the incubation. (a) No further addition. (b) Respiration initiated by addition of tris succinate (4 mM). The effect of phosphate (2 mM) is shown as the dashed trace. (c) The effect of mersalyl (20 nmoles/mg protein) on respiring mitochondria. (d) The effect of acetate (5 mM) and acetate plus mersalyl (20 nmoles/mg) on mitochondria respiring in the KCl medium.

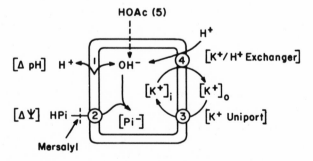

Fig. 5. Proposed model for influx and efflux of $K^+$ in mitochondria. See the text for explanation and details.

5. This model proposes that a relatively constant pool of matrix $K^+$ is maintained by a regulated interplay between the $K^+$ influx pathway ($K^+$ uniport) which responds to $\Delta\Psi$ and the $K^+$ efflux pathway ($K^+/H^+$ exchanger) which responds to $\Delta pH$.

In the absence of respiration, neither pathway is active and the only loss of $K^+$ is by a slow transmembrane leak. Respiration produces a combination of both $\Delta pH$ and $\Delta\Psi$,[3] so a limited turnover would be expected without net change in internal $K^+$. In the presence of Pi, conversion of $\Delta pH$ to $\Delta\Psi$ through the action of the phosphate transporter results in increased $K^+$ influx and decreased efflux, so that a net influx and swelling occur (Fig. 4b). After this swelling phase, the elevated turnover of $K^+$ which is observed could be due to either increased permeability (leaking back of accumulated $K^+$, as is seen in the presence of valinomycin; see ref. 42) or a change in the balance between $\Delta pH$ and $\Delta\Psi$ when large amounts of Pi are present in the matrix. This point remains to be clarified, but the $^{42}K^+$ efflux with mersalyl and acetate is readily explained by this model (Fig. 5). In the presence of sufficient mersalyl to block the phosphate transporter, conversion of $\Delta pH$ to $\Delta\Psi$ is prevented and the increased $\Delta pH$ is visible to a glass electrode.[51] This increase in $\Delta pH$ results in increased $K^+$ efflux via $K^+/H^+$ exchange and, since anion compensation does not occur in the chloride medium employed, the efflux of $K^+$ would produce a $\Delta\Psi$ which would activate influx through the uniport (Fig. 5). This would provide a rationale for increased turnover of $^{42}K^+$ in the absence of net $K^+$ alteration. With acetate also present, the $\Delta pH$ would now be converted quantitatively to an acetate gradient ($\Delta\Psi$), efflux on the exchanger would be abolished, and net influx without loss of matrix $^{42}K^+$ would be promoted.

In this model, in the presence of both Pi and ADP the proton motive force would be utilized for ATP formation,[3] and decreases in both uniport and exchange of $K^+$ would be expected. This is in line with the observation that both influx and efflux of $^{42}K^+$ are inhibited by ADP when Pi is present.[35-38] It also seems more likely that $Mg^{2+}$ interferes with the turnover via the uniport as opposed to the exchange process in this scheme.[38]

In the past we have called attention to the fact that mercurials and certain heavy metal cations (lead, zinc, cadmium, copper, and mercury) produce an activation of mitochondrial accumulation of monovalent cation acetates and pH transients in chloride salts which are very similar to the reactions induced by valinomycin.[10, 33] The reactions all discriminate in favor of $K^+$ over $Na^+$.[10, 33] We have considered that the increased cation uptake is a reflection of increased passive permeability to $K^+$ as a result of alteration of the protein portion of the membrane by the mercurial or heavy metal.[33] Southard *et al.*[52] and Blondin and Green[23] have interpreted similar observations in terms of release of a cryptic ionophore by the mercurial. The present studies suggest that the role of mercurials and heavy metals may be simply to shut off alternative pathways for conversion of $\Delta pH$ to $\Delta\Psi$ via the phosphate transporter and to divert all of the available proton motive force into acetate uptake. In this case, the existing uniport would be fully active for $K^+$ influx. The amount of mersalyl which produces maximum efflux of $^{42}K^+$ (Fig. 4c) is nearly the same as that required for a complete block of Pi transport (10 nmoles/mg). This and other evidence suggest that there is no direct effect of the mercurial on the uniport itself. Since the mitochondrial cation uptake mechanism (uniport) normally favors $K^+$ over $Na^+$, the existing cation discrimination would be maintained and accentuated in the presence of the mercurial. Further work will be necessary to clarify all aspects of this problem, but the use of $^{42}K^+$-labeled mitochondria should be very helpful in this regard.

Our studies of the turnover of $^{42}K^+$ in heart mitochondria support the conclusion of Diwan and Tedeschi[49] that mitochondrial $K^+$ is not maintained by a simple electrophoretic distribution of cation in response to membrane potential. The data also do not support the concept that mitochondrial $K^+$ levels are maintained by active influx working against passive efflux, as is seen at the level of the cell membrane. The experiments do suggest, however, that both influx and efflux of $^{42}K^+$ are energy dependent and that $K^+$ levels are maintained by a balance between the influx reaction which responds to $\Delta\Psi$ and the efflux pathway which is responsive to $\Delta pH$. Since the mitochondrion *in situ* is faced with about 140 mM external $K^+$, the influx pathway ($K_m$ 12 mM) should be saturated, and it seems possible that $K^+$ cycling may contribute to the maintenance and regulation of $\Delta\Psi$ and $\Delta pH$ under physiological conditions. Recent reports of an energy-dependent efflux of $Mg^{2+}$ [53] and of possible alternatives to $Ca^{2+}$ uniport, such as $Ca^{2+}/H^+$ exchange,[54, 55] suggest that complementary uniport and exchange pathways for divalent cations may also be provided in the mitochondrial membrane. It is obvious that many of the factors which contribute to the influx and efflux of cations in mitochondria remain to be clarified. In addition, at the present time we can only speculate as to the physiological significance of such cation fluxes.

ACKNOWLEDGMENTS

The studies presented here were supported in part by United States Public Health Services Grant HL09364 and by the Central Ohio Heart

Chapter and were done in collaboration with Drs. Edmundo Chávez, Dennis Jung, and Marianne Jurkowitz.

## References

1. Williamson, J. R. (1976) in *Mitochondria — Bioenergetics, Biogenesis, and Membrane Structure* (Packer, L., and Gomez-Puyou, A., eds.), pp. 79–107, Academic Press, New York
2. Meijer, A. J., and Van Dam, K. (1974) *Biochim. Biophys. Acta 36*, 213–244.
3. Mitchell, P. (1976) *Biochem. Soc. Trans. 4*, 399–430.
4. Williams, R. J. P. (1976) *T. I. B. S. 1*, N222–N224.
5. Brierley, G. P., Bachmann, E., and Green, D. E. (1962) *Proc. Natl. Acad. Sci. USA 48*, 1928–1935.
6. Brierley, G. P., Murer, E., and Green, D. E. (1963) *Science 140*, 60–62.
7. Brierley, G. P., Murer, E., Bachmann, E., and Green, D. E. (1963) *J. Biol. Chem. 238*, 3482–3489.
8. Lehninger, A. L., Carafoli, E., and Rossi, C. S. (1967) *Adv. Enzymol. 29*, 259–320.
9. Chappell, J. B., Cohn, M., and Greville, G. D. (1963) in *Energy-Linked Functions of Mitochondria* (Chance, B., ed.), pp. 219–231, Academic Press, New York.
10. Brierley, G. P. (1976) *Molec. Cell. Biochem. 10*, 41–62.
11. Chance, B. and Montal, M. (1971) *Curr. Topics Membranes Transport 2*, 99–156.
12. Pressman, B. C. (1976) *Ann. Rev. Biochem. 45*, 501–530.
13. Racker, E. (1976) *A New Look at Mechanisms in Bioenergetics*, Academic Press, New York.
14. Azzone, G. F., and Massari, S. (1973) *Biochim. Biophys. Acta 301*, 195–226.
15. Papa, S. (1976) *Biochim. Biophys. Acta 456*, 39–84.
16. Skulachev, V. P. (1975) in *Energy Transducing Mechanisms* (Racker, E., ed.), pp. 31–73, Butterworths, London.
17. Greville, G. D. (1969) *Curr. Topics Bioenergetics 3*, 1–78.
18. Harris, E. J. (1972) *Transport and Accumulation in Biological Systems*, Butterworths, London.
19. Klingenberg, M. (1970) *Essays Biochem. 6*, 119–159.
20. Packer, L., and Gomez-Puyou, A. (1976) *Mitochondria — Bioenergetics, Biogenesis and Membrane Structure*, Academic Press, New York.
21. Quagliariello, E., Papa, S., Palmeri, F., Slater, E. C., and Siliprandi, N. (eds.) (1975) *Electron Transfer Chains and Oxidative Phosphorylation*, North-Holland, Amsterdam.
22. Jagendorf, A. T. (1975) in *Bioenergetics of Photosynthesis* (Govindjee, ed.), pp. 414–492, Academic Press, New York.
23. Blondin, G. A., and Green, D. E. (1975) *Chem. Eng. News. 53*, 26–42.
24. Wilson, T. H., and Maloney, P. C. (1976) *Fed. Proc. 35*, 2174–2179.
25. Chappell, J. B. (1968) *Br. Med. Bull. 24*, 150–157.
26. Klingenberg, M., Durand, R., and Guerin, B. (1974) *Eur. J. Biochem. 42*, 135–150.
27. Brierley, G. P., and Stoner, C. D. (1970) *Biochemistry 9*, 708–713.
28. Brierley, G. P., Jurkowitz, M., Scott, K. M., and Merola, A. J. (1970) *J. Biol. Chem. 245*, 5405–5411.
29. Lehninger, A. L. (1974) *Circ. Res. Suppl. III, 34*, 83–88.
30. Carafoli, E. (1973) *Biochimie 55*, 755–762.
31. Rottenberg, H., and Scarpa, A. (1974) *Biochemistry 13*, 4811–4817.
32. Chappell, J. B., and Crofts, A. R. (1966) in *Regulation of Metabolic Processes in Mitochondria* (Tager, J. M., Papa, S., Quagliariello, E., and Slater, E. C., eds.), pp. 293–316, Elsevier, Amsterdam.

33. Brierley, G. P. (1974) *Ann. N.Y. Acad. Sci. 227*, 398–411.
34. Diwan, J. J., and Harrington, P. (1975) *Fed. Proc. 34*, 518 (abstr.).
36. Chávez, E., and Brierley, G. P. (1977) *Biophys. J. 17*, 253a.
36. Chávez, E., Jung, D. W., and Brierley, G. P. (1977) *Biochem. Biophys. Res. Commun. 75*, 69–75.
37. Jung, D. W., Chávez, E., and Brierley, G. P. (1977) *Arch. Biochem. Biophys. 183*, 452–459.
38. Chávez, E., Jung, D. W., and Brierley, G. P. (1977) *Arch. Biochem. Biophys. 183*, 460–470.
39. Brierley, G. P., Jurkowitz, M., Scott, K. M., and Merola, A. J. (1972) *Arch. Biochem. Biophys. 152*, 744–754.
40. Mueller, P. (1975) *Ann. N.Y. Acad. Sci. 264*, 247–264.
41. Harris, E. J., and Pressman, B. C. (1969) *Biochim. Biophys. Acta 172*, 66–70.
42. Rottenberg, H. (1972) *J. Membrane Biol. 11*, 117–137.
43. Brierley, G. P., and Jurkowitz, M. (1976) *Biochem. Biophys. Res. Commun. 68*, 82–88.
44. Brierley, G. P., Jurkowitz, M., and Chávez, E. (1977) *Biochem. Biophys. Res. Commun. 74*, 235–241.
45. Brierley, G. P., Jurkowitz, M., Chávez, E., and Jung, D. W. (1977) *J. Biol. Chem. 252*, 7932–7939.
46. Douglas, M. G., and Cockrell, R. S. (1974) *J. Biol. Chem. 249*, 5464–5471.
47. Reynafarje, B., Brand, M. D., and Lehninger, A. L. (1976) *J. Biol. Chem. 251*, 7442–7451.
48. Diwan, J. J., and Lehrer, P. H. (1975) *J. Cell Biol. 67*, 96a.
49. Diwan, J. J., and Tedeschi, H. (1975) *FEBS Lett. 60*, 176–179.
50. Diwan, J. J. (1973) *Biochem. Biophys. Res. Commun. 50*, 384–391.
51. Brierley, G. P., Jurkowitz, M., and Scott, K. M. (1973) *Arch. Biochem. Biophys. 159*, 742–756.
52. Southard, J. H., Blondin, G. A., and Green, D. E. (1974) *J. Biol. Chem. 249*, 678–681.
53. Crompton, M., Capano, M., and Carafoli, E. (1976) *Biochem. J. 154*, 735–742.
54. Cockrell, R. S. (1976) *Fed. Proc. 35*, 1759 (abstr.).
55. Puskin, J. S., Gunter, T. E., Gunter, K. K., and Russell, P. R. (1976) *Biochemistry 15*, 3834–3842.

*Chapter 15*

# Assembly of the Sarcoplasmic Reticulum

## David H. MacLennan, Elzbieta Zubrzycka, Annelise O. Jorgensen, and Vitauts I. Kalnins

### Experimental Systems for Membrane Biogenesis

The study of membrane biogenesis is greatly facilitated if systems are chosen which possess any of a set of unique features. First, the individual characteristics, localization, and orientation of each protein within the membrane should be known. For purposes of microanalysis, antibodies against single proteins should be available. Second, the synthesis of the membrane should be initiated, preferably *de novo*, under experimental conditions. Finally, genetic manipulation of the system should be possible. It is apparent from other papers in this volume that these features have been the major determinants of our extensive knowledge of the biogenesis of mitochondria in fungi (see Chapters 13, 16, and 17, this volume). In this chapter we review our work, showing that these same features are helping to advance understanding of the biogenesis of the sarcoplasmic reticulum in cultured muscle cells.

### Proteins of Sarcoplasmic Reticulum

The sarcoplasmic reticulum contains five well-defined proteins, three intrinsic and two extrinsic [1] (Fig. 1). The major intrinsic protein is

*David H. MacLennan and Elzbieta Zubrzycka* • Banting and Best Department of Medical Research, Charles H. Best Institute, University of Toronto, Toronto, Ontario, Canada M5G 1L6.    *Annelise O. Jorgensen and Vitauts I. Kalnins* • Department of Anatomy, University of Toronto, Toronto, Ontario, Canada M5G 1L6.

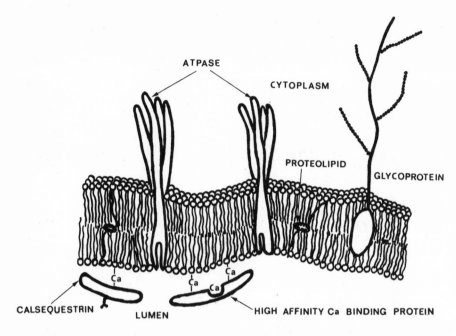

Fig. 1. Protein components of the sarcoplasmic reticulum. The ATPase is an intrinsic protein partially buried in the hydrophobic interior and partially extending into the cytoplasm. It probably associates as a tetramer. The proteolipid is probably buried within the lipid bilayer. The glycoprotein extends into the cytoplasm only in specialized regions of the membrane adjacent to the transverse tubular system. Calsequestrin and the high-affinity calcium-binding protein are extrinsic proteins found in the lumen of the membrane. They may be attached to the membrane by divalent metal bridges.

the $(Ca^{2+} + Mg^{2+})$-dependent ATPase, $M_r 100,000$, which constitutes about two-thirds of the total membrane protein.[2, 3] This protein is asymmetrically localized in the membrane with about one-half of the protein projecting on the cytoplasmic surface and one-half buried in the hydrophobic interior of the membrane.[4, 5] Buried portions of ATPase molecules may associate as tetramers.[6] A second intrinsic protein is the proteolipid, $M_r 12,000$, composed of 2 moles of fatty acid covalently linked to a peptide of relatively hydrophilic character.[7] This protein makes up only a few percent of the total membrane protein. The third intrinsic protein, $M_r 30,000$, is a glycoprotein which also makes up only a few percent of the total membrane protein mass.[8] The glycoprotein has been tentatively identified as the connecting link between the cisternal regions of the sarcoplasmic reticulum and the transverse tubules. Calsequestrin, $M_r 44,000$,[9] and the high-affinity calcium-binding protein, $M_r 55,000$,[10] are both extrinsic proteins localized on the interior of the sarcoplasmic reticulum membrane. It is not clear whether they are hydrophobically bonded to the membrane surface, bonded through divalent metal bridging, or merely free in the luminal space. They are both highly acidic and very water soluble, but differ in their $Ca^{2+}$-binding characteristics. Calsequestrin constitutes about 15% of the membrane protein, the high-affinity $Ca^{2+}$-binding protein about 10%.

Each of these proteins has been isolated in homogeneous form and antibodies specific against the ATPase, calsequestrin, and the high-affinity $Ca^{2+}$-binding protein have been prepared. Attempts to obtain antibodies against the proteolipid were unsuccessful. As yet, there are no reports of antibodies being raised against the glycoprotein. With the antibodies currently available, it has been possible to initiate studies of the synthesis and distribution of the major intrinsic protein, the ATPase, and the major extrinsic protein, calsequestrin. [11–13]

## Initiation of Sarcoplasmic Reticulum Synthesis

### Differentiation in Skeletal Muscle Cell Culture

Rat skeletal muscle cells can be grown in primary culture, [14] as well as in continuous cell lines. [15] In both cases, relatively undifferentiated, mononucleated myoblasts go through successive cell divisions, align themselves on the surface of the dish, and over a very short period fuse rampantly to form elongated multinucleated myotubes. [16] The myotubes gradually take on a transversely striated appearance as myofilaments form within. At a stage of advanced differentiation they pulsate spontaneously for short periods. With primary rat skeletal muscle cell cultures grown in standard medium, the period of cell growth, division, and alignment is about 50 hr, the period of rampant fusion begins at about 50 hr, and pulsation begins after about 120 hr in culture.

The differentiation process has been studied extensively in several laboratories. A common finding for a number of muscle-specific proteins, including myosin, actin, creatine, phosphokinase, aldolase, phosphorylase, and acetylcholine receptor, is that the synthesis is either increased five- to tenfold or turned on *de novo*. [17–26] There is some ambiguity in baseline synthetic rates since, in some cases, isozymes specific for adult muscle may be turned on *de novo*, but because of the low levels of another isozyme the *de novo* synthesis cannot readily be identified.

Morphological studies have been carried out on chicken muscle cells in culture. [27] These studies have shown that the smooth sarcoplasmic reticulum cannot be identified in myoblasts. At later stages, following fusion, smooth, somewhat swollen membranes can be observed to bud from the rough endoplasmic reticulum. These smooth membranes, with time, form the characteristic intracellular sarcoplasmic reticulum which surrounds each developing myofibril.

From these observations, the question arises whether the sarcoplasmic reticulum proteins are among those muscle-specific proteins whose synthesis is turned on coincident with fusion and other manifestations of differentiation to adult muscle.

*Biochemical Analysis of Synthesis*

In our initial studies of the development of sarcoplasmic reticulum, we decided to analyze the temporal pattern of synthesis of two major sarcoplasmic reticulum proteins, the ATPase and calsequestrin. The precise methodology has been published elsewhere.[11, 12] In summary, we plate primary myoblasts in Dulbecco's modified Eagle's medium, fortified with 0.1 vol of horse serum and 0.005 vol of chick embryo extract. At intervals over a period of 5 days the medium is replaced with a leucine-free medium containing 0.5 $\mu$Ci of [$^3$H] leucine per milliliter, and the cells are incubated for 2 hr. The radioactive medium is washed away, after which the cells are released from the plate by trypsinization and washed in phosphate-buffered saline containing protease inhibitors. Cells are then disrupted in 0.5% Triton, and insoluble material, including most of the DNA, is removed by centrifugation. The supernatant is made 0.5% in deoxycholate and 0.2% in sodium dodecylsulfate. About 10–20 $\mu$g of carrier protein (either ATPase or calsequestrin) is added to the supernatant, together with a twofold excess of antiserum specific for either protein. After 72 hr at 0°C, the precipitate is recovered by centrifugation, washed thoroughly in detergent-containing solution, and applied quantitatively to the top of disk gels. The carrier proteins are both identifiable in the disk gels in spite of the presence of large amounts of heavy and light chains of immunoglobulins. The gels are sliced, and areas corresponding to the protein under study are noted. Background counts are obtained from the bulk of the gel, and radioactivity in the specific protein band is defined as that part which exceeds the average background. The method does not define absolute synthetic rates but provides acceptable measurement of relative synthetic rates.

Data obtained on the synthesis of the ATPase[11] were compatible with observations made on other muscle-specific proteins. Within experimental error, there was no synthesis of the ATPase during the growth and division period prior to myoblast fusion at about 50 hr (Fig. 2). Synthesis rates increased dramatically after about 50 hr, remained

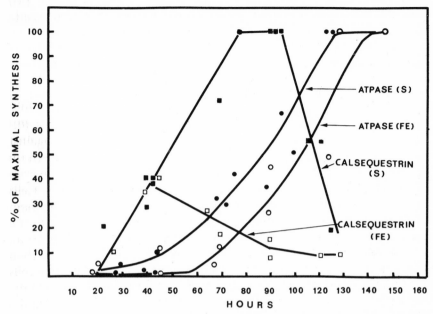

Fig. 2. Temporal patterns of synthesis of calsequestrin and ATPase during differentiation of rat skeletal muscle myoblasts in standard (S) medium and enriched (FE) medium. Cultures were pulse-labeled for 2 hr with tritiated leucine starting at the indicated times after plating. Incorporation of radioactive leucine into the two proteins was analyzed by immunoprecipitation, disk gel electrophoresis, and counting of gel slices. The data were collected from three experiments. ■, Calsequestrin, S medium; □, calsequestrin, FE medium; ●, ATPase, S medium; ○, ATPase, FE medium.

high for an extended period, and then declined as the cultures deteriorated. If the growth medium was enriched with fetal calf serum and chick embryo extract, myoblast fusion was delayed for about 20 hr (approximately the period of one cycle of cell division). In this case, ATPase synthesis was also delayed. It became clear, however, that cell fusion and synthesis of the ATPase were merely coincident events and not interdependent events, since growth of the cells in standard medium depleted of $Ca^{2+}$ inhibited cell fusion but not synthesis of the ATPase. [11]

A similar study of calsequestrin synthesis [12] gave surprising results. Calsequestrin synthesis was very low at 20–30 hr after plating, but by 40–45 hr, well before fusion was initiated and before ATPase synthesis was detectable, the rate of calsequestrin synthesis was already 40% of maximal (Fig. 2). The rate of synthesis increased to maximal just following fusion and then declined, well before the decline in the rate of ATPase synthesis. If the cells were transferred to low-$Ca^{2+}$ medium or to the enriched medium at about 45 hr, the rate of calsequestrin synthesis declined immediately, further enhancing the difference in the rates of synthesis between the ATPase and calsequestrin. Clearly, the two proteins, which ultimately reside in the same membrane, are not made coordinately.

### Immunofluorescence Studies of Protein Distribution

Confirmation of the biochemical observations and new insights into the assembly process were obtained through the use of indirect immunofluorescent staining of the cultures with antibodies specific against the ATPase and calsequestrin.[13] In this procedure, cells are grown on glass slides, but somewhat more sparsely, so that the time intervals cannot be precisely equated with those of the biochemical study. Cells adherent to the slide are fixed in formaldehyde, air-dried to make them permeable, and stored desiccated. Specific antibodies against the ATPase and calsequestrin are isolated by absorption to, and elution from, insoluble forms of the two proteins. The specific antibodies are then infused into the permeable cells. After incubation and washing, the cells are treated with a second, fluorescein isothiocyanate labeled antibody, this one specific against the immunoglobulin of the species in which the first antibody is made. The cells are washed again and examined under a fluorescent microscope. The presence or absence, and the distribution, of the sarcoplasmic reticulum proteins can now be detected by fluorescence.

As expected, there was no evidence for the presence of the ATPase prior to about 50 hr in culture (Fig. 3). After that time, the ATPase could be detected in an increasingly larger proportion of the cells. Surprisingly, the ATPase was found not only in bi- and multinucleated cells but also in mononucleated cells, showing that even under normal conditions of growth there is no obligatory linkage between fusion and initiation of ATPase synthesis. At the earliest periods, the ATPase was localized in foci throughout the cytoplasm, giving a granular appearance to the stained cells. As the myofibrillar structure became organized, the ATPase fluorescence became aligned in a direction parallel to the developing myofibrils.

Similarly, calsequestrin staining was absent from 24-hr myoblasts (Fig. 3). As predicted from biochemical studies, it could be found in myoblasts which had not begun to fuse. In contrast to the multiple foci characteristic of ATPase distribution, calsequestrin was found initially only in a sharply demarcated zone adjacent to, and sometimes indenting, the nucleus. This is the region where the Golgi apparatus is commonly found. With time, the fluorescence moved out of the perinuclear Golgi region into the surrounding cytoplasm until all of the myoblast was

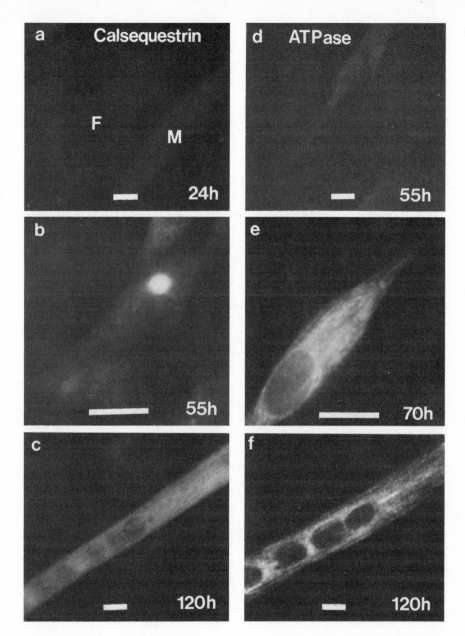

Fig. 3. Rat skeletal muscle cells grown in standard medium and treated at various times with antibodies to either calsequestrin (a,b,c) or ATPase (d,e,f). Bar = 10 $\mu$m. (a,d) Specific calsequestrin and ATPase staining are absent in both myoblasts (M) and fibroblasts (F). (b) Specific calsequestrin staining is present in a small region of the cytoplasm near to and indenting the nucleus. (e) Initial ATPase staining in myoblasts is present throughout the cytoplasm. The pattern is granular, possibly in longitudinal arrangement. (c,f) Staining patterns running longitudinally can be distinguished in multinucleated myotubes with both calsequestrin and ATPase antibodies.

stained. At later stages, in multinucleated cells, the intracellular distribution of calsequestrin and that of the ATPase were comparable.

*Postulated Assembly Process*

How can these data be interepreted? From first principles we can assume that the basic structure of the sarcoplasmic reticulum membrane is composed of the ATPase and phospholipid. Calsequestrin and the high-affinity $Ca^{2+}$-binding protein are not part of this intrinsic structure, the glycoprotein is probably located in a specialized region, and the proteolipid comprises less than 5% of the total intrinsic protein mass. Thus, for all intents and purposes, the appearance of the sarcoplasmic reticulum as a morphological structure is synonymous with the synthesis and incorporation of the ATPase into a phospholipid bilayer. This process apparently occurs at multiple foci in the cytoplasm, in agreement with electron microscopic studies which indicate that smooth membranes bud off from rough endoplasmic reticulum at this time.[27] Our first postulate, then, is that the ATPase is synthesized on bound polyribosomes at a "growing point" between rough and smooth reticulum (Fig. 4). The amino-terminal end of the protein penetrates only into the hydrophobic portion of the phospholipid bilayer and remains embedded within this bilayer. Its lateral displacement by newly forming peptide chains and the continued accretion of new phospholipid account for the growth of the smooth membrane.

For calsequestrin, the process might be more complex. Our second postulate is that calsequestrin is also formed on bound polyribosomes (Fig. 4). Unlike the ATPase, the amino-terminal end crosses the membrane and the completed protein is deposited in the lumen of the rough endoplasmic reticulum and is transported through luminal spaces to the Golgi apparatus in a process similar to that which is operative in the synthesis of secreted proteins.[28] Like many secreted proteins, calsequestrin is a glycoprotein.[13] However, it contains only the "core sugars" which are common to virtually all glycoproteins, a chain of two glucosamines linked to a chain of three mannoses through the center mannose.[29] The rough endoplasmic reticulum, in some cells at least, has the enzymic capacity for incorporation of these sugars into protein, and it may not be essential for calsequestrin to enter the Golgi for incorporation of these sugars.

Calsequestrin leaves the lumen of the Golgi apparatus and enters the lumen of the newly forming sarcoplasmic reticulum. As part of our postulate, then, we suggest that calsequestrin does not leave luminal compartments of the cell. Either it leaves the Golgi by a reverse pathway through the rough endoplasmic reticulum to the lumen of the newly

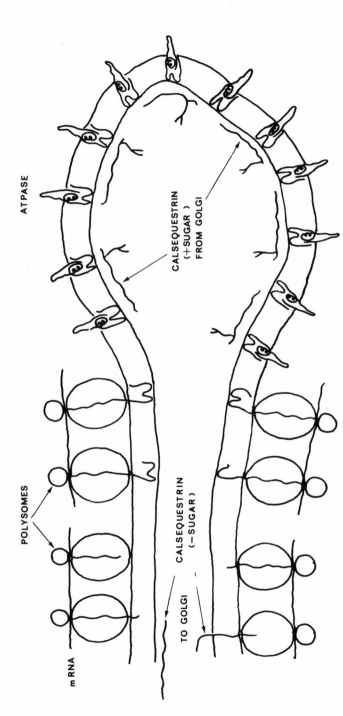

Fig. 4. Hypothesis for sarcoplasmic reticulum assembly. The ATPase is synthesized on bound polyribosomes. The amino-terminal end does not cross the phospholipid bilayer and the protein folds within the hydrophobic interior of the bilayer. Lateral displacement creates a "growing point" and the smooth sarcoplasmic reticulum appears to "bud" from the rough endoplasmic reticulum. Calsequestrin is also synthesized on rough endoplasmic reticulum. The amino-terminal end crosses the bilayer and calsequestrin enters the lumen of the membrane and is transported to the Golgi apparatus. Sugar is attached in luminal compartments. Calsequestrin returns to the newly forming sarcoplasmic reticulum through luminal spaces either by a reversal of the pathway leading to the Golgi or by fusion of Golgi-derived vesicles with the newly forming sarcoplasmic reticulum.

forming sarcoplasmic reticulum, or it enters the lumen of the sarcoplasmic reticulum by a fusion between Golgi-derived vesicles and sarcoplasmic reticulum. The processing of calsequestrin and the transport of
the protein through intracellular membranes may explain why the
synthesis of calsequestrin precedes the synthesis of the ATPase.

Clearly, there are many experiments still to be done to test these
postulates. Current studies in our laboratory are aimed at many of the
outstanding problems. We are analyzing the ultrastructure of the
developing membrane system more completely. It must be confirmed
that the perinuclear site of the initial calsequestrin accumulation is the
Golgi apparatus. Continuity between Golgi, rough endoplasmic reticulum, and sarcoplasmic reticulum must be analyzed in detail. Evidence for or against fusion of Golgi-derived vesicles and sarcoplasmic
reticulum must be obtained. This will involve the use of peroxidase- or
ferritin-conjugated, rather than FITC-conjugated, antibodies so that the
studies can be carried out at the resolution of the electron microscope.
Evidence must be obtained that the two proteins are indeed formed by
membrane-bound polyribosomes and that they are released as predicted. In biochemical studies, we are looking for kinetic evidence for
the movement of calsequestrin among different membrane fractions.
These combined studies should provide a rather clear picture of the
intricacy of assembly of this mammalian membrane system.

### Genetic Analysis

The potential for obtaining mutations in the sarcoplasmic reticulum
proteins does not, at the moment, seem very great. Nevertheless, it
should not be ruled out as a possibility. There are established cell lines of
rat skeletal muscle and mutations in a number of functions, such as RNA
synthesis,[30] glycosyltransferase,[31] and fusion,[30, 31] have been readily
obtained. Unfortunately, it is not easy to devise a selection procedure for
a fully differentiated function such as sarcoplasmic reticulum synthesis.
However, it will be interesting to examine cell lines mutant in glycoprotein synthesis for their ability to synthesize calsequestrin.

A second approach is to look for potential sarcoplasmic reticulum
mutations in the several muscle diseases which have been reported in a
variety of inbred mammalian and avian strains. We examined chicken
muscular dystrophic muscle for defects in sarcoplasmic recticulum.[32]
The ATPase and calsequestrin were normal in all respects analyzed, and
the high-affinity calcium-binding protein was present in the preparation,
although its function was not tested.

A more plausible candidate for a sarcoplasmic reticulum defect was the dysgenic mouse.[33] Homozygous mice manifest such a severe form of muscle disease that they do not survive *ex utero* and the strain is carried by heterozygotes. Nevertheless, the cells of *mdg/mdg* fetuses can be maintained in tissue culture, and they will fuse normally.[34] They do not undergo spontaneous, rhythmic contractions, although they are capable of induced contraction. This would indicate a possible defect in the system controlling intracellular $Ca^{2+}$ concentrations.

In collaboration with Dr. Francine Bowden-Essien of Rutgers University, we have analyzed the synthetic pattern of sarcoplasmic reticulum proteins in cultured cells from dysgenic mice.[35] We were able to show, from both biochemical and immunofluorescence studies, that the ATPase and calsequestrin develop normally. Moreover, these two proteins and the high-affinity $Ca^{2+}$-binding protein could be detected in homogenates of embryonic *mdg/mdg* tissue. It appears, therefore, that we can assay the presence or absence of the proteins even in severely diseased tissue by the combined use of tissue culture and biochemical and immunological analysis. Screening of other diseased tissues should be possible using immunological analysis of extracts combined with immunofluorescent studies of whole cells. Were a sarcoplasmic reticulum mutant uncovered, cell lines might be established which carried the mutation. It might also be feasible to use such a screening technique to analyze induced variants of established cell lines.

A nonlethal mutation in sarcoplasmic reticulum has already been described.[36] In the normal rabbit population, calsequestrin exists in two forms which differ in molecular weight. The forms appear to be the result of the interaction of two genes, possibly the calsequestrin gene and a processing gene. The usefulness of this mutant in developmental studies has yet to be demonstrated.

ACKNOWLEDGMENTS

Research reviewed in this chapter was supported by grant MT-3399 from the Medical Research Council of Canada, by a grant from the Muscular Dystrophy Association of Canada to D. H. MacLennan, by Grant MT-3302 from the Medical Research Council, and by a grant from the Ontario Heart Foundation to Vitauts I. Kalnins. Elzbieta Zubrzycka is a Fellow of the Muscular Dystrophy Association of Canada on leave of absence from the Department of Biochemistry of Nervous System and Muscle, Nencki Institute of Experimental Biology, Warsaw, Poland. Annelise O. Jorgensen is a Fellow of the Canadian Heart Foundation. The expert technical assistance of E. Zachwieja and M. Wassman is gratefully acknowledged.

## References

1. MacLennan, D. H., and Holland, P. C. (1975) *Ann. Rev. Biophys. Bioen. 4*, 377–404.
2. MacLennan, D. H. (1970) *J. Biol. Chem. 245*, 4508–4518.
3. MacLennan, D. H., Seeman, P., Iles, G. H., and Yip, C. C. (1971) *J. Biol. Chem. 246*, 2702–2710.
4. Stewart, P. S., and MacLennan, D. H. (1974) *J. Biol. Chem. 249*, 985–993.
5. Thorley-Lawson, D. A., and Green, N. M. (1973) *Eur. J. Biochem. 40*, 403–413.
6. Murphy, A. J. (1976) *Biochem. Biophys. Res. Commun. 70*, 160–166.
7. MacLennan, D. H., Yip, C. C., Iles, G. H., and Seeman, P. (1972) *Cold Spring Harbor Symp. Quant. Biol. 37*, 469–478.
8. Ikemoto, N., Cucchiaro, J., and Garcia, A. M. (1976) First International Congress on Cell Biology, Boston, September 5–10, abstr. 869.
9. MacLennan, D. H., and Wong, P. T. S. (1971) *Proc. Natl. Acad. Sci. USA 68*, 1231–1235.
10. Ostwald, T. J., and MacLennan, D. H. (1974) *J. Biol. Chem. 249*, 974–979.
11. Holland, P. C., and MacLennan, D. H. (1976) *J. Biol. Chem. 251*, 2030–2036.
12. Zubrzycka, E., and MacLennan, D. H. (1976) *J. Biol. Chem. 251*, 7733–7738.
13. Jorgensen, A. O., Kalnins, V. I., Zubrzycka, E., and MacLennan, D. H. (1977) *J. Cell Biol. 74*, 287–298.
14. Yaffe, D. (1973) in *Tissue Culture: Methods and Applications* (Kruse, P., and Patterson, M. K., eds.), pp. 106–114, Academic Press, New York.
15. Yaffe, D. (1968) *Proc. Natl. Acad. Sci. USA 61*, 477–483.
16. Yaffe, D. (1969) *Curr. Topics Dev. Biol. 4*, 37–77.
17. Coleman, J. R., and Coleman, A. W. (1968) *J. Cell. Physiol. 72*, 19–34.
18. Shainberg, A., Yagil, G., and Yaffe, D. (1971) *Biol. 25*, 1–29.
19. Paterson, B., and Strohman, R. C. (1972) *Dev. Biol. 29*, 113–138.
20. Delain, D., Meienhofer, M. C., Proux, D., and Schapira, F. (1973) *Differentiation 1*, 349–354.
21. Paterson, B., and Prives, J. (1973) *J. Cell Biol. 59*, 241–245
22. Morris, D. E., Cooke, A. and Coleman, R. J. (1972) *Exp. Cell Res. 74*, 582–585.
23. Turner, D. C., Maier, V., and Eppenberger, H. M. (1974) *Dev. Biol. 37*, 63–89.
24. Rubinstein, N., Chi, J., and Holtzer, H. (1976) *Exp. Cell Res. 97*, 387–393.
25. Chi, J. C. H., Rubinstein, N., Strahs, K., and Holtzer, H. (1975) *J. Cell Biol. 67*, 523–537.
26. Merlie, J. P., Sobel, A., Changeux, J. P., and Gros, F. (1975) *Proc. Natl. Acad. Sci. USA. 72*, 4028–4032.
27. Ezerman, E. B., and Ishikawa, H. (1967) *J. Cell Biol. 35*, 405–420.
28. Palade, G. (1975) *Science 189*, 347–358.
29. Montreuil, J. (1975) *Pure Appl. Chem. 42*, 431–477.
30. Somers, D. G., Pearson, M. L., and Ingles, C. J. (1975) *J. Biol. Chem. 250*, 4825–4831.
31. Whatley, R., Ng., S. K-C., Rogers, J., McMurray, W. C., and Sanwal, B. D. (1976) *Biochem. Biophys. Res. Commun. 70*, 180–186.
32. Yap, J. L., and MacLennan, D. H. (1976) *Can. J. Biochem. 54*, 670–673.
33. Gluecksohn-Waelsch, S. (1963) *Science 142*, 1269–1276.
34. Bowden-Essien, F. (1972) *Dev. Biol. 27*, 351–364.
35. Bowden-Essien, F., Jorgensen, A. O., Kalnins, V. I., Zubrzycka, E., and MacLennan, D. H. (1977) *Lab. Invest. 37*, 562–568.
36. MacLennan, D. H. (1974) *J. Biol. Chem. 249*, 980–984.

*Chapter 16*

# Mitochondrial Biogenesis: The Identification of Mitochondrial Genes and Gene Products

## Anthony W. Linnane and Ruth M. Hall

### Introduction

Elucidation of the biogenesis of subcellular organelles may be considered as one of the significant challenges in modern cell biology. Brief reflection on the problem of the biogenesis of mitochondria illustrates this thesis. The mitochondrion is a complex organelle with inner and outer membrane systems of still largely unknown composition; these membranes enclose a matrix space which is comprised of some hundreds of enzymes and small molecules of great functional and molecular diversity.

The great majority of the proteins and lipids composing functional mitochondria are under the control of the nuclear genetic system and are synthesized outside the mitochondrion. The mitochondrial genetic system codes for the tRNAs and rRNAs of the mitochondrial protein synthetic system but appears to direct the synthesis of only a small number of protein products, which are generally hydrophobic components of the inner membrane respiratory complexes. However, both the mitochondrial ribosome and the individual respiratory complexes also contain components which are encoded by the nuclear genetic system: and the biosynthesis of the macromolecular components of the

*Anthony W. Linnane and Ruth M. Hall*  •  Department of Biochemistry, Monash University, Clayton, Victoria, 3168, Australia.

mitochondrion is thus clearly a cooperative venture directed by two separate genetic systems, one maintained within the mitochondrion itself and the other residing in the nucleus.

The problems involved in the study of mitochondrial biogenesis are thus extensive and diverse. Major problems include (a) the identification of both mitochondrial and nuclear gene products required for mitochondrial function, (b) the determination of what controls the synthesis of individual components of the mitochondrion and what the interactions are between the two contributing genetic systems, (c) the determination of how extramitochondrially synthesized components are transported into the mitochondrion, and (d) the determination of how proteins are assembled into enzyme complexes and proteins and lipids into membrane arrays.

As a first step toward the solution of these problems, investigators over the past decade have concentrated principally on the contribution made by the mitochondrion to its own synthesis. This in itself is an enormous problem, involving the study of the characteristics of the largely unexplored mitochondrial genetic system and the identification of mitochondrial genes and gene products. The development of this particular area has, however, proceeded rapidly from the seminal investigations in a small number of laboratories in the mid-1960s into a major aspect of contemporary molecular biology with some hundreds of investigators now contributing. The purpose of this chapter is to review briefly the development of this area as it occurred in our laboratory from the first isolation of mitochondrial mutants of *Saccharomyces cerevisiae* which affect a single mitochondrial function to the development of a genetic and physical map of the mitochondrial genome and the use of mutants to identify mitochondrial gene products.

### Mitochondrial Mutants and Mapping the Mitochondrial Genome

The first group of mitochondrial mutants studied were the respiratory-deficient "petite" mutants of *Saccharomyces cerevisiae*, which were discovered by Ephrussi and co-workers in the late 1940s. Genetic studies indicated that the mutations did not behave in a classical Mendelian fashion (for review, see ref. 1) and it was concluded that a cytoplasmic genetic factor $(\rho)$ whose expression is essential for respiratory competence was involved. This conclusion was eventually confirmed when a specific species of mtDNA associated with mitochondria was detected in yeast,[2] and it was found that the petite mutation was frequently accompanied by a shift in the buoyant density of mtDNA.[3]

Biochemical studies on petites have, however, revealed that these mutants are defective in a large number of mitochondrial functions; cytochromes $aa_3$, $b$, and $c_1$ are not detectable and the oligomycin-sensitive ATPase, protein synthesis, and respiratory activities are also absent. Indeed, it is now known that petite mutants form a class in which the mtDNA has suffered large deletions resulting in the loss of genes coding for more than one mitochondrial product and that, since at least one gene required for mitochondrial protein synthetic activity is always absent, no mitochondrially coded proteins are synthesized. However, the absence of a particular protein from petite cells is not a sufficient condition to indicate that the protein is a mitochondrial gene product. Since the mitochondrial enzyme complexes are composed of both nuclear and mitochondrially coded proteins, the possibility that nuclear-coded proteins may also be absent from petites as a result of coordinate regulatory processes must be considered.

To identify the mitochondrial gene products, it is therefore necessary to study mutants which are either altered or defective in a single mitochondrial function as a result of a single small lesion in the mtDNA. The first isolation of such mutants followed shortly after the discovery, in this laboratory, that certain antibiotics such as chloramphenicol, erythromycin, and lincomycin specifically affect mitochondrial protein synthetic activity.[4-7] It was then possible to use a positive selection for antibiotic-resistant mutants by selecting for mutants which were able to grow on nonfermentable carbon sources in the presence of the antibiotic, and indeed among these mutants a number were detected which showed the same patterns of inheritance as described by Ephrussi for the cytoplasmic genetic factor.[8-10] Furthermore, a mitochondrial erythromycin resistance determinant was shown to directly affect the mitochondrial protein synthetic system, which exhibited strong *in vitro* resistance to erythromycin.[11]

Subsequently, a large number of mitochondrial mutants which show resistance to antibiotics such as the protein synthesis inhibitors (chloramphenicol, erythromycin, spiramycin, and paromomycin), inhibitors of the mtATPase (oligomycin and venturocidin), and inhibitors of complex III function (antimycin A and funiculosin) have been isolated in a number of laboratories (for a complete list, see ref. 12), and in many cases *in vitro* antibiotic resistance has also been demonstrated. More recently, mitochondrial respiratory-deficient mutants which differ from petites in that they have very small lesions in the mtDNA and retain mitochondrial protein synthetic activity have been identified.[13, 14]

Following the isolation of antibiotic-resistant mutants, recombination between different mitochondrial genomes was soon demonstrated in crosses between two haploid yeast strains,[15, 16] making possible formal analysis of the mitochondrial genetic system. However, although much effort has been addressed to the study of the recombination and transmission of mitochondrial markers, such analysis has provided little help in generating a map of the mitochondrial genome.

Classical three-point recombination analysis cannot be used to order markers, and to date in only one case where all the markers were linked (i.e., confined to a small segment of the mitochondrial genome) has a map order consistent with the physical map order (see below) been derived by recombination analysis.[17] Analysis of the frequency of transmission of markers and of recombination polarities has been used to order markers in a short segment of the genome affected by the polarity locus ω.[18, 19] However, in general, recombination analysis can be used only to demonstrate linkage between individual pairs of markers (see ref. 20). Indeed, such linkage analysis is of considerable importance for testing the allelism of mutations which confer the same phenotype, as when markers are sufficiently separated a minimum value for the number of independent loci which affect a particular function may be obtained.

Since a complete map of the mitochondrial genome cannot be derived by classical genetic procedures, two new procedures were developed in this laboratory. The rationale for the first procedure was based on the fact that petite mutants compose a set of mutants which have lost or retained different segments of the mitochondrial genome. The procedure involves analysis of the frequency of coretention and codeletion of markers in petite clones (petite deletion analysis). Such analyses will generate a unique map order only if the petites analyzed arise as the result of a single deletion event (i.e., have lost a single continuous segment of mtDNA), and this condition was found to hold only in the case of freshly arisen (primary) petite clones.[21] A unique circular map order was derived for five antibiotic resistance loci (Fig. 1a),[21] and this represented the first genetic map of the mitochondrial genome, as well as the first evidence for genetic circularity of the mitochondrial genome.

The second mapping procedure developed in our laboratory[22] generates a physical map of the mitochondrial genome. It has also led to a simple procedure for mapping any new mitochondrial marker. The principles of this procedure have been described in detail elsewhere[22, 20] and will be only briefly summarized here. For this analysis, extensively subcloned stable petite clones are used, and these petites form a reference library which is available for further genetic analysis. Initially the petite strains were characterized for the loss or retention of the five antibiotic resistance loci used in the petite deletion analysis. Quantitative DNA-DNA hybridization was then used to measure the size of the

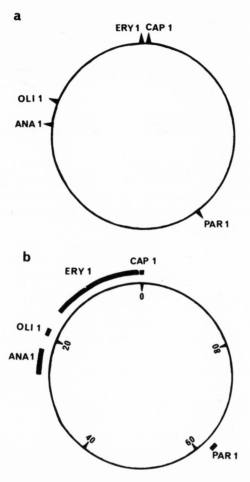

Fig. 1. Comparison of the first maps of the mitochondrial genome of *S. cerevisiae* obtained by (a) petite deletion analysis (after Molloy *et al.*[21]) and (b) physical mapping procedures (after Sriprakash *et al.*[22]). Markers represented are mitochondrial mutations conferring resistance to chloramphenicol (*cap1*), erythromycin (*ery1*), oligomycin (*oli1*), antimycin A (*ana1*), and paromomycin (*par1*).

mtDNA segment retained in petite strains, and to measure the size of the sequences common to pairs of petite strains. The circular map order generated by petite deletion analysis was used to deduce the relative positions of the mtDNA segments retained in the petite strains. Positions of markers can then be defined as lying within the mtDNA segment retained by all petites which retain the particular marker (inclusion) or in the segment outside the mtDNA segments retained in petites which have lost the marker (exclusion). Map positions generated by this procedure represent the physical location of the marker on the mitochondrial chromosome (Fig. 1b).

The library of petite strains used to generate the physical map of the mitochondrial genome can be used to derive the physical map position of any new mitochondrial marker by simple genetic analysis to determine whether the wild-type allele of the marker is retained or lost in each of the petite strains which form the library. We have now mapped a large number of mitochondrial mutations in this way.[17, 23-25] The map positions of the genes for the large and small rRNA species of the mitochondrial ribosome have also been determined by DNA-RNA hybridization.[26] Map positions can be refined as the petite library is enlarged.[27] The most recent version of the physical map of the mitochondrial genome is shown in Fig. 2.

## Identification of Mitochondrial Gene Products and Gene Product Relationships

After a long period of comparative stagnation, the process of identification of the specific proteins and RNA species coded by mtDNA is now beginning to make good progress. The RNA products of the mitochondrial genome can be identified simply by direct hybridization of the RNA species to mtDNA, and it is now well established that the small and large rRNAs associated with the mitochondrial ribosomes as well as a number of tRNA species (possibly a full complement) are products of the mtDNA, both in metazoa and in lower eukaryotes (for review, see ref. 28). However, identification of the protein products of the mitochondrial genome cannot be achieved directly, and the major approach used in this case has involved identification of the proteins which are synthesized by the mitochondrial protein synthetic system. The procedures used and the difficulties encountered with this approach, particularly in regard to the limitations of methods for specifically labeling mitochondrially synthesized proteins and for separating and characterizing the hydrophobic membrane proteins, have been discussed previously.[29] Proteins which can be shown to be synthesized by the mitochondrial protein synthetic system are putative gene products of the mtDNA, and it is now generally accepted that a number of components of the three inner mitochondrial membrane complexes (cytochrome $c$ oxidase, coenzyme $QH_2$-cytochrome $c$ reductase, and oligomycin-sensitive ATPase) are mitochondrially synthesized in *S. cerevisiae, Neurospora crassa,* and metazoa.

However, estimates from different laboratories of the number of mitochondrially synthesized proteins associated with a particular com-

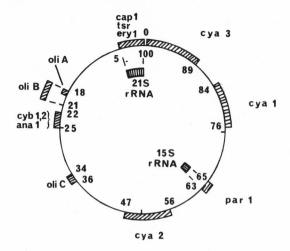

Fig. 2. Physical map of the mitochondrial genome of *S. cerevisiae*. Loci represented as follows: mutations conferring resistance to the antibiotics chloramphenicol (*cap1*), erythromycin (*ery1*), antimycin A (*ana1*), and paromomycin (*par1*) and three groups of mutations conferring resistance to oligomycin (*oliA, B,* and *C*); groups of respiratory-deficient (mit⁻) mutations with specific defects in complex III function (*cyb1* and *2*) or in cytochrome *c* oxidase function (*cya1, 2,* and *3*) and a mutation (*tsr*) which confers a respiratory-deficient phenotype at low temperature; and the large (21 S) and small (15 S) rRNA components of the mitochondrial ribosomes. For details of mapping procedures, see text.

plex are not always in agreement. For example, for *S. cerevisiae* estimates range from one to three proteins associated with complex III,[30, 31] for the ATPase complex estimates are three or four proteins,[32, 23] while for cytochrome oxidase estimates of three proteins are in agreement.[33, 34]

Although this approach has yielded a considerable amount of information, serious problems are encountered in determining functional relationships. The assignment of a particular protein to a specific enzyme complex is based on the final association of the protein with an isolated protein complex. However, in general the isolated complexes do not retain a full spectrum of enzymic activities. In particular, yeast complex III, which is currently isolated in the presence of antimycin A to prevent dissociation, cannot be assayed for enzyme activity,[30] and the oligomycin-sensitive ATPase complex has measurable activity as an ATPase but retains no ATP synthetase activity, which is its main biological function. Indeed, a functional protein of such a complex may be only weakly bound to other components and easily lost during isolation of the complex. Alternatively, it is possible that some proteins may be fortuitously copurified with a complex. It follows that the isolated complexes may or may not represent the functional complexes as they occur *in vivo*. As a consequence, definitive progress in the field of assignment of mitochondrial gene product relationships has been slow.

A more powerful approach to the problem of identification of the products of the mitochondrial genome involves the use of mitochondrial mutants. Biochemical analysis of mutants which have a single lesion in the mtDNA and are demonstrably altered in a specific mitochondrial function can be used to establish gene product relationships, and also to clearly assign a particular product to a specific role in the functioning of a mitochondrial enzyme complex. Furthermore, this approach will potentially lead to the identification of mitochondrially coded proteins which cannot be detected by other procedures.

The mitochondrial mutants available to date can be grouped into three major classes: (a) antibiotic-resistant mutants, (b) respiratory-deficient mutants which have small lesions in the mtDNA (mit⁻ and syn⁻), and (c) conditional respiratory-deficient mutants.

The availability of antibiotic-resistant mutants is determined by the availability of specific inhibitors of a mitochondrial function. Mutants resistant to antibiotics which inhibit mitochondrial protein synthesis (e.g., chloramphenicol, erythromycin, and paromomycin), the mtATPase (ventruicidin and oligomycin), and complex III function (antimycin A and funiculosin) have been reported (for a complete list, see ref. 12). These mutants have the great advantage of being conditional, and therefore they retain functional products albeit altered from wild type.

Respiratory-deficient mutants can potentially affect any mitochondrial gene product which is required for mitochondrial respiratory function. A considerable number of mutants of this type have now been described.[14, 23, 35–39] These mutants fall into two subgroups: (a) those which are defective in mitochondrial protein synthetic function (syn⁻) and (b) those which retain mitochondrial protein synthetic activity (mit⁻). Mit⁻ mutants which have specific defects in either cytochrome *c* oxidase, coenzyme QH₂-cytochrome *c* reductase, or ATPase function have been reported. Mutants with defects in more than one of these functions have in most cases been shown to contain more than one lesion in the mtDNA.

Conditional respiratory-deficient mutants (e.g., heat or cold sensitive) have also been isolated in a number of laboratories.[25, 40–43] The conditional mutants isolated in this laboratory have been shown to have defects in the ATPase complex[42] and in the large subunit of the mitochondrial ribosome.[44] For both genetic and biochemical studies, conditional respiratory-deficient mutants have considerable advantages over mit⁻ mutants. In particular, growth of conditional mutants under permissive conditions can be used for the maintenance of stocks free of petite mutants and for genetic manipulations (e.g., sporulation) for which respiratory competence is required. Also, gene products affected by a conditional mutation will be altered rather than completely absent from the cell, which should allow a more definitive assignment of gene product relationships. The importance of this aspect of gene product identification is emphasized by the studies of cytochrome oxidase mutants, where a single nuclear mutation has been shown to lead to the apparent loss of a number of products (both nuclear and mitochondrially coded) of the cytochrome oxidase complex.[45]

## Analysis of Mutants with Altered mtATPase

The mitochondrial oligomycin-sensitive ATPase complex has been extensively studied by Tzagoloff and co-workers (see ref. 32). Polyacrylamide gel analyses of immunoprecipitates of the mtATPase complex indicate that five protein bands are mitochondrially synthesized. One of these bands (approx. MW 45,000) has been shown to be a multimer of the low molecular weight (approx. 8000–9000) proteolipid subunit. Thus four mitochondrially synthesized proteins (subunits 5, 6, 8, and 9) are found associated with the mtATPase complex.

Similar analysis of the mtATPase complex performed in our laboratory has revealed three mitochondrially synthesized proteins (MW 29,000, subunit 5; MW 22,000, subunit 6; MW 8500 and 44,000, subunit 9) associated with this complex. One of these, a protein of MW 29,000 (subunit 5), is observed only when immunoprecipitates are not washed prior to gel analysis.[23] We conclude that this protein is only loosely associated with the isolated mtATPase complex and is therefore not necessarily associated with the complex *in vivo*.

Analysis of the mtATPase isolated from oligomycin-resistant mutants has allowed us to partially clarify this situation.

In an analysis of the linkage group which includes the known oligomycin resistance mutations and the antimycin resistance locus *ana1-r*,[17] three oligomycin resistance loci, *oliA, oliB,* and *oliC,* were distinguished by recombination analysis. The order *oliA–oliB–ana1–oliC* was derived from recombination analysis, petite deletion analysis, and physical mapping procedures. The *oliC* locus is clearly distinct from the *oliA* and *B* loci, indicating at least two different mitochondrial genes which affect the mtATPase complex. The *oliA* and *oliB* loci are separated at an average frequency of recombination of 1%, and thus represent two distinct genetic loci. Although current knowledge of the mitochondrial genetic system does not allow a distinction between the possibilities that these two genetic loci represent one gene or two genes, a study of the mitochondrially synthesized proteins of the ATPase in oligomycin-resistant mutants belonging to the classes *oliA* and *oliB* suggests that these two loci in fact reside in different genes.[23] In a number of mutants belonging to the class *oliB*, the proteolipid (subunit 9) of approximate molecular weight of 8500 does not aggregate to form the 45,000 MW peak which is routinely observed in wild-type strains and in mutants of the *A* and *C* classes. These data strongly suggest that subunit 9 of the ATPase complex is directly affected by the oligomycin resistance mutation. We have also shown that in one mutant of the *oliA* class, which is both oligomycin resistant and cold sensitive for respiratory function, an underproduction of subunit 6 (MW 22,000) of the mtATPase is observed at the restrictive growth temperature.[23] No alteration in any of the mtATPase proteins was detectable in the *oliC* mutant; thus the gene product affected by this mutation cannot yet be identified. These results constitute evidence that two of the mitochondrially synthesized proteins, subunits 6 and 9, which are associated with isolated mtATPase complex are indeed mitochondrial gene products and functionally associated with the ATPase.

### Analysis of Mutants Which Affect Complex III

Analysis of the mitochondrial respiratory complex III (coenzyme $QH_2$-cytochrome $c$ reductase) has been severely hampered by the difficulties encountered in isolating the complex. Using complex III preparations highly purified with respect to spectral $b$ and $c_1$, Katan and Groot[30] have identified six protein bands, one of which (the presumptive cytochrome $b$ protein) has been shown to be a mitochondrial translation product. Complex III preparations obtained by Marjanen and Ryrie[31] contain seven protein bands, and evidence that three of these bands are mitochondrially synthesized has been obtained. However, neither preparation is enzymically active and it is not possible to distinguish whether the preparations of Katan and Groot have lost polypeptides functionally associated with complex III *in vivo*, or whether the Marjanen and Ryrie preparation contains proteins which are not functionally associated with complex III *in vivo*. Thus in *S. cerevisiae* the number of mitochondrially coded components of complex III may be one, two, or three.

Two types of mutants which affect complex III function have been identified and studied in our laboratory: (a) antimycin A resistant mutants and (b) respiratory-deficient mutants (mit⁻) with specific defects in complex III function.

We have recently shown that a mitochondrial mutant originally isolated in this laboratory as resistant to mikamycin[46, 47] is in fact resistant to antimycin A.[23, 48] This mitochondrial mutation, now desig-

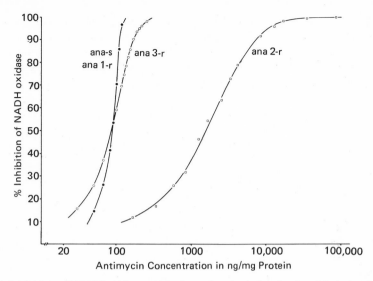

Fig. 3. Inhibition of NADH oxidase activity by antimycin A, in mitochondria isolated from wild type (*ana-s*) and antimycin A resistant mutants (*ana1-r*, *ana2-r*, and *ana3-r*) of *S. cerevisiae*.

nated *ana1-r* (formerly *mik1-r*), has been used extensively in both genetic and physical mapping analyses, and is located between the *oliA*,*B* and *oliC* loci (see Fig. 2). Although the *ana1* mutation confers a tenfold resistance to antimycin A *in vivo*, no detectable resistance to antimycin A could be demonstrated in assays of NADH oxidase activity using isolated mitochondria[48] (see also Fig. 3). However, we have now isolated a large number of mitochondrial antimycin A resistant mutants, and four of these mutants have been analyzed in detail.[49] Studies of NADH oxidase activity in isolated mitochondria reveal three classes of response to antimycin A *in vitro* (Fig. 3). One mutant is similar to the *ana1* strain and shows no detectable resistance to antimycin A. A second mutant (*ana3-r*) shows an altered response to antimycin A *in vitro* in that the shape of the inhibition curve is significantly different from that observed in the wild-type strain (*ana-s*), although the antimycin A concentration required for 50% inhibition of NADH oxidase activity is identical in both strains. Two mutants (*ana2-r* and *ana4-r*) exhibit a twentyfold resistance to antimycin A *in vitro*.

The *ana2-r*, *ana3-r*, and *ana4-r* mutations have been mapped by marker rescue analysis using the petite library, and are located within the mtDNA segment 22–25 units, which contains the *ana1* mutation (see Fig. 2). Thus we conclude that at least one protein associated with complex III is mitochondrially coded. Furthermore, the antimycin A resistant mutants may allow gene product identification by analysis of the mitochondrially synthesized proteins associated with complex III, and these studies are currently in progress.

We have also isolated a number of mit⁻ mutants which are specifically defective in complex III function, and the mitochondrial lesions (designated *cyb*) in these strains map in the same mtDNA segment (22–25 units) as the *ana* mutations.[23, 24, 38]

Analysis of the frequency of recombination in pairwise crosses between *cyb* mutants has allowed us to group the *cyb* mutations into two distinct loci, *cyb1* and *cyb2*, which are separated at a frequency of recombination of at least 10%. Mutations within each group are separated at frequencies of less than 0.1%.[50] Tzagoloff and co-workers have previously reported a similar analysis of mit⁻ mutants which affect complex III function. Two genetic loci, COB1 and COB2, which are located between the *oliB* and *oliC* loci and separated by 8–13% recombination, were distinguished.[35, 36] Although the high frequency of separation of the *cyb1* and *cyb2* loci suggests two genes, a decision as to whether these two loci indeed represent separate mitochondrial genes must await further biochemical analysis.

Although, for reasons outlined above, we consider that assignment of gene product relationships using mit⁻ strains will generally prove complicated, one of the *cyb* mutants isolated in this laboratory may be particularly suitable for such analysis. This mutant, Mn 37-16-6 (*cyb2*), retains spectrally detectable cytochrome *b* and a low but significant level of NADH-cytochrome *c* reductase activity. The absorption maximum of cytochrome *b* in the mutant is at a longer wavelength ($\lambda_{max}$ 560.5 nm) than in the parent strain ($\lambda_{max}$ 558 nm).[38] These properties suggest that all components of complex III are probably present and assembled in this mutant, and it may be possible to identify the protein altered by the *cyb2* mutation, by studying the mitochondrially synthesized proteins of isolated complex III. This analysis is currently in progress.

*Conclusion*

Detailed study of the mitochondrial genome of *S. cerevisiae* has been hampered for many years by the failure of traditional genetic procedures to allow the construction of a genome map. This problem has been overcome by the development in our laboratory of new mapping procedures, and a physical map of the mitochondrial genome has been constructed. Furthermore, the library of physically characterized petite strains used to construct the map can now be used to map rapidly any new mitochondrial mutation.

Significant progress has also been made in the area of mitochondrial gene product identification, using mitochondrial mutants. Genetic and biochemical analysis of oligomycin-resistant mutants has allowed us to identify three mitochondrial genes which directly affect the mtATPase complex, and the protein products of two of these genes have been identified. A similar approach is now being used to study the mitochondrially coded components of respiratory complex II.

The isolation and genetic and biochemical analysis of mitochondrial mutants have provided a powerful approach to understanding the nature of the gene products of mitochondrial DNA. Mitochondrial mutants should also prove useful in the study of the assembly and integration of the mitochondrial enzyme complexes and in the study of the control of mitochondrial gene expression.

## References

1. Ephrussi, B. (1953) *Nucleo-cytoplasmic Relationships in Microorganisms*, Clarendon Press, Oxford.
2. Schatz, G., Halsbrunner, E., and Tuppy, H. (1964) *Biochem. Biophys. Res. Commun. 15*, 127–132.
3. Mounolou, J. C., Jakob, H., and Slonimski, P. P. (1966) *Biochem. Biophys. Res. Commun. 24*, 218–224.
4. Clark-Walker, G. D., and Linnane, A. W. (1966) *Biochem. Biophys. Res. Commun. 25*, 8–13.
5. Clark-Walker, G. D., and Linnane, A. W. (1967) *J. Cell Biol. 34*, 1-14.
6. Huang, M., Biggs, D. R., Clark-Walker, G. D., and Linnane, A. W. (1966) *Biochim. Biophys. Acta 114*, 434–436.
7. Lamb, A.J., Clark-Walker, G. D., and Linnane, A. W. (1968) *Biochim. Biophys. Acta 161*, 415–427.
8. Linnane, A. W. (1968) in *Biochemical Aspects of the Biogenesis of Mitochondria* (Slater, E. C., Tager, J. M., Papa, S., and Quagliariello, E., eds.), pp. 333–353. Adriatica Editrice, Bari.
9. Linnane, A. W., Saunders, G. W., Gingold, E. B., and Lukins, H. B. (1968) *Proc. Natl. Acad. Sci. USA 59*, 903–910.
10. Thomas, D. Y., and Wilkie, D. (1968) *Genet. Res. 11*, 33–41.
11. Linnane, A. W., Lamb, A. J., Christodoulou, C., and Lukins, H. B. (1968) *Proc. Natl. Acad. Sci. USA 59*, 1288–1293.
12. Nagley, P., Sriprakash, K. S., and Linnane, A. W. (1977) in *Advances in Microbial Physiology*, Vol. 16 (Rose, A. H., and Morris, J. G., eds.), Academic Press, London.
13. Flury, U., Mahler, H. R., and Feldman, F. (1974) *J. Biol. Chem. 249*, 6130–6137.
14. Tzagoloff, A., Akai, A., Needleman, E. B., and Zulch, G. (1975) *J. Biol. Chem. 250*, 8236–8242.
15. Gingold, E. B., Saunders, G. W., Lukins, H. B., and Linnane, A. W. (1969) *Genetics 62*, 735–744.
16. Thomas, D. Y., and Wilkie, D. (1968) *Biochem. Biophys. Res. Commun. 30*, 368–372.
17. Trembath, M. K., Molloy, P. L., Sriprakash, K. S., Cutting, G. J., Linnane, A. W., and Lukins, H. B. (1976) *Mol. Gen. Genet. 145*, 43–52.

18. Linnane, A. W., Howell, N., and Lukins, H. B. (1974) in *The Biogenesis of Mitochondria: Transcriptional, Translational and Genetic Aspects* (Kroon, A. M., and Saccone, C., eds.), pp. 193–213. Academic Press, New York.

19. Netter, P., Petrochilo, E., Slonimski, P., Bolotin-Fukuhara, M., Coen, D., Deutsch, J., and Dujon, B. (1974) *Genetics 78*, 1063–1100.

20. Linnane, A. W., Lukins, H. B., Molloy, P. L., Nagley, P., Rytka, J., Sriprakash, K. S., and Trembath, M. K. (1976) *Proc. Natl. Acad. Sci. USA 73*, 2082–2085.

21. Molloy, P., Linnane, A. W., and Lukins, H. B. (1975) *J. Bacteriol. 122*, 7–18.

22. Sriprakash, K. S., Molloy, P. L., Nagley, P., Lukins, H. B., and Linnane, A. W. (1976) *J. Mol. Biol. 104*, 485–503.

23. Groot Obbink, D. J., Hall, R. M., Linnane, A. W., Lukins, H. B., Monk, B. C., Spithill, T. W., and Trembath, M. K. (1976) in *Genetic Function of Mitochondrial DNA* (Saccone, C., and Kroon, A. M., eds.), pp. 163–173. Elsevier/North-Holland Biomedical Press, Amsterdam.

24. Rytka, J., English, K. J., Hall, R. M., Linnane, A. W., and Lukins, H. B. (1976) in *Genetics and Biogenesis of Chloroplasts and Mitochondria* (Bucher, *et al.*, eds.), pp. 427–434, Elsevier/North-Holland Biomedical Press, Amsterdam.

25. Devenish, R., English, K., Hall, R. M., Linnane, A. W., and Lukins, H. B. (1977) *Mol. Gen. Genet.*, submitted.

26. Sriprakash, K. S., Choo, K. B., Nagley, P., and Linnane, A. W. (1976) *Biochem. Biophys. Res. Commun. 68*, 85–91.

27. Choo, K. B., Nagley, P., Lukins, H. B., and Linnane, A. W. (1977) *Mol. Gen. Genet. 153*, 279–288.

28. Hall, R. M., and Linnane, A. W. (1977) in *Cell Biology: A Comprehensive Treatise*, Vol. II (Goldstein, L., and Prescott, D. M., eds.), Academic Press, New York.

29. Schatz, G., and Mason, T. L. (1974) *Ann. Rev. Biochem. 43*, 51–87.

30. Katan, M. B., and Groot, G. S. P. (1975) in *Electron Transfer Chains and Oxidative Phosphorylation* (Quagliariello, E., *et al.*, eds.), pp. 127–132, North-Holland, Amsterdam.

31. Marjanen, L. A., and Ryrie, I. J. (1976) *Arch. Biochem. Biophys. 172*, 679–684.

32. Tzagoloff, A., and Meagher, P. (1972) *J. Biol. Chem. 247*, 594–603.

33. Rubin, M. S., and Tzagoloff, A. (1973) *J. Biol. Chem. 248*, 4275–4279.

34. Mason, T. L., and Schatz, G. (1973) *J. Biol. Chem. 248*, 1355–1360.

35. Tzagoloff, A., Foury, F., and Akai, A. (1976) in *The Genetic Function of Mitochondrial DNA* (Saccone, C., and Kroon, A. M., eds.), pp. 155–161. Elsevier/North-Holland Biomedical Press, Amsterdam.

36. Tzagoloff, A., Foury, F., and Akai, A. (1976) *Mol. Gen. Genet. 149*, 33–42.

37. Faye, G., Bolotin-Fukuhara, M., and Fukuhara, H. (1976) in *Genetics and Biogenesis of Chloroplasts and Mitochondria* (Bucher, T., *et al.*, eds.), pp. 547–555, Elsevier/North-Holland Biomedical Press, Amsterdam.

38. Cobon, G. S., Groot Obbink, D. J., Hall, R. M., Maxwell, R., Murphy, M., Rytka, J., and Linnane, A. W. (1976) in *Genetics and Biogenesis of Mitochondria and Chloroplasts* (Bucher, T. *et al.*, eds.), pp. 453–460, Elsevier/North-Holland Biomedical Press, Amsterdam.

39. Pajot, P., Wambier-Kluppel, M. L., Kotylak, Z., and Slonimski, P. P. (1976) in *Genetics and Biogenesis of Chloroplasts and Mitochondria* (Bucher, T., *et al.*, eds.), pp. 443–452, Elsevier/North-Holland Biomedical Press, Amsterdam.

40. Storm, E. M., and Marmur, J. (1975) *Biochem. Biophys. Res. Commun. 64*, 752–759.

41. Handwerker, A., Schweyen, R. J., Wolf, K., and Kaudewitz, F. (1973) *J. Bacteriol. 113*, 1307–1310.

42. Trembath, M. K., Monk, B. C., Kellerman, G. M., and Linnane, A. W. (1975) *Mol. Gen. Genet. 141*, 9–22.

43. Lancashire, W. (1976) in *Genetics and Biogenesis of Chloroplasts and Mitochondria* (Bucher, T., *et al.*, eds.), pp. 481–490., Elsevier/North-Holland Biomedical Press, Amsterdam.

44. Lukins, H. B., English, K. J., Spithill, T. W., Devenish, R. J., Hall, R. M., Nagley, P., and Linnane, A. W. (1977) in *Genetics and Biogenesis of Mitochondria* (Kauclewitz, F., Schweyen, R. J., Bandlow, W., and Wolf, K., eds.), De Gruyter, Berlin, in press.

45. Schatz, G. (1975) in *Molecular Biology of Nucleo-cytoplasmic Relationships* (Piuseaux-Dao, S., ed.), pp. 157–169, Elsevier, Amsterdam.

46. Bunn, C. L., Mitchell, C. H., Lukins, H. B., and Linnane, A. W. (1970) *Proc. Natl. Acad. Sci. USA 67*, 1233–1240.

47. Howell, N., Molloy, P. L., Linnane, A. W., and Lukins, H. B. (1974) *Mol. Gen. Genet. 128*, 43–54.

48. Groot Obbink, D. J., Spithill, T. W., Maxwell, R. J., and Linnane, A. W. (1977) *Mol. Gen. Genet. 151*, 127–136.

49. Hall, R. M., Harris, M., Maxwell, R., and Linnane, A. W., in preparation.

50. Hall, R. M., Astin, A., Devenish, R., Marzuki, S., and Linnane, A. W., in preparation.

*Chapter 17*

# Mitochondrial Genome of Saccharomyces cerevisiae

## Alexander Tzagoloff

### Introduction

New methods of screening for mitochondrial mutants in yeast have given fresh impetus to the developing field of mitochondrial genetics. The immediate aim in this area is to arrive at as complete a picture as possible of the genetic content of mitochondrial DNA in terms of the number of genes and the identity of their products. Because of the small genome size, there is good reason to think that these are realistic goals, achievable within the next few years. A detailed map of mitochondrial DNA will at once define the extent of autonomy of the organelle and perhaps reveal new organizational and regulatory features in a eukaryotic genome.

In this chapter, I would like to review some of the approaches currently being used to identify mitochondrial genes and to study their distribution on the genome.

### Mitochondrial Genes

A good starting point is to consider the two broad classes of genes present on mitochondrial DNA. It is now known that some of the electron transfer complexes of the respiratory chain (coenzyme $QH_2$-

*Alexander Tzagoloff* • Department of Biological Sciences, Columbia University, New York, New York 10032.

Table I. Properties of Cytoplasmic Mutants of Yeast

| Mutant | Nature of mutation | Growth on nonfermentable substrates | Mitochondrial protein synthesis |
|---|---|---|---|
| $\rho^-$ | Long deletions | − | − |
| Antiobiotic resistant | Point mutations | + | + |
| Mit⁻ | Point mutations | − | + |
| Syn⁻ | Point mutations | − | − |

cytochrome *c* reductase and cytochrome oxidase) and the oligomycin-sensitive ATPase each contain a set of subunit polypeptides that are synthesized in mitochondria.[1-3] There is also good evidence that these proteins are encoded in mitochondrial DNA. We will define as "mit genes" those genetic elements on mitochondrial DNA, whether structural or regulatory, that code for products involved in the biosynthesis of the above three enzymes. Genes coding for subunits of the ATPase complex or for proteins that may control their transcription or translation are examples of mit genes. In addition to mit genes, mitochondria are also known to contain genetic information essential for the production of a functional system of mitochondrial protein synthesis. The two ribosomal RNAs as well as all the transfer RNAs are gene products of mitochondrial DNA. This second class of genes, necessary for the biogenesis of the mitochondrial translational machinery, will be referred to as "syn genes."

## Consequences of Mutations in Mit and Syn Genes

A variety of mutants of *Saccharomyces cerevisiae* have been isolated with lesions in mit and syn genes. Based on their phenotypes, they can be divided into two types. Some mutations cause a loss of function (mit⁻ or syn⁻ mutants) as a result of which the cell is no longer capable of growing on nonfermentable substrates. In mit⁻ mutants, the lack of growth on a substrate such as glycerol is due to a specific deficiency in a respiratory complex or in the ATPase. Mit⁻ mutants, however, are still competent in carrying out mitochondrial protein synthesis. In contrast, syn⁻ mutants have a defective system of mitochondrial protein synthesis and consequently show pleiotropic deficiencies in all the enzymes which contain one or more mitochondrial translation products.

The second type of mutant is recognized by its resistance to an antibiotic or drug which normally inhibits growth of the wild type on a nonfermentable substrate. Such antibiotic-resistant strains, however, do not show any gross impairment of mitochondrial functions. Antibiotic-resistant mutants may have lesions in either mit or syn genes.

In addition to mit$^-$, syn$^-$, and antibiotic-resistant mutants, *S. cerevisiae* can sustain large deletions in mitochondrial DNA resulting in the loss of either mit, syn, or both types of genes. Such strains are known as $\rho^-$ mutants and in the special case where the entire genome is lost, as $\rho^0$. In Table I are summarized the different types of mutants of *S. cerevisiae* along with their observed phenotypes.

## Specific Mit$^-$, Syn$^-$, and Antibiotic-Resistant Mutants

Up to a few years ago, mitochondrial mutants were relatively rare and confined to antibiotic-resistant strains. This situation has changed in the last few years, and there are presently new antibiotic-resistant strains and many mit$^-$ mutants of *S. cerevisiae* available for genetic studies. Some salient properties of the antibiotic-resistant, mit$^-$, and syn$^-$ mutants are presented in Table II. Although only a few syn$^-$ mutants have been isolated, several laboratories are engaged in obtaining such strains and it may be anticipated that more will be found and will further enlarge the number of mitochondrial genetic markers.

## Selection Methods

### Antibiotic-Resistant Strains

Mutants resistant to antibiotics are by far the easiest to obtain since a positive selection is used. Usually, a heavy lawn of wild-type strain is spread on solid medium containing a nonfermentable substrate and the antibiotic. Depending on the antibiotic used, spontaneous mutants appear as single colonies after several days to several weeks of incubation.[4-6]

### Mit$^-$ Strains

Several methods can be used to isolate mit$^-$ mutants. In the original method, a haploid wild-type strain was mutagenized with $Mn^{2+}$ and spread for single colonies on a solid medium containing a low concentration of glucose and a high concentration of a nonfermentable substrate such as glycerol. On this medium, strains with defects in mitochondrial respiratory functions ($\rho^-$, mit$^-$, and syn$^-$) grow only to the extent that glucose is available and appear as small colonies, while the wild-type cells

Table II. Known Antibiotic-Resistant, Mit⁻, and Syn⁻ Mutants of Yeast

| Mutant | Phenotype | Affected gene |
|---|---|---|
| 1. Antibiotic resistant | | |
| cap | Resistance to chloramphenicol | syn |
| ery | Resistance to erythromycin | syn |
| par | Resistance to paromomycin | syn |
| oli | Resistance to oligomycin | mit |
| ana | Resistance to antimycin | mit |
| 2. Mit⁻ | | |
| oxi | Defect in cytochrome oxidase | mit |
| cob | Defect in cytochrome b | mit |
| pho | Defect in oligomycin-sensitive ATPase | mit |
| 3. Syn⁻ | | |
| asp | Defective tRNA$_{asp}$ | syn |
| thr1 | Defective tRNA$_{thr}$ | syn |

continue to utilize glycerol and form large colonies. The glycerol-negative strains are collected and assayed for mitochondrial protein synthesis. This assay eliminates the preponderant class of $\rho^-$ strains (greater than 90%) induced by the mutagenic treatment. The glycerol-negative mutants that are capable of mitochondrial protein synthesis consist of a class highly enriched in mit⁻ mutants.[7] Another method involves the use of a strain carrying a nuclear mutation ($op_1$) which causes the $\rho^-$ condition to be lethal. After mutagenesis with $Mn^{2+}$, all the $\rho^-$ mutants are killed and the survivors are again enriched in mit⁻ mutants.[8] In the third method, glycerol-negative mutants are crossed to a set of $\rho^-$ testers which contain different segments of mitochondrial DNA and the diploids are checked for growth on glycerol.[9] This test does two things. First, it eliminates $\rho^-$ mutants since diploids issued from a cross of two $\rho^-$ strains have never been found to be capable of growth on nonfermentable substrates. Second, presumptive mit⁻ mutants whose growth on glycerol is restored by the testers can be localized in the segment of DNA retained by the tester.

*Syn⁻ Strains*

There are several ways in which it is possible to select for syn⁻ mutants. Protein synthesis negative strains can be checked for spontaneous or mutagen-induced revertants. Strains that show a measurable reversion frequency should have point mutations in syn genes. An alternative method is to use $\rho^-$ testers with segments of mitochondrial DNA known to contain syn genes. Preferably the testers should have

DNA in which the mit genes have been deleted. Such $\rho^-$ strains can be used to enrich for mutations in particular syn genes. Advantage can also be taken of some special properties of mutations in syn genes. For example, *S. cerevisiae* contains isoaccepting species of transfer RNAs. Mutations in transfer RNAs which have isoaccepting species may reduce but not completely abolish growth on nonfermentable substrates. An initial selection of slow-growing mutants may therefore be used to enrich for mutations in transfer RNAs.

## Mapping of Mitochondrial Mutations

When two haploid strains of yeast are mated, the resultant zygotes form a pool of mitochondrial DNA consisting of multiple copies of each of the two parental types. Recombination of genetic markers ensues shortly after zygote formation and continues until pure clones are formed as a result of the segregation of DNA molecules during growth. The segregation of pure recombinant types is a rapid process requiring 10–20 generations.

At present all genetic mapping procedures for mitochondrial mutations are based on recombination. Mutated alleles or sites can therefore be assigned to genetic loci but not genes. Operationally, the term "locus" is given the following definition: two mutations are considered to be in the same locus if they exhibit similar phenotypes (e.g., absence of an enzyme activity or resistance to an antibiotic) and if they can be shown to be close to each other by recombination or other criteria (recombination frequencies of 1% or less). This definition implies that there must exist at least one gene per locus but does not exclude more than one. The other point to be noted is that the two genetic loci may be included in the same gene.

The assignment of mutations to loci and their localization relative to different markers on the genome have been possible with three kinds of genetic analyses: (a) the frequency of wild-type recombinants issued from crosses of mit$^-$ and/or syn$^-$ mutants with each other, (b) the production of wild-type recombinants in crosses of mit$^-$ or syn$^-$ mutants with $\rho^-$ testers having different retained segments of mitochondrial DNA, and (c) recombination of syn$^-$ or mit$^-$ mutations with antibiotic resistance markers.

### $Mit_1^- \times Mit_2^-$ Crosses

Pairwise crosses of mit$^-$ or syn$^-$ strains allow the mutations to be placed in different loci. If the production of wild-type recombinants is

*Table III. Pairwise Crosses of Cytochrome b Mutants* [a]

|  | Cob1 | | | | | | | | Cob2 | | | | |
|---|---|---|---|---|---|---|---|---|---|---|---|---|---|
|  | M6-200 | M7-40 | M8-53 | M8-181 | M13-101 | M15-207 | M17-231 | M24-241 | M10-152 | M17-162 | M18-68 | M33-119 | M21-71 |
| M6-200 | − | P | P | P | P | P | − | P | + | + | + | + | + |
| M7-40 |  | − | − | + | − | P | P | P | + | + | + | + | + |
| M8-53 |  |  | − | + | − | P | P | − | + | + | + | + | + |
| M8-181 |  |  |  | − | + | P | + | P | + | + | + | + | + |
| M13-101 |  |  |  |  | − | − | P | − | + | + | + | + | + |
| M15-207 |  |  |  |  |  | − | P | P | + | + | + | + | + |
| M17-231 |  |  |  |  |  |  | − | P | + | + | + | + | + |
| M24-241 |  |  |  |  |  |  |  | − | + | P | + | + | + |
| M10-152 |  |  |  |  |  |  |  |  | − | − | + | + | + |
| M17-162 |  |  |  |  |  |  |  |  |  | − | P | − | − |
| M18-68 |  |  |  |  |  |  |  |  |  |  | − | − | P |
| M33-119 |  |  |  |  |  |  |  |  |  |  |  | − | P |
| M21-71 |  |  |  |  |  |  |  |  |  |  |  |  | − |

[a] P, Papillae growth; +, confluent growth; −, absence of growth on glycerol.

low (less than 1%), the mutated alleles are genetically linked and said to be in the same locus. Crosses giving rise to a high percentage of recombinants (5–15%) indicate mutations that are unlinked and consequently represent different loci. The results of crosses done with a group of mutants deficient in cytochrome *b* (Table III) indicate that there are two separate loci, *cob1* and *cob2*, controlling the synthesis of this respiratory carrier. Similar criteria have been used to establish the existence of three distinct cytochrome oxidase loci[11] and two ATPase loci.[12, 13] The seven known mit loci are listed in Table IV.

In addition to being useful for assigning mutations to loci, mit⁻ × mit⁻ crosses can be used to map intralocus mutations relative to each other. A large number of mutations in the *oxi3* locus have been used to construct a deletion map of this region (Fig. 1). Although in theory it should be possible to obtain map distances from the observed recombination values, this has proven to be difficult since many crosses do not give additive distances. The reason for this is not clear but may be related to the nature of the mutations (insertions, deletions, frameshifts) which would tend to give anomalous results in such recombinational analysis.

## Deletion Mapping with ρ⁻ Testers

A ρ⁻ mutant which retains as little as a few percent of the genome is still capable of recombination. This fact has been taken advantage of in developing techniques for mapping mutations on mitochondrial DNA using $\rho^-$ clones.[14-16] Crosses of mit$^-$ or syn$^-$ mutants to $\rho^-$ testers can be used to assign mutations to genetic loci as well as to order them on the genome. The use of testers for the purpose of determining the locus of a mutated allele is illustrated in Fig. 2. Here ten different mit$^-$ or syn$^-$ strains carrying mutations in the alleles designated by the letters $A-H$ are crossed to four different $\rho^-$ testers covering different regions of the genome. In these crosses, respiratory-competent diploids will be formed only in those cases where the DNA of the tester has retained the wild-type allele of the mutant (*viz.* tester 1 $\times A, B, C,$ or $D$; tester 3 $\times F$; etc.). The restoration pattern shown in Fig. 2 indicates a clustering of mutations $A-D$ and $G-H$ and suggests that $E$ and $F$ are separate from these two groups. The results of crosses of mit$^-$ mutants to a large number of independent $\rho^-$ clones have corroborated the existence of the seven loci determined by the mit$^-$ $\times$ mit$^-$ crosses.

The application of $\rho^-$ clones for positioning mit$^-$ mutations relative to each other and to other mitochondrial markers is shown in Fig. 3. The 125 testers used in this experiment were independent clones isolated for their ability to restore wild-type growth to one of two different cytochrome $b$ mutants. When the clones were tested against other cytochrome $b$ deficient strains, 105 were found to be nondiscriminating and restored all the cytochrome $b$ mutants, while 20 clones discriminated in their ability to restore only some of the mutations. The results of the crosses with the discriminating clones allowed an ordering of the mutations as shown in Fig. 3. Furthermore, the mutations could be

*Table IV. Gene Products of Mitochondrial Loci*

|  | Gene product |
| --- | --- |
| Antibiotic resistance loci | |
| *cap* | 21 S RNA |
| *ery* | 21 S RNA |
| *par* | 16 S RNA |
| *oli1* | Subunit 9 of ATPase |
| *oli2* | ? |
| *ana1* | Cytochrome *b* |
| *ana2* | Cytochrome *b* |
| Mit loci | |
| *oxi1* | Subunit 2 of cytochrome oxidase |
| *oxi2* | Probably subunit 3 of cytochrome oxidase |
| *oxi3* | Subunit 1 of cytochrome oxidase |
| *cob1* | ? |
| *cob2* | Cytochrome *b* |
| *pho1* | ? |
| *pho2* | Subunit 9 of ATPase |

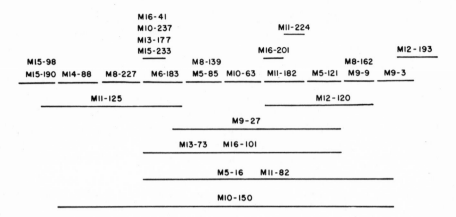

Fig. 1. Order of mutations in the *oxi3* locus. The order was established from pairwise crosses of all the mutants shown in the map. Overlapping lines indicate crosses in which the diploids did not grow on glycerol. From Slonimski and Tzagoloff.[11]

oriented relative to the antibiotic resistance loci, *oli1* and *oil2*, by checking the retention of these markers in the discriminating testers.

A similar approach has been used to map the *oxi* and *pho* loci.[11, 13, 17]

*Recombination of Mit⁻ and Syn⁻ Mutations with Antibiotic Resistance Markers*

There are presently six antiobiotic resistance loci on mitochondrial DNA that are essentially unlinked and therefore serve as useful genetic markers for different regions of the genome. Several methods have been adopted to study the position of mit⁻ and syn⁻ mutations relative to these markers. The simplest approach has been to determine the extent of recombination with the antibiotic resistance alleles. The existence of genetic linkage gives a first approximation of the localization of the mutation. For example, *pho1* and *pho2* mutations have been shown to be very tightly linked to *oli2* and *oli1*, respectively.[12,113]

Slonimski and Tzagoloff [11] have described a means of localizing mit⁻ mutations in spans of mitochondrial DNA that are bordered by two different antibiotic resistance markers. This analysis involves crossing a strain that contains a mit⁻ mutation and is antibiotic sensitive to a respiratory-competent tester with three different antibiotic-resistant markers. The frequency of the different recombinant classes issued from the cross permits the mit⁻ mutation to be located in one of the three spans defined by the resistance markers. This technique was used to localize the *oxi1* and *oxi2* loci in the *cap–par* span and the *oxi3* locus in the *par–oli2* span.[11]

The distribution of the antibiotic resistance and of the known mit and syn loci is shown in Fig. 4.

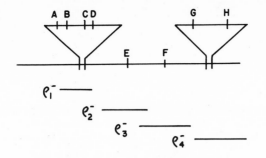

RESTORATION OF RESPIRATION

|         | A | B | C | D | E | F | G | H |
|---------|---|---|---|---|---|---|---|---|
| $\rho_1^-$ | + | + | + | + | − | − | − | − |
| $\rho_2^-$ | − | − | − | − | + | − | − | − |
| $\rho_3^-$ | − | − | − | − | − | + | − | − |
| $\rho_4^-$ | − | − | − | − | − | − | + | + |

Fig. 2. Use of $\rho^-$ clones in recognizing closely clustered mutations.

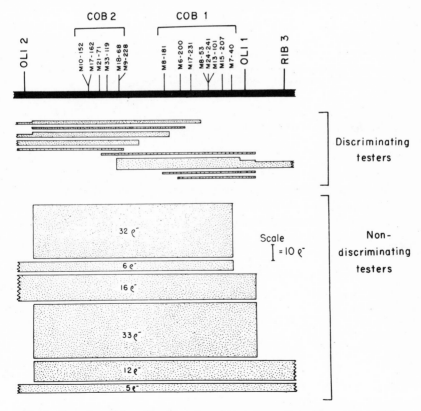

Fig. 3. Dissection of the *cob1 cob2* region with $\rho^-$ clones. From Tzagolóff *et al.*[10]

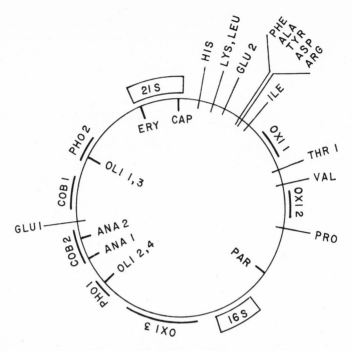

Fig. 4. Map of the mitochondrial genome of *S. cerevisiae*. The designation of genes and loci is the same as that of Tables II and IV. The location of the different tRNAs (with the exception of *thr1*) is based on the published map of Faye *et al.*[9]

## Identification of Mitochondrial Gene Products

Since there is no genetic complementation test for mitochondrial mutations, it is not possible at present to specify the number of genes that are represented by the seven mit loci. Some progress, however, has been made in identifying some of the structural genes of cytochrome oxidase and ATPase subunits and of cytochrome *b*.

### Oxi Loci

An indirect approach has been used to determine whether the *oxi* loci code for structural components of cytochrome oxidase. Spontaneous or mutagen-induced temperature-sensitive revertants have been obtained of different mit⁻ strains with lesions in *oxi1*, *oxi2*, and *oxi3*. Such revertants are capable of growth on glycerol at the permissive temperatures (25°–40°C). Some of the revertants in each of the three loci have been found to express the temperature sensitivity *in vitro* when cytochrome oxidase is assayed in isolated mitochondria (Trembath and Tzagoloff, unpublished studies). This finding suggests that *oxi1, 2,* and *3* each contain at least one structural gene of cytochrome oxidase. Additional evidence for the identity of the gene products of the *oxi* loci comes from studies of Cabral *et al.,*[18], who found that some mutations in *oxi1* lead to an altered form of subunit 2 of cytochrome oxidase. The presence of the structural gene of subunit 1 in the *oxi3* locus is indicated by the absence of this protein in many mutants carrying mutations in this locus.[7]

## Cob Loci

Two lines of evidence suggest that the *cob2* locus contains the structural gene of cytochrome *b*. Several mutants in *cob2* fail to synthesize a normal cytochrome *b*—instead, these strains produce a new peptide which migrates with an apparent lower molecular weight on SDS-polyacrylamide gels. Since these mutants revert to the wild type, it is probable that they have nonsense mutations in the structural gene of cytochrome *b*. Pratje and Michaelis[19] have recently shown that the antimycin resistance locus *(ana2)* is genetically linked to one of the nonsense mutations in cytochrome *b*. The resistant strain requires higher concentrations of antimycin to inhibit respiration in isolated mitochondria, indicating that the resistance allele must be in a structural gene of coenzyme $QH_2$-cytochrome *c* reductase. This allele is likely to be in the structural gene of cytochrome *b*.

## Pho Loci

The oligomycin-sensitive ATPase complex of yeast is a ten-subunit enzyme of which four proteins are translated on mitochondrial ribosomes.[3] One of the mitochondrial products (subunit 9) is a highly lipophilic 7800 dalton polypeptide which exists as a polymer of 45,000 daltons in the native enzyme.[20] Studies on the primary structure of subunit 9 of the wild type and oligomycin-resistant and *pho2* mutants have conclusively shown that the *oli1* and *pho2* loci are in the structural gene of this ATPase component (Sebald and Tzagoloff, unpublished studies). The identity of the gene products(s) of the *oli2* and *pho1* loci is not known at present.

The gene products currently thought to be represented by the various genetic loci on mitochondrial DNA are summarized in Table IV.

## References

1. Mason, T. L., and Schatz, G. (1973) *J. Biol. Chem. 248,* 1355–1360.
2. Weiss, H., and Ziganke, B. (1974) *Eur. J. Biochem. 41,* 63–71.
3. Tzagoloff, A., and Meagher, P. (1972) *J. Biol. Chem. 247,* 594–603.
4. Linnane, A. W., Saunders, G. W., Gingold, E. B., and Lukins, H. B. (1968) *Proc. Natl. Acad. Sci. USA 59,* 903–910.

5. Coen, D., Deutsch, J., Netter, P., Petrochilo, E., and Slonimski, P. P. (1970) in *Control of Organelle Development*, pp. 449–496, Cambridge University Press, Cambridge.
6. Avner, P. R., and Griffiths, D. E. (1973) *Eur. J. Biochem. 32*, 312–319.
7. Tzagoloff, A., Akai, A., Needleman, R. B., and Zulch, G. (1975) *J. Biol. Chem. 250*, 8236–8242.
8. Kotylak, Z., and Slonimski, P. P. (1976) in *The Genetic Function of Mitochondrial DNA* (Saccone, C. and Kroon, A. M., eds.,) pp. 143–154, North-Holland Press, Amsterdam.
9. Faye, G., Bolotin-Fukuhara, M., and Fukuhara, H. (1976) in *The Genetics and Biogenesis of Chloroplasts and Mitochondria* (Bücher, T., *et al.*, eds.), North-Holland, Amsterdam.
10. Tzagoloff, A., Foury, F., and Akai, A. (1976) *Mol. Gen. Genet. 149*, 33–42.
11. Slonimski, P. P., and Tzagoloff, A. (1976) *Eur. J. Biochem. 61*, 27–41.
12. Foury, F., and Tzagoloff, A. (1976) *Eur. J. Biochem. 68*, 113–119.
13. Coruzzi, G., Trembath, M. K., and Tzagoloff, A. (1977), submitted.
14. Bolotin-Fukuhara, M., Faye, G., and Fukuhara, H. (1976) in *Function of Mitochondrial DNA* (Saccone, C., and Kroon, A. M., eds.), pp. 243–250, North-Holland, Amsterdam.
15. Nagley, P., Sriprakash, K. S., Rytka, J., Choo, K. B., Trembath, M. K., Lukins, H. B., and Linnane, A. W. (1976) in *The Genetic Function of Mitochondrial DNA* (Saccone, C., and Kroon, A. M., eds.), pp. 231–242, North-Holland, Amsterdam.
16. Schweyen, R. J., Weiss-Brummer, B., Backhaus, B., and Kaudewitz, F. (1976) in *The Genetic Function of Mitochondrial DNA*, pp. 251–258, North-Holland, Amsterdam.
17. Trembath, M. K., Macino, G., and Tzagoloff, A. (1977) *Mol. Gen. Genet. 158*, 35–45.
18. Cabral, F., Rudin, Y., Solioz, M., Schatz, G., Clavilier, L., and Slonimski, P. P. (1978) *J. Biol. Chem. 253*, 297–304.
19. Pratje, E., and Michaelis, G. (1977) *Mol. Gen. Genet. 152*, 167–174.
20. Sierra, M. F., and Tzagoloff, A. (1973) *Proc. Natl. Acad. Sci. USA 70*, 3155–3159.

# Index